MW00710565

ANTENNA THEORY AND PRACTICE

Antenna Theory and Practice

Rajeswari Chatterjee

Department of Electrical Communication Engineering
Indian Institute of Science
Bangalore, India

JOHN WILEY & SONS

New York   Chichester   Brisbane   Toronto   Singapore

First published in 1988 by
WILEY EASTERN LIMITED
4835/24 Ansari Road, Daryaganj
New Delhi 110 002, India

Distributors:

Australia and New Zealand:
Jacaranda-Wiley Ltd., Jacaranda Press
JOHN WILEY & SONS, INC.
GPO Box 859, Brisbane, Queensland 4001, Australia

Canada:
JOHN WILEY & SONS CANADA LIMITED
22 Worcester Road, Rexdale, Ontario, Canada

Europe and Africa:
JOHN WILEY & SONS LIMITED
Baffins Lane, Chichester, West Sussex, England

South East Asia:
JOHN WILEY & SONS, INC.
05–05 Block B, Union Industrial Building
37 Jalan Pemimpin, Singapore 2057

Africa and South Asia:
WILEY EASTERN LIMITED
4835/24 Ansari Road, Daryaganj
New Delhi 110 002, India

North and South America and rest of the world:
JOHN WILEY & SONS, INC.
605 Third Avenue, New York, N.Y. 10158, USA

Copyright © 1988, WILEY EASTERN LIMITED
New Delhi, India

Library of Congress Cataloging in Publication Data

Chatterjee, Rajeswari, 1922—
    Antenna theory and practice.

    Includes bibliographies.
    1. Antennas (Electronics) I. Titles.

TK7871.6.C438   1988   621.38′028′3   87–23156

ISBN 0-470-20957-7 John Wiley & Sons, Inc.
ISBN 81-224-0005-1 Wiley Eastern Limited

Printed in India at Urvashi Press, Meerut, India.

PREFACE

The aim of this book is to present the basic theory of antennas and their applications. The antenna is a very important component of a communication system, and without it no communication over long distances is possible. Innumerable types of antennas have been designed for use at different frequency bands. They are of different shapes and are made of both conductors and dielectrics. The theory underlying the behaviour and design of these antennas are based on electromagnetic theory. Therefore a chapter on electromagnetic theory has been included in the beginning of the book.

Chapters on different types of antennas like the thin linear antenna, cylindrical antenna, biconical antenna, loop antenna, helical antenna, slot and micro-strip antennas, horn antenna, reflector-type antennas, lens antennas, leaky-wave and surface-wave antennas including dielectric and dielectric-loaded antennas, and wide-band antennas are included. Also antenna arrays, antenna synthesis, antenna measurements and electromagnetic wave propagation are treated in other chapters. Applications of antennas in different frequency ranges and for special purposes are treated in a chapter on 'Antenna Practice'.

The book is an outgrowth of lectures given at the under-graduate and post-graduate level at the Indian Institute of Science, Bangalore, India. The material can be used for teaching at both these levels as well as a reference book for practising engineers. Problems are given at the end of most of the chapters and other books and references are suggested for further reading at the end of each chapter.

I wish to thank the University Grants Commission, New Delhi, for sponsoring a project for writing this text and the Indian Institute of Science for using their library and other facilities. I should also like to express my gratitude to Mr. Y. S. Ramaswamiah, Ms. Ranganayaki and Mr. G. Jois for their help in the preparation of the manuscript.

R. CHATTERJEE

CONTENTS

1

INTRODUCTION

1.1 General

Electrical communication is the transmission of speech, music, pictures and other information by means of electrical signals. Speech and music are transmitted directly from their sources to the listeners across short distances by means of sound or acoustic waves. A picture is similarly transmitted directly by light or optical waves across a short distance. Over longer distances, wire communication and radio communication are used to transmit the signals. It is the special circuit component called the 'antenna' which has made radio communication possible in practice. An antenna creates sufficiently strong electromagnetic fields at large distances and, reciprocally, it is very sensitive to the electromagnetic fields impressed on it externally. However, the coupling between a transmitting antenna and a receiving antenna is so small that the signals have to be amplified both at the transmitting and receiving stations. It is impossible to attempt radio communication without antennas.

Antennas are usually obtained by modifying ordinary circuits or transmission lines to work effectively, and new physical and mathematical ideas are required to understand their behaviour.

The antennas required for radio communication, radio broadcasting and receiving, radar, satellite communication, etc. depend on the band of operating wavelengths (from tens of kilometers to a few millimeters), the amount of power radiated or received, and the transmission range. For long-wave (LW) transmission, large horizontal grids of bimetallic, often steel-aluminium wires of diameter about 20 mm or more, are used. Medium wave (MW) antennas are usually mast or tower antennas. For short-wave (SW) transmission, arrays of wire dipoles, single and double rhomboid antennas suspended from towers of heights 60 m are used. For receiving purposes travelling wave antenna and loop antennas are used. In the microwave and millimeter-wave range of wavelengths varying from a few metres to millimeters, dipole antennas, short antennas, parabolic and other reflector antennas, horn antennas, periscopic antennas, helical antennas, spiral antennas, surface-wave and leaky-wave antennas including dielectric and dielectric-loaded antennas, microstrip antennas, etc. are used. All these different types of antennas have been treated in different chapters. A chapter on 'Antenna Practice' is also included.

Since antenna theory is based on classical electromagnetic theory as described by Maxwell's equations, a review of this subject has been included. Also chapters on 'Electromagnetic Wave Propagation' and on 'Antenna Measurements' are included.

1.2   Antenna Fundamentals and Definitions

Some definitions and fundamental ideas concerning antennas are presented in this section. Other terms will be defined in the text. The reader is also referred to IEEE (formerly IRE) standards.

An *antenna* used as a radiator or receiver of electromagnetic energy may be defined as the transition region between free-space and a guiding structure like a transmission line. The guiding structure transports electromagnetic energy to or from the antenna. In the former case it is a 'transmitting antenna' and in the latter case it is a 'receiving antenna'.

The *radiation pattern* is a very important property of the antenna. The power received at a point by a receiving antenna is a function of the position of the receiving antenna with respect to the transmitting antenna. The graph of the received power at a constant radius from the transmitting antenna is called the 'power pattern' of the antenna, which is a spatial pattern. The spatial pattern of the electric (or magnetic field) is called a field pattern. A cross-section of this field pattern in any particular plane is called the 'radiation pattern' in that plane.

The *power density* is defined as the power per unit area in the field of the antenna. The power density multiplied by the square of the radial distance from the antenna gives the *radiation intensity* which is the power per unit solid angle.

The directivity of an antenna is defined as the ratio of the radiation intensity of the antenna in the direction of maximum radiation to the average radiation intensity of the antenna. If $U(\theta, \phi)$ is the radiation intensity in the direction (θ, ϕ), then the directivity D is given by

$$D = \frac{4\pi U_{\max}(\theta, \phi)}{\int U(\theta, \phi)\, d\Omega} \tag{1.1}$$

where Ω is the solid angle and $d\Omega = \sin\theta\, d\theta\, d\phi$.

The *gain* of an antenna is defined as the ratio of the maximum radiation intensity of the antenna to the maximum radiation intensity of a reference antenna with the same power input. If the reference antenna is an isotropic source then this ratio is the *absolute* gain of the antenna. If any other reference antenna like a dipole or a horn is used, then it should be clearly stated. The antenna gain is the product of the directivity and the efficiency. The directivity can be calculated, but the gain is usually measured.

The *polarization* of an antenna is the orientation of the electric field of an antenna. With reference to the earth's surface, the principal polarizations are horizontal and vertical. Many antennas are designed to radiate or receive either vertical or horizontal polarization or some other orientation of the electric field. These antennas are called *linearly polarized* antennas. An antenna is called *elliptically polarized* if it responds to two orthogonal field components with some phase difference. If the magnitudes of the orthogonal field components are equal and if the phase difference is $\pm 90°$, then the antenna is said to be *circularly polarized*. If the electric field rotates in the clockwise direction, it is called right-circular polarization, and if it rotates in the counter-clockwise direction it is called left-circular polarization.

The *impedance* of an antenna is the impedance at the antenna terminals with no load attached, where impedance may be defined as the ratio of voltage to current at the terminals or the ratio of the appropriate components of electric and magnetic fields at a point. In general, the antenna impedance is complex.

Any antenna when used as a receiving antenna collects a certain amount of energy from the incident electromagnetic wave. If W is the power in watts delivered to a load impedance at the antenna terminals and if P_i is the power density in watts per square meter incident on the antennas, then $A_e = W/P_i$ is defined as the *effective aperture* of the antenna. The effective aperture depends on the load impedance and the polarization and direction of the incident wave. In the special case where polarization and impedance are matched, and the incoming wave is in the direction of maximum directivity of the antenna, then A_e is given by

$$A_e = \frac{\lambda^2 D}{4\pi}$$

where D is the directivity and λ is the wavelength. For large aperture antennas the effective aperture is of the same order as the physical aperture. For pyramidal horns the effective aperture is about 50 to 80 per cent of the physical aperture, and for parabolic reflector antennas it is about 50 to 65 per cent of the physical aperture.

2

ELECTROMAGNETIC FIELDS AND
THEORY OF RADIATION

This book deals with the theory of antennas which radiate and receive energy and their applications. Antenna theory is based on classical electromagnetic theory as described by Maxwell's equations. Therefore, a review of electromagnetic phenomena is useful in order to have a proper understanding of antenna theory. This is discussed in this chapter. It is assumed that the reader is familiar with the more elementary aspects of electromagnetic fields and their properties.

The electric and magnetic fields are vector fields which are dependent on spatial coordinates like the rectangular coordinates x, y, z, and of time t. The important vector and scalar quantities are:

  (i) $\mathbf{E}$, the electric field intensity in volts/meter
 (ii) $\mathbf{H}$, the magnetic field intensity in amperes/meter
(iii) $\mathbf{D}$, the electric displacement density in Coulombs/meter²
 (iv) $\mathbf{B}$, the magnetic flux density in Webers/meter²
  (v) $\mathbf{A}$, the magnetic vector potential in amperes/meter²
 (vi) $\mathbf{F}$, the electric vector potential in volts/meter²
(vii) $\mathbf{P}$, the Poynting vector in watts/meter²
(viii) V, the electric scalar potential in volts
 (ix) U, the magnetic scalar potential in amperes
  (x) ϵ, the permittivity of the medium in farads/meter
 (xi) μ, the permeability of the medium in henrys/meter
(xii) σ, the conductivity of the medium in mhos/meter
(xiii) $\mathbf{M}$, the magnetic current density in volts/meter²
(xiv) $\mathbf{J}$, the electric current density in amperes/meter²
 (xv) ρ, the electric charge density in Coulombs/meter²
(xvi) $\boldsymbol{v}$, velocity in meters/second.

2.1 The Two Fundamental Electromagnetic Equations or Maxwell's Equations

The first and second laws of electromagnetic induction may be called the two fundamental electromagnetic (or Maxwell's) equations. These equations are

$$\text{curl } \mathbf{E} - \text{curl } (\mathbf{v} \times \mathbf{B}) = \frac{-\partial \mathbf{B}}{\partial t} \tag{2.1}$$

$$\text{curl } \mathbf{H} = \sigma \mathbf{E} + \frac{\partial \mathbf{D}}{\partial t} \tag{2.2}$$

In most problems of microwave engineering it is not necessary to include the term curl $(\mathbf{v} \times \mathbf{B})$ which is due to motion. Hence, Eqs. 2.1 and 2.2 may be rewritten as

$$\text{curl } \mathbf{E} = -\frac{\partial \mathbf{B}}{\partial t} \tag{2.3}$$

$$\text{curl } \mathbf{H} = \sigma \mathbf{E} + \frac{\partial \mathbf{D}}{\partial t} \tag{2.4}$$

In Eqs. (2.3) and (2.4), the sources of the electromagnetic field, namely, the electric and magnetic currents, have been omitted. However, if the sources are included, the two fundamental electromagnetic equations may be written as

$$\text{curl } \mathbf{E} = \frac{-\partial \mathbf{B}}{\partial t} - \mathbf{M}^i \text{ volts/metre}^2, \tag{2.5}$$

$$\text{curl } \mathbf{H} = \sigma \mathbf{E} + \frac{\partial \mathbf{D}}{\partial t} + \mathbf{J}^i \text{ amperes/metre}^2, \tag{2.6}$$

where $\mathbf{M}^i$ and $\mathbf{J}^i$ may be called respectively the source or impressed magnetic and source or impressed electric current densities.

In integral form, the two equations may be written as

$$\oint_c \mathbf{E} \cdot ds = -\int_s \left(\frac{\partial \mathbf{B}_n}{\partial t} + \mathbf{M}_n^i \right) \cdot d\mathbf{S}, \tag{2.7}$$

$$\oint_c \mathbf{H} \cdot ds = \int_s \left(\sigma \mathbf{E}_n + \frac{\partial \mathbf{D}_n}{\partial t} + \mathbf{J}_n^i \right) d\mathbf{S}. \tag{2.8}$$

In practical situations, we are mainly concerned with sources and fields varying harmonically with time. In such circumstances, Eqs. 2.7 and 2.8 become

$$\oint_c \mathbf{E} \cdot d\mathbf{s} = -\int_s (j\omega\mu\mathbf{H}_n + \mathbf{M}_n^i) \cdot d\mathbf{S} \text{ volts}, \tag{2.9}$$

$$\oint_c \mathbf{H} \cdot ds = \int_s [(\sigma + j\omega\mu)\mathbf{E}_n + \mathbf{J}_n^i] \cdot d\mathbf{S} \text{ amperes}, \tag{2.10}$$

using the relations $\mathbf{B} = \mu\mathbf{H}$ and $\mathbf{D} = \epsilon\mathbf{E}$ and assuming that all quantities vary as exp $(j\omega t)$.

When Maxwell's equations are expressed in the form of Eqs. 2.9 and 2.10, they possess considerable symmetry: $\mathbf{E}$ and $\mathbf{H}$ correspond to each other, being expressed in volts per metre and amperes per metre, respectively; $\mathbf{D}$ and $\mathbf{B}$ correspond to each other, being expressed in coulombs or ampere-seconds per square metre and webers or volt-seconds per square metre, respectively; electric and magnetic currents correspond to each other, being measured in amperes and volts, respectively. It is argued often, that **E** (force per unit positive charge) and **B** (force per unit electric current element) form one pair and **D** and **H** another. **D** and **H** respectively are determined from the charges and currents in the field, and represent lines of force resulting from these charges and currents. However, since electric and magnetic quantities are physically different, it is better to consider the similarity that arises from Maxwell's equations.

2.2 Currents Across Closed Surface

We shall now show that the total electric and magnetic currents across a closed surface vanish. Applying the two fundamental electromagnetic equations

$$\oint_c \mathbf{E}_s \cdot ds = -\int_s \left(\frac{\partial \mathbf{B}_n}{\partial t} + \mathbf{M}_n^i\right) \cdot d\mathbf{S}, \qquad (2.11)$$

$$\oint_c \mathbf{H}_s \cdot ds = \int_s \left(\sigma \mathbf{E}_n + \frac{\partial \mathbf{D}_n}{\partial t} + \mathbf{J}_n^i\right) \cdot d\mathbf{S} \qquad (2.12)$$

to the two parts S_1 and S_2 of a closed surface S, which has been divided into two parts by a closed curve C drawn on the surface (see Fig. 2.1), we obtain

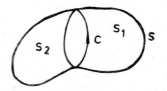

$$-\int_{s_1} \left(\frac{\partial \mathbf{B}_n}{\partial t} + \mathbf{M}_n^i\right) \cdot d\mathbf{S} - \int_{s_2} \left(\frac{\partial \mathbf{B}_n}{\partial t} + \mathbf{M}_n^i\right) \cdot d\mathbf{S}$$

$$= \oint_c \mathbf{E}_s \cdot ds - \oint_c \mathbf{E}_s \cdot ds = 0 \qquad (2.13)$$

Fig. 2.1 Illustration of total currents across closed surface S.

or

$$\int_s \left(\frac{\partial \mathbf{B}_n}{\partial t} + \mathbf{M}_n^i\right) \cdot d\mathbf{S} = 0. \qquad (2.14)$$

Similarly,

$$\int_s \left(\sigma \mathbf{E}_n + \frac{\partial \mathbf{D}_n}{\partial t} + \mathbf{J}_n^i\right) \cdot d\mathbf{S} = 0. \qquad (2.15)$$

Equations 2.14 and 2.15 prove that the total electric and magnetic currents across a closed surface vanish. These equations can be rewritten as

$$\oint_s \frac{\partial \mathbf{B}_n}{\partial t} \cdot d\mathbf{S} = -\oint_s \mathbf{M}_n^i \cdot d\mathbf{S} = -K, \qquad (2.16)$$

$$\oint_s \left(\sigma \mathbf{E}_n + \frac{\sigma \mathbf{D}_n}{\sigma t}\right) \cdot d\mathbf{S} = -\oint_s \mathbf{J}_n^i \cdot d\mathbf{S} = -I, \qquad (2.17)$$

where **K** and **I** are respectively the source magnetic and source electric currents flowing out of the closed surface S.

In perfect dielectrics, $\sigma = 0$. Substituting in Eq. 2.17 and integrating with respect to t, we obtain,

$$\int_{-\infty}^{t} dt \oint_{s} \frac{\partial \mathbf{D}_n}{\partial t} \cdot d\mathbf{S} = -\int_{-\infty}^{t} I \, dt \qquad (2.18)$$

or

$$\oint_{s} \mathbf{D}_n \cdot d\mathbf{S} = \oint_{s} \epsilon \mathbf{E}_n \cdot d\mathbf{S} = -\int_{-\infty}^{t} I \, dt = q, \qquad (2.19)$$

where q is the total electric charge enclosed by the closed surface S. Equation 2.19 is called Gauss' theorem. Similarly, integrating Eq. 2.16 with respect to time, we get,

$$\oint_{s} \mathbf{B}_n \cdot d\mathbf{S} = \oint_{s} \mu \mathbf{H}_n \cdot d\mathbf{S} = -\int_{-\infty}^{t} K \, dt = m, \qquad (2.20)$$

where m is the total magnetic pole enclosed by S. But since positive and negative magnetic poles cannot be separated, m is always equal to zero. Hence,

$$\oint_{s} \mathbf{B}_n \cdot d\mathbf{S} = 0. \qquad (2.21)$$

It may be noted that the concept of existence of $\mathbf{M}^i$ or source magnetic current is questionable. However, this is not true of source electric current $\mathbf{J}^i$ which can exist in the form of current in various types of generators of electric energy such as batteries, electric generators, and electron devices. Despite the uncertainty about the existence of source magnetic current, the concept of a fictitious source of magnetic current is very useful in practice. Besides, this concept makes the fundamental electromagnetic (or Maxwell's) equations mathematically symmetric and helps to solve such practical problems as radiation from antennas.

2.3 Boundary Conditions in Electromagnetic Field

Applying the fundamental equations

$$\oint_{c} \mathbf{E}_s \cdot d\mathbf{S} = -\int_{s} \left(\mu \frac{\partial \mathbf{H}_n}{\partial t} + \mathbf{M}_n^i \right) d\mathbf{S}, \qquad (2.22)$$

$$\oint_{c} \mathbf{H}_s \cdot d\mathbf{S} = \int_{s} \left(\sigma \mathbf{E}_n + \epsilon \frac{\partial \mathbf{E}_n}{\partial t} + \mathbf{J}_n^i \right) \cdot d\mathbf{S} \qquad (2.23)$$

to a rectangle $ABB'A'$ across a boundary S' between two media, as shown in Fig. 2.2, we get

$$E_t \cdot AB + E_n' \cdot BB' - E_t' \cdot A'B' - E_n \cdot AA' = \int_{s} \left(\frac{\mu \partial \mathbf{H}_n}{\partial t} + \mathbf{M}_n^i \right) \cdot d\mathbf{S} \quad (2.24)$$

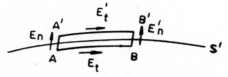

Fig. 2.2 Cross-section of boundary between two media and rectangle having two sides parallel to boundary and other two sides vanishingly small.

and E'_n are the normal components of E along AA' and BB', respectively. As $AA' \to 0$ and $BB' \to 0$,

$$E_t - E'_t = 0. \qquad (2.25)$$

Similarly,

$$H_t - H'_t = 0. \qquad (2.26)$$

Hence, the tangential components of E and H at the interface between the two media are continuous.

Since the circulation of the tangential component of **H** per unit area is the normal component of the electric current density **J**, we can say that the normal component of **J** is also continuous across the boundary, or

$$J_n = J'_n. \qquad (2.27)$$

Similarly,

$$M_n = M'_n \qquad (2.28)$$

For fields varying harmonically with time, that is, as $e^{j\omega t}$, the boundary conditions given in Eqs. 2.27 and 2.28 become

$$(\sigma + j\omega\epsilon)E_n = (\sigma' + j\omega\epsilon')E'_n \qquad (2.29)$$

$$\mu H_n = \mu' H'_n. \qquad (2.30)$$

For static fields,

$$\sigma E_n = \sigma' E'_n, \quad \mu H_n = \mu' H'_n \qquad (2.31)$$

In perfect dielectrics ($\sigma = 0$), Eqs. 2.29 and 2.30 become

$$\epsilon E_n = \epsilon' E'_n$$

or

$$D_n = D'_n \qquad (2.32)$$

and

$$\mu H_n = \mu' H'_n$$

or

$$B_n = B'_n \qquad (2.33)$$

This means that the normal components of **D** and **B** are continuous across the boundary.

In perfect conductors ($\sigma = \infty$), the electric intensity is zero for finite currents, and hence at the interface between a perfect conductor and a perfect dielectric,

$$E_t = 0$$

or

$$H_n = 0 \qquad (2.34)$$

The concept of perfect conductors is valuable chiefly because it helps in simplifying mathematical calculations and in providing approximation to solutions of problems involving good conductors.

2.4   Boundary Conditions in the Vicinity of Current Sheet

A current sheet is defined as an infinitely thin sheet carrying finite current (electric or magnetic) per unit length normal to the lines of flow. Figure 2.3 shows the cross-section of an electric current sheet whose linear current density J_s in amperes per metre is normal to the plane of the figure and is directed towards the reader.

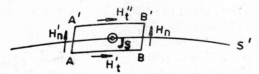

Fig. 2.3 Cross-section of boundary between two media.

Applying the equation,

$$\oint_c \mathbf{H}_s \cdot ds = \int_s \left(\sigma \mathbf{E}_n + \epsilon \frac{\partial \mathbf{E}_n}{\partial t} \right) \cdot d\mathbf{S} + \int_s J_n^i \cdot d\mathbf{S} \qquad (2.35)$$

to the rectangle $ABB'A'$, we obtain

$$H_t' \cdot AB + H_n \cdot BB' - H_t'' A'B' - H_n' \cdot AA' = \int_s \left(\sigma \mathbf{E}_n + \epsilon \frac{\partial \mathbf{E}_n}{\partial t} \right) \cdot d\mathbf{S} + \int_s J_n^i \cdot d\mathbf{S} \qquad (2.36)$$

As $AA' \to 0$ and $BB' \to 0$, relation (2.36) becomes

$$(H_t' - H_t'')AB = 0 + AB \cdot \lim_{AA' \to 0} AA' J_n^i \qquad (2.37)$$

If

$$\lim_{AA' \to 0} AA' \cdot J_n^i = J_s,$$

then

$$H_t' - H_t'' = J_s. \qquad (2.38)$$

In relation (2.38), the positive direction of the current density, the tangential component of H, and the normal to the sheet form a right-handed triad. Since

$$\lim_{AA' \to 0} AA' M_n^i = 0,$$

applying the equation

$$\oint_c \mathbf{E}_s \cdot ds = \int_s \left(\frac{\mu \partial \mathbf{H}_n}{\partial t} + \mathbf{M}_n^i \right) \cdot d\mathbf{S} \qquad (2.39)$$

to the same rectangle $ABB'A'$, we get

$$E_t' = E_t''. \qquad (2.40)$$

Therefore, we can say that the tangential components of the electric field intensity in the vicinity of an electric current sheet of density J_s are continuous, whereas the tangential components of the magnetic field intensity are discontinuous by an amount equal to the linear current density J_s.

Similarly, for a magnetic current sheet density M_s volts per metre, we can show that

$$E_t' - E_t'' = -M_s, \qquad (2.41)$$

$$H'_t = H''_t \qquad (2.42)$$

In other words, there is a discontinuity in the tangential components of electric field intensity equal to the linear magnetic current density M_s, and the tangential components of the magnetic field intensity are continuous. The discontinuities in the tangential components of the field intensities imply discontinuities in the normal components of the current densities. Imagine a pill box with its broad faces infinitely close and parallel to the electric current sheet on its opposite sides, as illustrated in Fig. 2.4. Let A be the cross-section and dl the height of this pill box. If J'_n and J''_n are the normal components of the electric current density $\mathbf{J}$ on either side of the electric current sheet, and if $\mathbf{J}_s$ is the surface current density on the electric current sheet, then using the theorem that the total electric current across the closed surface of the pill box is zero, we obtain,

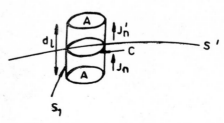

Fig. 2.4 Discontinuity in normal components of electric current densities in the vicinity of electric current sheet.

$$(J'_n - J''_n)A + \int_{S_1} \mathbf{J}_n \cdot d\mathbf{S} + \int_{A+A} \mathbf{J}^n \cdot d\mathbf{S} + \int_{S_1} \mathbf{J}^i_n \cdot d\mathbf{S} = 0, \qquad (2.43)$$

where $\mathbf{J}^i$ is the source (or impressed) current inside the pill box. As $dl \to 0$,

$$\int_s \mathbf{J}_n \cdot d\mathbf{S} = 0, \quad \int_{A+A} \mathbf{J}^i_n \cdot d\mathbf{S} = 0, \qquad (2.44)$$

$$\int_{S_1} \mathbf{J}^i_n \cdot d\mathbf{S} = \lim_{dl \to 0} \oint_c (\mathbf{J}^i_n \cdot dl)\, d\mathbf{S}, \qquad (2.45)$$

where c is the perimeter of the cross-section of the pill box. As $dl \to 0$ and $J^i_n \to \infty$, we have

$$\lim_{dl \to 0} \mathbf{J}^i_n \cdot d\mathbf{l} = J_s = \text{surface current density}, \qquad (2.46)$$

and hence

$$(J'_n - J''_n)A = \oint_c \frac{\mathbf{J}_s \cdot ds}{A} \qquad (2.47)$$

By definition, as $A \to 0$,

$$\frac{\oint_c \mathbf{J}_s \cdot ds}{A} = \text{div}' \, \mathbf{J}_S = \text{surface divergence of } \mathbf{J}_S \qquad (2.48)$$

Therefore,

$$J'_n - J''_n = \text{div}' \, \mathbf{J}_S = \text{surface divergence of } \mathbf{J}_S \qquad (2.49)$$

Thus, the discontinuity in the normal component of the electric current density is equal to the surface divergence of the surface electric current density of the electric current sheet. Similarly, the discontinuity in the normal component of the magnetic current density is equal to the surface divergence of the magnetic surface current density of the magnetic current sheet.

2.5 Conditions in the Vicinity of Infinitely Thin Linear Current Filaments

For an infinitely thin electric current filament carrying current I amperes per metre of length in the z-direction, we have

$$H_\phi = \frac{I}{2\pi\rho},\tag{2.50}$$

and for an infinitely thin magnetic current filament carrying current K amperes per metre of length in the z-direction,

$$E_\phi = \frac{K}{2\pi\rho}.\tag{2.51}$$

2.6 Energy Theorems, Poynting Vector

Starting with the fundamental equations of electromagnetic induction given by

$$\text{curl } \mathbf{E} = -\mu\,\frac{\partial \mathbf{H}}{\partial t} - \mathbf{M}^i,\tag{2.52}$$

$$\text{curl } \mathbf{H} = \sigma\mathbf{E} + \epsilon\,\frac{\partial \mathbf{E}}{\partial t} + \mathbf{J}^i\tag{2.53}$$

and taking the scalar product of the first equation and $\mathbf{H}$ and subtracting from it the scalar product of the second equation and $\mathbf{E}$, we obtain

$$\mathbf{H}\cdot\text{curl } \mathbf{E} - \mathbf{E}\cdot\text{curl } \mathbf{H} = -\mathbf{M}^i\cdot\mathbf{H} - \mathbf{E}\cdot\mathbf{J}^i - \sigma E^2 - \mu H\frac{\partial \mathbf{H}}{\partial t} - \epsilon\mathbf{E}\cdot\frac{\partial \mathbf{E}}{\partial t}\tag{2.54}$$

Integrating over a volume V bounded by a closed surface S, and using the vector identity

$$\text{div } [\mathbf{F}\times\mathbf{G}] = \mathbf{G}\cdot\text{curl } \mathbf{F} - \mathbf{F}\cdot\text{curl } \mathbf{G}\tag{2.55}$$

and Gauss' theorem, we get,

$$-\int_v \mathbf{E}\cdot\mathbf{J}^i\,dV - \int_v \mathbf{M}^i\cdot\mathbf{H}\,dV = \int_v \sigma E^2\,dV + \frac{\partial}{\partial t}\int_v \tfrac{1}{2}\epsilon E^2\,dV + \frac{\partial}{\partial t}\int_v \tfrac{1}{2}\mu H^2\,dV$$

$$+\int_s (\mathbf{E}\times\mathbf{H})_n\cdot d\mathbf{S}\tag{2.56}$$

Integrating Eq. 2.56 with respect to t in the interval $(-\infty, t)$ and assuming that originally the space was field free, we obtain,

$$-\int_{-\infty}^{t} dt \int_{v} (\mathbf{E}\cdot\mathbf{J}^i + \mathbf{M}^i\cdot\mathbf{H})\, dV = \int_{-\infty}^{t} dt \int_{v} \sigma E^2\, dV + \int_{v} (\tfrac{1}{2}\epsilon E^2 + \tfrac{1}{2}\mu H^2)\, dV$$
$$+ \int_{-\infty}^{t} dt \int_{s} (\mathbf{E}\times\mathbf{H})_n\cdot d\mathbf{S} \qquad (2.57)$$

The left-hand side of Eq. 2.57 represents the total work performed by the impressed forces up to the instant t against the forces of the field in maintaining the impressed or source currents $\mathbf{J}^i$ and $\mathbf{M}^i$. In accordance with the principle of conservation of energy, we say that this work appears as electromagnetic energy, and we explain the various terms on the right-hand side of Eq. 2.57 as follows:

$$\int_{-\infty}^{t} dt \int_{v} \sigma E^2\, dV$$

is the total energy converted into heat,

$$\int_{v} \sigma E^2\, dV$$

being the rate at which the total energy is converted into heat, and

$$\int_{v} (\tfrac{1}{2}\epsilon E^2 + \tfrac{1}{2}\mu H^2)\, dV$$

is the total stored electric and magnetic energies within S,

$$\frac{\partial}{\partial t} \int_{v} (\tfrac{1}{2}\epsilon E^2 + \tfrac{1}{2}\mu H^2)\, dV$$

being the rate at which this total stored energy is changing with time. The last term in Eq. 2.57, namely,

$$\int_{-\infty}^{t} dt \int_{s} (\mathbf{E}\times\mathbf{H})_n\cdot d\mathbf{S}$$

can be interpreted as the total flow of energy across S up to time t, and hence $\int_{s} (\mathbf{E}\times\mathbf{H})_n\cdot d\mathbf{S}$ can be treated as the time rate of flow of energy or power across S. Hence, the vector $\mathbf{E}\times\mathbf{H} = \mathbf{P}$, called the Poynting vector, can be considered as a vector representing the time rate of energy flow or power per unit area. Its unit is watt per square metre.

The surface integral of the vector $\mathbf{P} = \mathbf{E}\times\mathbf{H}$ over a closed surface represents the difference between the energy contributed to the field inside S and the energy accounted for within S. On the other hand, it is also true that the value of this integral remains the same if the curl of an arbitrary vector $\mathbf{P}'$ is added to $\mathbf{P}$.

When the fields vary harmonically with time, multiplying Eq. 2.52 by $\mathbf{H}^*$ and the conjugate of Eq. 2.53 by $\mathbf{E}$ and subtracting, we obtain

$$\mathbf{H}^*\cdot \text{curl } \mathbf{E} - \mathbf{E}\cdot \text{curl } \mathbf{H}^* = -\mathbf{M}^i\cdot\mathbf{H}^* - \mathbf{E}\cdot\mathbf{J}^{i*} - \sigma\mathbf{E}\cdot\mathbf{E}^*$$
$$+ j\omega\epsilon\mathbf{E}\cdot\mathbf{E}^* - j\omega\mu\mathbf{H}\cdot\mathbf{H}^* \qquad (2.58)$$

Integrating Eq. 2.58 over a volume V bounded by a closed surface S, we get

$$-\tfrac{1}{2}\int_v (\mathbf{E}\cdot\mathbf{J}^{i*} + \mathbf{M}^i\cdot\mathbf{H}^*)\,dV = \tfrac{1}{2}\int_v \sigma\mathbf{E}\cdot\mathbf{E}^*dV + \tfrac{1}{2}j\omega\int_v \mu\mathbf{H}\cdot\mathbf{H}^*dV$$

$$-\tfrac{1}{2}j\omega\int_v \epsilon\mathbf{E}\cdot\mathbf{E}^*dV + \tfrac{1}{2}\int_s (\mathbf{E}\times\mathbf{H}^*)_n\cdot d\mathbf{S}$$

$$(2.59)$$

Let us take the real part of Eq. 2.59. The real part of the left-hand side of this equation is the average power spent by the impressed forces in sustaining the electromagnetic field. Some of this power is transformed into heat and this amount is given by

$$\tfrac{1}{2}\int_v \sigma\mathbf{E}\cdot\mathbf{E}^*dV$$

The remaining power flows out of the volume across S, and the amount is represented by the real part of the term

$$\tfrac{1}{2}\int_s (\mathbf{E}\times\mathbf{H}^*)_n\cdot d$$

The term,

$$\tfrac{1}{2}\int_v (\mu\mathbf{H}\cdot\mathbf{H}^* - \epsilon\mathbf{E}\cdot\mathbf{E}^*)\,dV$$

represents the difference between the average magnetic and electric power stored inside S, and it is equal to the imaginary part of the left-hand side of the equation. The term

$$\psi = \tfrac{1}{2}\int_s (\mathbf{E}\times\mathbf{H}^*)_n\cdot d\mathbf{S} \qquad (2.60)$$

is called the complex power flow across S. The vector $\mathbf{P} = \tfrac{1}{2}\mathbf{E}\times\mathbf{H}^*$ is the complex Poynting vector, and its real part is the average power flow per unit area.

If S is a perfect conductor, the tangential component of $\mathbf{E}$ on S vanishes, and hence there is no flow of energy across S and it is parallel to the surface. A perfectly conducting closed sheet separates space into two electromagnetically independent regions. A similar complete separation is provided by a closed surface on which the tangential component of H is zero. In the physical world, metals are good approximations to perfect conductors, but there are no good approximations to sheets on which H_t is zero except at zero (or nearly zero) frequency when substances with extremely high permeability have this property.

2.7   Normal and Surface Impedances of Surfaces

Only the tangential components of $\mathbf{E}$ and $\mathbf{H}$ contribute to the complex power flow ψ across a surface S. If u and v are orthogonal coordinates on S and if u, v and n form a right-handed triplet of directions, then from

Eq. 2.60,

$$\psi = \tfrac{1}{2} \int_s (E_u H_v^* - E_v H_u^*) \, dS$$

Putting

$$Z_{uv} = \frac{E_u}{H_v}, \quad Z_{vu} = \frac{-E_v}{H_u}, \tag{2.61}$$

we obtain

$$\psi = \tfrac{1}{2} \int_s (Z_{vu} H_u H_u^* + Z_{uv} H_v H_v^*) \, dS \tag{2.62}$$

If $Z_{uv} = Z_{vu} = Z_n$, then

$$\psi = \tfrac{1}{2} \int_s Z_n (H_u H_u^* + H_v H_v^*) \, dS. \tag{2.63}$$

Here, Z_n is called the impedance normal to the surface S.

Consider now a conducting surface of thickness t. The surface current density $\mathbf{J}_s$ in amperes per metre is equal to $\mathbf{J}_t$, where $\mathbf{J}$ is the volume current density in amperes per square metre. If σ is the conductivity, then $\mathbf{J} = \sigma\mathbf{E}$, and hence $\mathbf{J}_s = \sigma t\mathbf{E}$. If t approaches zero and σ increases to a very large value so that the product $G = \sigma t$ remains constant, we have

$$\mathbf{J}_s = G\mathbf{E}, \quad \mathbf{E} = R\mathbf{J}_s, \tag{2.64}$$

where G and R are called respectively the surface conductance (in mhos) and surface resistance (in ohms) of the sheet.

In any medium, if the displacement current density is considered in addition to the conduction current density, then for time-harmonic fields,

$$\mathbf{J} = j\omega\epsilon\mathbf{E} + \sigma\mathbf{E}, \tag{2.65}$$

and hence

$$\mathbf{J}_s = (j\omega\epsilon + \sigma)t\mathbf{E} \tag{2.66}$$

As $t \to 0$ and σ and $\omega\epsilon$ become very large,

$$\mathbf{J}_s = (G_s + jB_s)\mathbf{E} = Y_s\mathbf{E} \tag{2.67}$$

$$\mathbf{E} = (R_s + jX_s)\mathbf{J}_s = Z_s\mathbf{J}_s \tag{2.68}$$

where Y_s is called the surface admittance, G_s the surface conductance, B_s the surface susceptance, Z_s the surface impedance, R_s the surface resistance, and X_s the surface reactance.

A perfect conducting surface has zero normal impedance because the tangential component of $\mathbf{E}$ is zero, and hence it is called a surface of zero impedance. A surface of infinite impedance may be defined as one with zero tangential component of $\mathbf{H}$, and it can be approximated by a surface of high permeability only at zero frequency.

2.8   Primary and Secondary Electromagnetic Constants of a Medium

Conductivity σ, permittivity ϵ, and permeability μ are called the primary electromagnetic constants of a medium, in the sense that they appear directly in the formulation of the electromagnetic equations

$$\text{curl } \mathbf{E} = -\mu \frac{\partial \mathbf{H}}{\partial t} - \mathbf{M}^i, \tag{2.69}$$

$$\text{curl } \mathbf{H} = \sigma \mathbf{E} + \epsilon \frac{\partial \mathbf{E}}{\partial t} + \mathbf{J}^i \tag{2.70}$$

For time-harmonic fields,

$$\text{curl } \mathbf{E} = -j\omega\mu\mathbf{H} - \mathbf{M}^i \tag{2.71}$$

$$\text{curl } \mathbf{H} = \sigma\mathbf{E} + j\omega\epsilon\mathbf{E} + \mathbf{J}^i \tag{2.72}$$

Compare Maxwell's Eqs. 2.69 and 2.70 with the transmission line equations

$$\frac{dV}{dx} = -(R + j\omega L)I + E(x) \tag{2.73}$$

$$\frac{dI}{dx} = -(G + j\omega C)V \tag{2.74}$$

where $E(x)$ is the voltage per unit length impressed along the line and in series with the line, V and I are the instantaneous transverse voltage across the line and the longitudinal electric current in the line, and R, L, G and C are respectively the series resistance, series inductance, shunt conductance and shunt capacitance per unit length. We then observe that the transmission line equations form a special one-dimensional case of Maxwell's equations, and the terminology of the transmission line may be extended to Maxwell's equations. Thus, we may call $j\omega\mu$ the distributed series impedance, σ the distributed shunt conductance, and $j\omega\epsilon$ the distributed shunt susceptance, and $(\sigma + j\omega\epsilon)$ the distributed shunt admittance per metre of the medium. The constants μ, σ, and ϵ may be called respectively the distributed series inductance, distributed shunt conductance and distributed shunt capacitance per metre of the medium. In transmission line theory, two secondary constants are introduced, namely, the propagation constant

$$\gamma = [(R + j\omega L)(G + j\omega C)]^{1/2}$$

and the characteristic impedance

$$Z_0 = [(R + j\omega L)/(G + j\omega C)]^{1/2}$$

Similarly, in the three-dimensional Maxwell's equations, we may introduce the two secondary electromagnetic constants, namely, the intrinsic propagation constant γ and the intrinsic impedance $Z_0 = \eta$, given by

$$\gamma = [j\omega\mu(\sigma + j\omega\epsilon)]^{1/2}, \tag{2.75}$$

$$\eta = Z_0 = \left(\frac{j\omega\mu}{\sigma + j\omega\epsilon}\right)^{1/2} \tag{2.76}$$

The constants γ and η are independent of the geometry of the wave, and hence the adjective 'intrinsic' or 'characteristic' of the medium. The characteristic impedances of different types of wave contain η as a factor.

The primary constants are positive except when the frequencies are very high (at optical frequencies ϵ may be negative), and hence $\gamma = \alpha + j\beta$ and $\eta = \mathscr{R} + j\mathscr{X}$ are either in the first quadrant or in the third quadrant. The definitions of γ and η are made unambiguous if it is agreed that they always lie in the first quadrant or on its boundaries. For perfect conductors, $\sigma = \infty$, and hence both γ and η lie on the bisector of the first quadrant. For perfect dielectrics, $\sigma = 0$, and hence γ is on the positive imaginary and η on the positive real axis. In general, both γ and η are complex quantities:

$$\gamma = \alpha + j\beta \tag{2.77}$$

$$\eta = \mathscr{R} + j\mathscr{X} \tag{2.78}$$

where α, β, $\mathscr{R}$, $\mathscr{X}$ are called respectively the intrinsic attenuation constant, intrinsic phase constant, intrinsic resistance, and intrinsic reactance of the medium.

For perfect dielectrics,

$$\gamma = j\beta, \ \beta = \omega\sqrt{\mu\epsilon} = \frac{\omega}{v} = \frac{2\pi}{\lambda},$$

$$\eta = \sqrt{\frac{\mu}{\epsilon}}, f\lambda = v, \tag{2.79}$$

$$v = \frac{1}{\sqrt{\mu\epsilon}}, \ \lambda = \frac{2\pi}{\beta}, \ \mu = \frac{\eta}{v}, \ \epsilon = \frac{1}{\eta v},$$

where v and λ are called the intrinsic or characteristic velocity and wavelength in the medium. In general, the velocity and wavelength for different types of waves may have respectively the characteristic velocity and characteristic wavelength as factors.

For free space, the numerical values for various constants are

$$\eta_0 = \sqrt{\frac{\mu_0}{\epsilon_0}} = 376.7 \approx 377 \approx 120\pi \text{ ohms,} \tag{2.80}$$

$$v_0 = \frac{1}{\sqrt{\mu_0\epsilon_0}} = 2.998 \times 10^8 \approx 3 \times 10^8 \text{ metres/second,} \tag{2.81}$$

$$\eta_0^{-1} = \sqrt{\frac{\epsilon_0}{\mu_0}} \approx \frac{1}{120\pi} \text{ mho,} \tag{2.82}$$

$$\mu_0 = 4\pi \times 10^{-7} = 1.257 \times 10^{-6} \text{ henry/metre,} \tag{2.83}$$

$$\epsilon_0 = 8.854 \times 10^{-12} \approx \frac{1}{36\pi} \times 10^{-9} \text{ farad/metre.} \tag{2.84}$$

For good conductors, $\sigma \gg \omega\epsilon$, and hence

$$\gamma = \sqrt{j\omega\mu\sigma}, \ \eta = \sqrt{\frac{j\omega\mu}{\sigma}}, \tag{2.85}$$

$$\alpha = \beta = \sqrt{\frac{\omega\mu\sigma}{2}}, \tag{2.86}$$

$$\mathscr{R} = \mathscr{X} = \sqrt{\frac{\omega\mu}{2\sigma}} \tag{2.87}$$

The Q of a medium is defined as the ratio of the displacement current density to the conduction current density, and hence

$$Q = \frac{\omega \epsilon}{\sigma} = \frac{1}{D},$$ (2.88)

where D is called the dissipation factor. For good dielectrics, $Q \gg 1$, and for good conductors, $Q \ll 1$. Table 2.1 gives the values of ϵ_r for the various commonly used materials.

2.9   Electromagnetic Wave Equation in Dielectrics and Conductors

In Maxwell's equations

$$\text{curl } \mathbf{E} = -\mu \frac{\partial \mathbf{H}}{\partial t} - \mathbf{M}^i$$ (2.89)

$$\text{curl } \mathbf{H} = \sigma \mathbf{E} + \epsilon \frac{\partial \mathbf{E}}{\partial t} + \mathbf{J}^i$$ (2.90)

If $\mathbf{M}^i$ and $\mathbf{J}^i$ are differentiable, then $\mathbf{E}$ and $\mathbf{H}$, which satisfy the nonhomogeneous wave equations, can be eliminated. In practice, however, $\mathbf{J}^i$ and $\mathbf{M}^i$ are usually discontinuous, and hence nondifferentiable; in such cases, $\mathbf{E}$ and $\mathbf{H}$ cannot be eliminated without introducing the auxiliary functions which are also called potential functions.

In source-free regions, where $\mathbf{J}^i = \mathbf{M}^i = 0$, on the other hand, we can always eliminate either $\mathbf{E}$ or $\mathbf{H}$. In homogeneous source-free regions, we obtain the wave equations

$$\nabla^2 \mathbf{E} = \mu \epsilon \frac{\partial^2 \mathbf{E}}{\partial t^2} + \mu \sigma \frac{\partial \mathbf{E}}{\partial t}$$ (2.91)

$$\nabla^2 \mathbf{H} = \mu \epsilon \frac{\partial^2 \mathbf{H}}{\partial t^2} + \mu \sigma \frac{\partial \mathbf{H}}{\partial t}$$ (2.92)

In perfect dielectrics, $\sigma = 0$, and Eqs. 2.91 and 2.92 become

$$\nabla^2 \mathbf{E} = \mu \epsilon \frac{\partial^2 \mathbf{E}}{\partial t^2},$$

$$\nabla^2 \mathbf{H} = \mu \epsilon \frac{\partial^2 \mathbf{H}}{\partial t^2}$$ (2.93)

For time-harmonic fields, the wave equations are

$$\nabla^2 \mathbf{E} = \gamma^2 \mathbf{E},$$ (2.94)

$$\nabla^2 \mathbf{H} = \gamma^2 \mathbf{H}.$$ (2.95)

2.10   Solution of Wave Equation in Cartesian Coordinates in Homogeneous Dissipative and Nondissipative Regions: Uniform Plane Waves

In cartesian coordinates, the vector wave equation

$$\nabla^2 \mathbf{E} = \gamma^2 \mathbf{E}$$ (2.96)

Table 2.1   Relative Permittivity of Materials

Material	Relative permittivity (ϵ_r)
Vacuum	1
Air (°C)	1
40 atmospheres	1.00059
80 atmospheres	1.0218
Carbon dioxide (0°C)	1.000985
Hydrogen (0°C)	1.000264
Water vapour (145°C)	1.00705
Styrofoam	1.03
Air (−191°C)	1.43
Paraffin	2.1
Plywood	2.1
Polystyrene	2.7
Paper	2.0–2.5
Rubber	2.3–4.0
Amber	2.7
Asphalt	2.68
Benzene	2.29
Petroleum	2.13
Amber	3.0
Linseed oil	3.35
Plexiglas	3.4
Dry sandy soil	3.4
Nylon (hard)	3.8
Wood	2.5–7.7
Sulphur (amorphous)	3.98
Sulphur (cast, fresh)	4.92
Guttapercha	3.3–4.9
Castor oil	4.67
Quartz	5.0
Bakelite	5.0
Mica	5.6–6.0
Lead glass	5.4–8.0
Formica	6.0
Flint glass	6.6–9.9
Alcohol	
amyl	16.0
ethyl	25.8
methyl	31.2
Ammonia	22.0
Glycerin (15°C)	56.2
Water (distilled)	81.1
Rutile (TiO_2)	89–173
Barium titanate ($BaTiO_3$)	1200
Barium strontium titanate ($2BaTiO_3 : 1SrTiO_3$)	10,000
Barium titanate zirconate ($4BaTiO_3 : 1BaZrO_3$)	13,000
Barium titanate stannate ($9BaTiO_3 : 1BaSnO_3$)	20,000

reduces to three scalar wave equations, namely

$$\nabla^2 E_x = \gamma^2 E_x,$$

$$\nabla^2 E_y = \gamma^2 E_y, \qquad (2.97)$$

$$\nabla^2 E_z = \gamma^2 E_z.$$

A solution of any of the Eq. 2.97 is of the form

$$E_x = \exp\,(\pm \Gamma_x x \pm \Gamma_y y \pm \Gamma_z z), \qquad (2.98)$$

where the propagation constants Γ_x, Γ_y, Γ_z in the x-, y-, z-coordinate axes respectively satisfy the condition

$$\Gamma_x^2 + \Gamma_y^2 + \Gamma_z^2 = j\omega\mu(\sigma + j\omega\epsilon). \qquad (2.99)$$

In nondissipative media, we have $\sigma = 0$, and hence

$$\Gamma_x^2 + \Gamma_y^2 + \Gamma_z^2 = -\omega^2\mu\epsilon = -\beta^2 = -\frac{4\pi^2}{\lambda^2} \qquad (2.100)$$

Because of the relation given in Eq. 2.100 only two of the propagation constants Γ_x, Γ_y, Γ_z are independent, but they depend on the distribution of sources producing the field. If the distribution of sources is uniform in planes parallel to the xy-plane, the field is uniform in these planes, and $\Gamma_x = \Gamma_y = 0$, then the propagation constant Γ_z in the z-direction is equal to the intrinsic propagation constant γ. Such a wave is a uniform plane wave whose equiphase and equiamplitude planes are parallel to the xy-plane, and the expression for any of the components of the field is of the form

$$E_x = \exp\,(\pm \Gamma_z z)e^{j\omega t} = e^{\pm\gamma z}e^{j\omega t} = e^{\pm(\alpha + j\beta)z}e^{j\omega t}. \qquad (2.101)$$

The positive signs in Eq. 2.101 give a wave with increasing amplitude as z increases. This is not possible in a physical situation unless there is a conversion of energy of some other form into electromagnetic energy (as seen in an electron device where the kinetic or potential energy of the charge carriers is converted into electromagnetic energy). Hence, it is customary to use the negative sign, and therefore,

$$E_x = e^{-\alpha z}e^{j(\omega t - \beta z)} \qquad (2.102)$$

which represents an attenuated uniform plane travelling in the positive z-direction with phase velocity v equal to ω/β. In a nondissipative or perfect dielectric medium, $\alpha = 0$, and hence there is no attenuation.

For a uniform plane wave, whose equiamplitude and equiphase planes are parallel to a plane

$$x\cos A + y\cos B + z\cos C = \text{constant},$$

where $\cos A$, $\cos B$, $\cos C$ are the direction cosines of a normal to the plane, the expression for any of the field components is of the form

$$E_x = \exp\{-\gamma(x \cos A + y \cos B + z \cos C)\}$$
$$= \exp(-\Gamma_x x - \Gamma_y y - \Gamma_z z) \tag{2.103}$$

so that the propagation constants along the three axes are

$$\Gamma_x = \gamma \cos A, \ \Gamma_y = \gamma \cos B, \ \Gamma_z = \gamma \cos C \tag{2.104}$$

If the medium is nondissipative, $\sigma = 0$, and hence $\gamma = j\beta$. Therefore

$$\Gamma_x = j\beta_x = j\beta \cos A \quad \text{or} \quad \beta_x = \beta \cos A,$$
$$\beta_y = \beta \cos B, \quad \beta_z = \beta \cos C. \tag{2.105}$$

Hence

$$\lambda_x = \frac{2\pi}{\beta_x} = \frac{2\pi}{\beta \cos A} = \frac{\lambda}{\cos A}$$
$$\lambda_y = \frac{\lambda}{\cos B}, \quad \lambda_z = \frac{\lambda}{\cos C} \tag{2.106}$$

$$v_x = \frac{\omega}{\beta_x} = \frac{\omega}{\beta \cos A} = \frac{v}{\cos A}$$
$$v_y = \frac{v}{\cos B}, \quad v_z = \frac{v}{\cos C} \tag{2.107}$$

$$\beta_x^2 + \beta_y^2 + \beta_z^2 = \beta^2 \tag{2.108}$$

$$\frac{1}{\lambda_x^2} + \frac{1}{\lambda_y^2} + \frac{1}{\lambda_z^2} = \frac{1}{\lambda^2} \tag{2.109}$$

$$\frac{1}{v_x^2} + \frac{1}{v_y^2} + \frac{1}{v_z^2} = \frac{1}{v^2} \tag{2.110}$$

Thus, for uniform plane waves in a nondissipative medium, the phase velocities in various directions are always greater than the characteristic phase velocity, and the phase constants β_x, β_y, β_z are always less than the characteristic phase constant β of the medium. This property is true for a rectangular metal waveguide.

There are some types of physical situations, for example, for surface waves supported by surface-wave structures or slow-wave structures in nondissipative media, where the phase constant in some direction, for example, the z-direction, is greater than β or, in other words, $\beta_z > \beta$. Then

$$\Gamma_x^2 + \Gamma_y^2 = -\beta^2 + \beta_z^2 \tag{2.111}$$

If $\Gamma_y = 0$ or there is no variation of the field in the y-direction, then

$$\Gamma_x^2 = -\beta^2 + \beta_z^2 > 0 \tag{2.112}$$

or Γ_x is a real quantity $\pm\alpha_x$.

Taking the negative value $-\alpha_x$ for a real physical situation, we find that there is attenuation in the x-direction. As $\beta_z > \beta$, $v_z < v$, the wave travels in the z-direction with a velocity less than the characteristic velocity of the medium, and is attenuated exponentially in the x-direction or, in other words, is evanescent in the x-direction. Such a wave is called a surface (or slow) wave. The attenuation is due to the boundary conditions, and not due to the loss in the medium. Any component of the field varies as $\exp(-\alpha_x x) \exp[j(\omega t - \beta_z z)]$ in a nondissipative medium. On the other hand, if $\beta_z < \beta$ and if $\Gamma_y = 0$, then

$$\Gamma_x^2 = \beta_z^2 - \beta^2 < 0 \tag{2.113}$$

or

$$\Gamma_x = j\beta_x \tag{2.114}$$

and

$$v_z > v.$$

In this case, the wave travels in a direction making an angle $\theta = \tan^{-1}(\beta_z/\beta_x)$ with the z-axis and with a velocity v_z which is greater than the characteristic velocity v of the medium. Such a wave is called a leaky (or fast) wave, and any field component varies as $\exp[j(\omega t - \beta_x x - \beta_z z)]$.

In general, in a nondissipative medium, if the propagation constants Γ_x, Γ_y, Γ_z in the three directions are complex, then,

$$\Gamma_x^2 + \Gamma_y^2 + \Gamma_z^2 = -\beta^2 \tag{2.115}$$

becomes

$$(\alpha_x + j\beta_x)^2 + (\alpha_y + j\beta_y)^2 + (\alpha_z + j\beta_z)^2 = -\beta^2 \tag{2.116}$$

or

$$\alpha_x^2 + \alpha_y^2 + \alpha_z^2 - \beta_x^2 - \beta_y^2 - \beta_z^2 = -\beta^2 \tag{2.117}$$

and

$$\alpha_x \beta_x + \alpha_y \beta_y + \alpha_z \beta_z = 0 \tag{2.118}$$

Equation 2.118 shows that the equiamplitude planes

$$\alpha_x x + \alpha_y y + \alpha_z z = \text{constant}$$

are perpendicular to the equiphase planes

$$\beta_x x + \beta_y y + \beta_z z = \text{constant}.$$

Thus, in nondissipative media, equiamplitude and equiphase planes either coincide with each other (as for uniform plane waves) or are orthogonal to each other. In the first instance, the waves are uniform on equiphase planes in the sense that E and H have constant values at all points of a given equiphase plane at a given instant; in the second case, the amplitude varies exponentially, the fastest variation being in the direction given by the direction components α_x, α_y, α_z. For example, for a surface (or slow) wave any field component varies as

$$\exp(-\alpha_x x) \exp[j(\omega t - \beta_z z)]$$

and hence the equiphase planes are $z =$ constant and the equiamplitude planes are $x =$ constant, and these two families of planes are orthogonal to each other. For a leaky (or fast) wave, in general,

$$\Gamma_x = \alpha_x + j\beta_x, \ \Gamma_z = \alpha_z + j\beta_z, \ \Gamma_y = 0,$$

and hence the expression for any field component varies as

$$[\exp - (\alpha_x x - \alpha_z z) \exp \{ j(\omega t - \beta_x x - \beta_z z) \}].$$

Therefore, the equiamplitude planes $\alpha_x x + \alpha_z z =$ constant are orthogonal to the equiphase planes $\beta_x x + \beta_z z =$ constant (see Fig. 2.5).

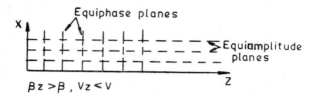

(a) Surface (or slow) waves

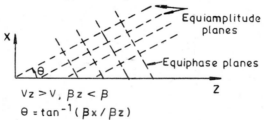

(b) Leaky (or fast) wave

Fig. 2.5   Illustration of surface (or slow) waves and leaky (or fast) waves.

In dissipative media, $\gamma^2 = (\alpha + j\beta)^2$, and hence

$$(\alpha_x + j\beta_x)^2 + (\alpha_y + j\beta_y)^2 + (\alpha_z + j\beta_z)^2 = (\alpha + j\beta)^2 \qquad (2.119)$$

or

$$\alpha_x^2 + \alpha_y^2 + \alpha_z^2 - \beta_x^2 - \beta_y^2 - \beta_z^2 = \alpha^2 - \beta^2 \qquad (2.120)$$

and

$$\alpha_x \beta_x + \alpha_y \beta_y + \alpha_z \beta_z = \alpha\beta = \tfrac{1}{2}\omega\mu\sigma \qquad (2.121)$$

Thus, equiamplitude planes are no longer perpendicular to equiphase planes.

The foregoing general conclusions on waves of exponential type of the form

$$E_x = \exp(-\Gamma_x x - \Gamma_y y - \Gamma_z z)e^{j\omega t} \qquad (2.122)$$

have a broader significance than appears at first sight. The constant Γ_x represents the relative rate of change of E_x in the x-direction, and we have

$$\Gamma_x = -\frac{1}{E_x}\frac{\partial E_x}{\partial x} \qquad (2.123)$$

$$\Gamma_x^2 = \frac{1}{E_x} \frac{\partial^2 E_x}{\partial x^2} \tag{2.124}$$

Equation 2.124 is also satisfied by $-\Gamma_x$. If the wave function is not exponential, we may still define Γ_x by Eq. 2.124 and, similarly, we may define Γ_y and Γ_z by

$$\Gamma_y^2 = \frac{1}{E_x} \frac{\partial^2 E_x}{\partial y^2} \tag{2.125}$$

$$\Gamma_z^2 = \frac{1}{E_x} \frac{\partial^2 E_x}{\partial z^2} \tag{2.126}$$

If the quantities Γ_x, Γ_y and Γ_z vary slowly from one point to another, the solution of the wave equation will be approximately exponential, and the properties of exponential waves just discussed will be applicable in sufficiently small regions.

2.11   Waves at Interface between Conductors and Dielectrics

Consider a plane interface (the xy-plane) between air (substantially free space) above the plane and a conductor below the plane, as shown in Fig. 2.6. For an exponential wave of the form

$$F = \exp(-\Gamma_x x - \Gamma_y y - \Gamma_z z), \tag{2.127}$$

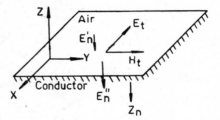

Fig. 2.6   Plane boundary between two semi-infinite media.

the propagation constants Γ_x and Γ_y in directions parallel to the boundary must be the same in both media so that the boundary conditions may be satisfied at all points of the air-conductor interface. This is obvious because F represents a component of either $\mathbf{E}$ or $\mathbf{H}$ parallel to the boundary; in this instance, F is continuous across the boundary and the continuity cannot be satisfied at all points unless Γ_x, Γ_y are the same on both sides of the boundary. The same condition applies to the normal components of the current densities $(\sigma + j\omega\epsilon)E_z$ and $j\omega\mu H_z$. Thus, we have

$$\Gamma_x^2 + \Gamma_y^2 + \Gamma_{z,0}^2 = -\beta_0^2 \quad \text{(in air)} \tag{2.128}$$

$$\Gamma_x^2 + \Gamma_y^2 + \Gamma_z^2 = \gamma^2 \quad \text{(in conductor)} \tag{2.129}$$

where β_0 is the characteristic phase constant $\omega\sqrt{\mu_0\epsilon_0}$ of air and γ is the characteristic propagation constant of the conductor. Subtracting Eq. 2.128 from Eq. 2.129, we obtain

$$\Gamma_z^2 = \gamma^2 + \beta_0^2 + \Gamma_{z,0}^2. \tag{2.130}$$

The propagation constant γ in the conductor is much larger than that in free space (or air). Hence, $\gamma^2 \gg \beta_0^2$. Also, $\Gamma_{z,0}$ which is the propagation constant in free space in the direction normal to the interface, is comparable to β_0 if the wave direction is nearly normal to the interface or it is much smaller than β_0 if the wave direction is nearly parallel to the interface. In other words, $\Gamma_{z,0} \approx \beta_0$ or $\Gamma_{z,0} \ll \beta_0$. Hence $\Gamma_{z,0} \ll \gamma$. Therefore, from Eq. 2.130,

$$\Gamma_z \approx \gamma \qquad (2.131)$$

or, in the conductor, the propagation constant normal to the interface is substantially equal to the intrinsic propagation constant of the conductor.

Since the current density normal to the interface is continuous, we have

$$\sigma E_n'' = j\omega\epsilon_0 E_n', \qquad (2.132)$$

or

$$\sigma E_n'' = \frac{j\omega\epsilon_0}{\sigma} E_n', \qquad (2.133)$$

where E_n' and E_n'' are the normal components of E in air and the conductor, respectively. As σ for the conductor is very large, $E_n'' \ll E_n'$ or the normal component of $\mathbf{E}$ in the conductor is much smaller than the normal component of $\mathbf{E}$ in air.

Even at moderately high frequencies, the attenuation constant in the conductor is large, and the field becomes quite small at rather short distances from the interface. Except at very low frequencies, the fields are confined largely to thin skins of conductors, and hence this effect is known as the skin effect.

The current density at the surface of the conductor is σE_t, where E_t is the tangential component of E at the interface. Elsewhere inside the conductor it is $\sigma E_t e^{-\gamma z}$, where z is the normal distance from the surface. The total current per unit length normal to the lines of flow in a conductor is

$$J_s = \int_0^\infty \sigma E_t e^{-\gamma z}\, dz = \frac{\sigma}{\gamma} E_t = \frac{1}{\eta} E_t, \qquad (2.134)$$

where $\eta = \left(\dfrac{j\omega\mu}{\sigma}\right)^{1/2}$.

Also, $\gamma = \sqrt{j\omega\mu\sigma}$. On the other hand, at the surface of the conductor, the tangential component of $\mathbf{H}$, namely, H_t, is equal to J_s, and is given by

$$H_t = J_s = \frac{1}{\eta} E_t \qquad (2.135)$$

or $E_t/H_t = Z_n =$ the normal impedance of the surface $= \eta$, which is the intrinsic impedance of the conductor.

By definition, E_t/J_s is the surface impedance Z_s of the interface. Therefore, Z_s of an interface between a good conductor and a good dielectric is equal to the normal impedance Z_n. The conductor may be replaced by a sheet, whose surface impedance is η adjacent to a sheet of infinite impedance, which would effectively exclude the space previously occupied by the conductor. Z_s and γ are given by

$$Z_s = \mathscr{R} + j\mathscr{X} = \eta = (j\omega\mu/2\sigma)^{1/2} = (\omega\mu/2\sigma)^{1/2} + j(\omega\mu/2\sigma)^{1/2} \quad (2.136)$$

$$\gamma = \alpha + j\beta = \sqrt{j\omega\mu\sigma} = (\omega\mu\sigma/2)^{1/2} + j(\omega\mu\sigma/2)^{1/2} \quad (2.137)$$

Therefore,

$$\mathscr{R} = \left(\frac{\omega\mu}{2\sigma}\right)^{1/2} = \frac{\alpha}{\sigma} = \frac{1}{\sigma t}, \quad (2.138)$$

where

$$t = \frac{1}{\alpha} = \left(\frac{2}{\omega\mu\sigma}\right)^{1/2} \quad (2.139)$$

Hence, the surface resistance $\mathscr{R}$ is equal to the dc resistance of a plate of thickness t, defined by the reciprocal of the attenuation constant. The thickness t is called the skin depth or depth of penetration but the term should not be interpreted to mean that the rest of the conductor could be removed without changing its ac resistance. The field reduces to $1/e$ of its value at the surface at a distance equal to the skin depth, and the attenuation through the skin depth is only one neper.

2.12   Special Forms of Maxwell's Equations in Source-Free Regions in Different Systems of Coordinates

Fundamental electromagnetic (or Maxwell's) equations in source-free regions for time-harmonic fields are

$$\text{curl } \mathbf{E} = -j\omega\mu\mathbf{H} \quad (2.140)$$

$$\text{curl } \mathbf{H} = j\omega\epsilon\mathbf{E} + \sigma\mathbf{E} \quad (2.141)$$

In cartesian coordinates, they become

$$\frac{\partial E_z}{\partial y} - \frac{\partial E_y}{\partial z} = -j\omega\mu H_x$$

$$\frac{\partial E_x}{\partial z} - \frac{\partial E_z}{\partial x} = -j\omega\mu H_y \quad (2.142)$$

$$\frac{\partial E_y}{\partial x} - \frac{\partial E_x}{\partial y} = -j\omega\mu H_z$$

$$\frac{\partial H_z}{\partial y} - \frac{\partial H_y}{\partial z} = (\sigma + j\omega\epsilon)E_x$$

$$\frac{\partial H_x}{\partial z} - \frac{\partial H_z}{\partial x} = (\sigma + j\omega\epsilon)E_y \quad (2.143)$$

$$\frac{\partial H_y}{\partial x} - \frac{\partial H_x}{\partial y} = (\sigma + j\omega\epsilon)E_z$$

In cylindrical coordinates, they are of the form

$$\frac{\partial E_z}{\partial \phi} - \frac{\partial E_\phi}{\partial z} = -j\omega\mu\rho H_\rho$$

$$\frac{\partial E_\rho}{\partial z} - \frac{\partial E_z}{\partial \rho} = -j\omega\mu H_\phi \tag{2.144}$$

$$\frac{\partial}{\partial \rho}(\rho E_\phi) - \frac{\partial E_\rho}{\partial \phi} = -j\omega\mu\rho H_z$$

$$\frac{\partial H_z}{\partial \phi} - \rho\,\frac{\partial H_\phi}{\partial z} = (\sigma + j\omega\epsilon)\rho E_\rho$$

$$\frac{\partial H_\rho}{\partial z} - \frac{\partial H_z}{\partial \rho} = (\sigma + j\omega\epsilon)E_\phi \tag{2.145}$$

$$\frac{\partial}{\partial \rho}(\rho H_\phi) - \frac{\partial H_\rho}{\partial \phi} = (\sigma + j\omega\epsilon)E_z$$

In spherical coordinates, they take the form

$$\frac{\partial}{\partial \theta}(\sin\theta E_\phi) - \frac{\partial E_\theta}{\partial \phi} = -j\omega\mu r \sin\theta H_r$$

$$\frac{\partial}{\partial r}(rE_\theta) - \frac{\partial E_r}{\partial \theta} = -j\omega\mu r H_\phi \tag{2.146}$$

$$\frac{\partial E_r}{\partial \phi} - \sin\theta\,\frac{\partial}{\partial r}r(E_\phi) = -j\omega\mu r \sin\theta H_\theta$$

$$\frac{\partial}{\partial \theta}(\sin\theta H_\phi) - \frac{\partial H_\theta}{\partial \phi} = (\sigma + j\omega\epsilon)r \sin\theta E_r$$

$$\frac{\partial}{\partial r}(rH_\theta) - \frac{\partial H_r}{\partial \theta} = (\sigma + j\omega\epsilon)rE_\phi \tag{2.147}$$

$$\frac{\partial H_r}{\partial \phi} - \sin\theta\,\frac{\partial}{\partial r}(H_\phi r) = (\sigma + j\omega\epsilon)\sin\theta E_\theta$$

2.13 Field Produced by Given Distribution of Currents in Infinite Homogeneous Medium: Vector and Scalar Wave Potentials

When the source (or impressed) currents are known throughout an infinite homogeneous medium, the field can be calculated fairly easily. If we obtain the field of a current element, then by using the principle of superposition the field due to any distribution may be calculated.

The electromagnetic equations for harmonic fields are

$$\text{curl } \mathbf{E} = -j\omega\mu\mathbf{H} - \mathbf{M}^i \tag{2.148}$$

$$\text{curl } \mathbf{H} = (\sigma + j\omega\epsilon)\mathbf{E} + \mathbf{J}^i \qquad (2.149)$$

If these equations are solved, the solution of the most general case can then be expressed in the form of a contour integral in the oscillation constant plane. If $\mathbf{M}^i$ is a continuous and differentiable function, then it is possible to eliminate $\mathbf{E}$ from Eqs. 2.148 and 2.149 and obtain a second-order differential equation for $\mathbf{H}$. Similarly, if $\mathbf{J}^i$ is a continuous and differentiable function, we can eliminate $\mathbf{H}$ from Eqs. 2.148 and 2.149 and derive a second-order differential equation for $\mathbf{E}$. However, in practical problems, $\mathbf{J}^i$ and $\mathbf{M}^i$ are localized and the regions occupied by them have sharp boundaries. When this happens, the equations can be solved by introducing a set of auxiliary functions, generally called potential functions. Let

$$\mathbf{E} = \mathbf{E}' + \mathbf{E}'', \qquad \mathbf{H} = \mathbf{H}' + \mathbf{H}'' \qquad (2.150)$$

where $(\mathbf{E}', \mathbf{H}')$ and $(\mathbf{E}'', \mathbf{H}'')$ respectively are the solutions of

$$\text{curl } \mathbf{E}' = -j\omega\mu\mathbf{H}' \qquad (2.151)$$

$$\text{curl } \mathbf{H}' = \mathbf{J}^i + (\sigma + j\omega\epsilon)\mathbf{E}' \qquad (2.152)$$

and

$$\text{curl } \mathbf{E}'' = -\mathbf{M}^i - j\omega\mu\mathbf{H}'' \qquad (2.153)$$

$$\text{curl } \mathbf{H}'' = (\sigma + j\omega\epsilon)\mathbf{E}'' \qquad (2.154)$$

The field $(\mathbf{E}', \mathbf{H}')$ is produced by the electric current $\mathbf{J}^i$, and the field $(\mathbf{E}'', \mathbf{H}'')$ by the magnetic current $\mathbf{M}^i$. The sum of these fields satisfies Eqs. 2.148 and 2.149 and hence it is produced by $\mathbf{J}^i$ and $\mathbf{M}^i$.

Taking the divergence of each of the equations 2.151–2.154, we obtain

$$\text{div } \mathbf{H}' = 0 \qquad (2.155)$$

$$\text{div } \mathbf{E}' = \frac{-\text{div } \mathbf{J}^i}{\sigma + j\omega\epsilon} \qquad (2.156)$$

$$\text{div } \mathbf{H}'' = \frac{-\text{div } \mathbf{M}^i}{j\omega\mu} \qquad (2.157)$$

$$\text{div } \mathbf{E}'' = 0. \qquad (2.158)$$

Equations 2.156 and 2.157 require $\mathbf{J}^i$ and $\mathbf{M}^i$ to be continuous and differentiable. But one form of the solution to our problem is obtained without using these equations. In the other form of the solution which depends on them, to begin with we may assume $\mathbf{J}^i$ and $\mathbf{M}^i$ to be differentiable and then extend the results to include discontinuous distributions.

Equations 2.155 and 2.158 show that $\mathbf{H}'$ and $\mathbf{E}''$ can be represented as the curls of some vector point functions:

$$\mathbf{H}' = \text{curl } \mathbf{A} \qquad (2.159)$$

$$\mathbf{E}'' = -\text{curl } \mathbf{F} \qquad (2.160)$$

Substituting Eqs. 2.159 and 2.160 in Eqs. 2.151 and 2.154, we obtain

$$\mathbf{E'} = -j\omega\mu\mathbf{A} - \text{grad } V, \tag{2.161}$$

$$\mathbf{H''} = -(\sigma + j\omega\epsilon)\mathbf{F} - \text{grad } U, \tag{2.162}$$

where V and U are two new scalar point functions which are introduced because the equality of the curls of the two vectors does not imply that the vectors are identical. From Eqs. 2.161 and 2.162 and using Eqs. 2.152 and 2.153, we obtain

$$\text{curl curl } \mathbf{A} = \mathbf{J}^i - \gamma^2\mathbf{A} - (\sigma + j\omega\epsilon)\text{ grad } V, \tag{2.163}$$

$$\text{curl curl } \mathbf{F} = \mathbf{M}^i - \gamma^2\mathbf{F} - j\omega\mu\text{ grad } U. \tag{2.164}$$

Using the vector identity

$$\text{curl curl } \mathbf{A} = -\nabla^2\mathbf{A} + \text{grad div } \mathbf{A}, \tag{2.165}$$

we obtain

$$\nabla^2\mathbf{A} - \text{grad div } \mathbf{A} = -\mathbf{J}^i + \gamma^2\mathbf{A} + (\sigma + j\omega\epsilon)\text{ grad } V, \tag{2.166}$$

$$\nabla^2\mathbf{F} - \text{grad div } \mathbf{F} = -\mathbf{M}^i + \gamma^2\mathbf{F} + j\omega\mu\text{ grad } U. \tag{2.167}$$

Thus, we have been able to express $\mathbf{E}$ and $\mathbf{H}$ in terms of the vectors $\mathbf{A}$ and $\mathbf{F}$ and the scalars V and U by means of Eqs. 2.166 and 2.167. The new functions, called auxiliary functions, are related to each other by Eqs. 2.159–2.162. If we add the gradients of arbitrary functions to $\mathbf{A}$ and $\mathbf{F}$, then Eqs. 2.159 and 2.160 remain unchanged. Hence, both $\mathbf{A}$ and $\mathbf{F}$ and V and U are arbitrary. Therefore, to suit our convenience, it is possible to impose further conditions on these functions. For instance, by putting

$$V = -\frac{\text{div } \mathbf{A}}{\sigma + j\omega\epsilon}, \tag{2.168}$$

$$U = -\frac{\text{div } \mathbf{F}}{j\omega\mu}. \tag{2.169}$$

Equations 2.166 and 2.167 become

$$\nabla^2\mathbf{A} = \gamma^2\mathbf{A} - \mathbf{J}^i, \tag{2.170}$$

$$\nabla^2\mathbf{F} = \gamma^2\mathbf{F} - \mathbf{M}^i. \tag{2.171}$$

When the auxiliary functions $\mathbf{A}$, $\mathbf{F}$, V and U are specified as in Eqs. 2.168–2.171, they are called wave potentials. $\mathbf{A}$ is called the magnetic vector potential, $\mathbf{F}$ the electric vector potential, V the electric scalar potential, and U the magnetic scalar potential. H. A. Lorentz, while considering electromagnetic wave propagation in nondissipative media, introduced these wave potentials for the first time and called them retarded potentials. These wave potentials are not only retarded but also attenuated, and, therefore, the term wave potentials is appropriate.

In terms of the wave potentials $\mathbf{A}$, $\mathbf{F}$, V and U the field vectors $\mathbf{E}$ and $\mathbf{H}$ can be expressed as

$$\mathbf{E} = \mathbf{E'} + \mathbf{E''} = -j\omega\mu\mathbf{A} - \text{grad } V - \text{curl } \mathbf{F}, \tag{2.172}$$

$$\mathbf{H} = \mathbf{H}' + \mathbf{H}'' = \text{curl } \mathbf{A} - \text{grad } U - (\sigma + j\omega\epsilon)\mathbf{F}, \tag{2.173}$$

where V and U are defined by Eqs. 2.168 and 2.169 and $\mathbf{A}$ and $\mathbf{F}$ are the solutions of Eqs. 2.170 and 2.171.

If $\mathbf{J}^i$ and $\mathbf{M}^i$ are differentiable functions, V and U satisfy equations similar to those satisfied by $\mathbf{A}$ and $\mathbf{F}$. Taking the divergence of Eqs. 2.170 and 2.171 and substituting from Eqs. 2.168 and 2.169, we obtain

$$\nabla^2 V = \gamma^2 V + \frac{\text{div } \mathbf{J}^i}{\sigma + j\omega\epsilon}, \tag{2.174}$$

$$\nabla^2 U = \gamma^2 U + \frac{\text{div } \mathbf{M}^i}{j\omega\mu}. \tag{2.175}$$

In nondissipative media,

$$\text{div } \mathbf{J}^i = -\frac{\partial\rho}{\partial t} = -j\omega\rho, \tag{2.176}$$

$$\text{div } \mathbf{M}^i = -\frac{\partial m}{\partial t} = -j\omega m, \tag{2.177}$$

where ρ and m are the electric and magnetic volume charge densities. In fact, m is equal to zero. Consequently,

$$\nabla^2 V = -\beta^2 V - \frac{\rho}{\epsilon}, \tag{2.178}$$

$$\nabla^2 U = -\beta^2 U - \frac{m}{\mu} = -\beta^2 U. \tag{2.179}$$

For static fields, $\beta = \omega\sqrt{\mu\epsilon} = 0$. Hence

$$\nabla^2 V = -\frac{\rho}{\epsilon}, \tag{2.180}$$

$$\nabla^2 U = 0 \tag{2.181}$$

which are Poisson's and Laplace's equations, respectively.

2.14 Field of Electric Current Element

Let Il be an electric current element (see Fig. 2.7). Assume that the current I is uniform and steady between the terminals A and B so that the entire current is forced to flow out of B into the external medium and then back into A. If the external medium is a perfect dielectric medium, then there would be concentration of positive electric charge at B and negative electric charge at A at the rate of I amperes or I coulombs per second. This results in an ever increasing electric field around the filament. The moment of the electric current element

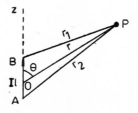

Fig. 2.7 Electric current element.

is defined as Il, the product of the current and length of the filament. If the current is along the z-axis and is centred at the origin of the coordinates, then the current density $\mathbf{J}$ at a point P at a radial distance r from the origin is given by

$$\mathbf{J} = \mathbf{u}_r J_r = \mathbf{u}_r \frac{I}{4\pi r^2} = \mathbf{u}_r \frac{\partial}{\partial r}\left(-\frac{I}{4\pi r}\right). \tag{2.182}$$

Assuming that the current element consists of two point sources, namely, $+I$ at B and $-I$ at A, the current density at a distant point $P(r \gg 1)$ is the gradient of the function

$$P = \frac{-I}{4\pi r_1} + \frac{I}{4\pi r_2} + \frac{+I}{4\pi}\left[\frac{-1}{r - (l/2)\cos\theta} + \frac{1}{r + (l/2)\cos\theta}\right] \approx \frac{-Il\cos\theta}{4\pi r^2} \tag{2.183}$$

Hence

$$J_r = \frac{\partial P}{\partial r} = \frac{Il\cos\theta}{2\pi r^3} \tag{2.184}$$

$$J_\theta = \frac{\partial P}{r\,\partial\theta} = \frac{Il\sin\theta}{4\pi r^3} \tag{2.185}$$

In Fig. 2.8, the dashed lines represent the current flow lines and the solid lines show the magnetic field lines which are coaxial with the element. The magnetomotive force around the circumference of a circle PP' coaxial with the element is equal to the total electric current $I(\theta)$ flowing through any surface S bounded by the circumference of this circle. If S is chosen as part of a sphere concentric with the origin, as shown in the figure, then we have

$$I(\theta) = \int_{\phi=0}^{2\pi}\int_{\theta=0}^{\theta} J_r r^2 \sin\theta\,d\theta\,d\phi$$

$$= \int_{\phi=0}^{2\pi}\int_{\theta=0}^{\theta} \frac{Il\cos\theta}{2\pi r^3} r^2 \sin\theta\,d\theta\,d\phi = \frac{Il\sin^2\theta}{2r} \tag{2.186}$$

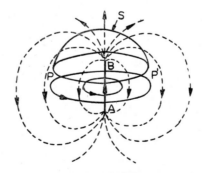

--- Current flow lines and electric field lines
——Magnetic field lines

Fig. 2.8   Current flow lines, electric field
lines, and magnetic field lines in
the vicinity of current element.

Therefore,

$$I(\theta) = \frac{Il \sin^2 \theta}{2r} = H_\phi 2\pi r \sin \theta \qquad (2.187)$$

or

$$H_\phi = \frac{Il \sin \theta}{4\pi r^2} \qquad (2.188)$$

But from Eq. 2.159,

$$\mathbf{H} = \text{curl } \mathbf{A},$$

and since

$$\nabla^2 \mathbf{A} = \gamma^2 \mathbf{A} - \mathbf{J}^i, \qquad (2.189)$$

the component A_z of $\mathbf{A}$ depends only on the z-component of the source current $\mathbf{J}^i$. Since the current element is in the z-direction, $\mathbf{A}$ has only a z-component. Hence

$$H_\phi = \frac{-\partial A_z}{\partial \rho} = \frac{Il \sin \theta}{4\pi r^2} \qquad (2.190)$$

Since

$$r^2 = \rho^2 + z^2, \quad \sin \theta = \rho/r = \frac{\rho}{\sqrt{\rho^2 + z^2}}$$

we have

$$\frac{\partial A_z}{\partial \rho} = \frac{-Il \sin \theta}{4\pi r^2} = \frac{-Il(\rho/r)}{\rho^2 + z^2} = \frac{-Il\rho}{(\rho^2 + z^2)^{3/2}}$$

Therefore

$$A_z = \frac{Il}{4\pi(\rho^2 + z^2)^{1/2}} = \frac{Il}{4\pi r} \qquad (2.191)$$

If the current is a harmonic function of time, or $I = I_0 e^{j\omega t}$ then as the frequency ω approaches zero, the field of the alternating current element $I_0 l e^{j\omega t}$, must approach the values given by Eqs. 2.190 and 2.191 for a steady current element $I\, dl$. The magnetic vector potential $\mathbf{A}$ satisfies the equation

$$V^2 \mathbf{A} = \gamma^2 \mathbf{A} \qquad (2.192)$$

at all points external to the current element. And from Eq. 2.191, $\mathbf{A} = \mathbf{u}_z A_z$ must be independent of θ and ϕ, and hence

$$\frac{\partial}{\partial r}\left(r^2 \frac{\partial A_z}{\partial r}\right) = \gamma^2 r^2 A_z \qquad (2.193)$$

whose solution is of the form

$$A_z = \frac{Pe^{-\gamma r}}{r} + \frac{Qe^{+\gamma r}}{r} \qquad (2.194)$$

In dissipative media ($\sigma \neq 0$), $[Q \exp(+\gamma r)]/r$ increases exponentially with the distance r from the element and it cannot therefore represent a physically realizable field. Hence, only the first term is retained, and

$$A_z = \frac{Pe^{-\gamma r}}{r} \qquad (2.195)$$

approaches the value given by Eq. 2.191 for a steady current element if $P = Il/(4\pi)$. Hence

$$A_z = \frac{Ile^{-\gamma r}}{4\pi r} \qquad (2.196)$$

In a nondissipative medium, $\gamma = j\beta$, and hence

$$A_z = \frac{Ile^{-j\beta r}}{4\pi r} \qquad (2.197)$$

Therefore,

$$\mathbf{A} = \mathbf{u}_z A_z = \mathbf{u}_z \frac{Ile^{-j\beta r}}{4\pi r} \qquad (2.198)$$

From Eq. 2.168,

$$V = \frac{-\operatorname{div}\mathbf{A}}{\sigma + j\omega\epsilon}, \qquad (2.199)$$

$$V = -\frac{1}{\sigma + j\omega\epsilon}\frac{\partial A_z}{\partial z} = \frac{\eta Il}{4\pi r}\left(1 + \frac{1}{\gamma r}\right)e^{-\gamma r}\cos\theta \qquad (2.200)$$

where

$$\eta = \left(\frac{j\omega\mu}{\sigma + j\omega\epsilon}\right)^{1/2}$$

E and **H** are now obtained from

$$\mathbf{E} = -j\omega\mu\mathbf{A} - \operatorname{grad}V, \qquad (2.201)$$

$$\mathbf{H} = \operatorname{curl}\mathbf{A} \qquad (2.202)$$

and have the components E_r E_θ, H_ϕ given by

$$E_r = \frac{\eta Il}{2\pi r^2}\left(1 + \frac{1}{\gamma r}\right)e^{-\gamma r}\cos\theta \qquad (2.203)$$

$$E_\theta = \frac{j\omega\mu Il}{4\pi r}\left(1 + \frac{1}{\gamma r} + \frac{1}{\gamma^2 r^2}\right)e^{-\gamma r}\sin\theta \qquad (2.204)$$

$$H_\phi = \frac{\gamma Il}{4\pi r}\left(1 + \frac{1}{\gamma r}\right)e^{-\gamma r}\sin\theta \qquad (2.205)$$

In Eqs. 2.203–2.205, E_r has a $1/r^2$ term and a $1/r^3$ term; E_θ has a $1/r$ term, $1/r^2$ term, and $1/r^3$ term; and H_ϕ has a $1/r$ term and a $1/r^2$ term. The $1/r$ term is called the radiation term, the $1/r^2$ term the induction term, and the $1/r^3$ term the electrostatic term. At a large distance r, ony the $1/r$ term predominates, and hence

$$H_\phi = \frac{jIl}{2\pi r}e^{-j\beta r}\sin\theta, \quad E_\theta = \eta H_\phi, \quad E_r = 0$$

Similarly, the field of a magnetic current element Kl can be obtained from the electric vector potential

$$\mathbf{F} = \mathbf{u}_z F_z = \mathbf{u}_z\frac{Kle^{-\gamma r}}{4\pi r} \qquad (2.206)$$

2.15   Radiation from Electric Current Element

The total power W radiated from an electric current element is calculated as the total flow of power across a large sphere concentric with the element.

$$W = \int_{\text{sphere}} P_r dS = \tfrac{1}{2} \int_{\phi=0}^{2\pi} \int_{\theta=0}^{\pi} E_\theta H_\phi^* r^2 \sin \theta \; d\theta \; d\phi \qquad (2.207)$$

where P_r is the radial component of the Poynting vector. W is given by

$$W = \frac{\eta I^2 l^2}{8\lambda^2} \int_0^{2\pi} \int_0^\pi \sin^3 \theta \; d\theta \; d\phi = \tfrac{1}{3}\pi\eta(Il/\lambda)^2 \qquad (2.208)$$

which, in free space, becomes equal to

$$40\pi^2(Il/\lambda_0)^2 = \frac{80\pi^2 l^2 I^2}{2\lambda_0^2} \qquad (2.209)$$

It should be noted here that, while integrating, only the $1/r$ terms in E_θ and H_ϕ contribute to this power. The other terms, namely, the $1/r^2$ and $1/r^3$ terms, contribute to the reactive power which is exchanged back and forth between the source and the field.

The radiated power can also be calculated from the work done by the electromotive force impressed on the current element. The electric intensity on the axis of the element in a nondissipative medium is given by

$$E_z = E_r = \frac{\eta Il}{2\pi}\left(\frac{1}{j\beta r^3} - \frac{j\beta}{r} - \frac{1}{3}\beta^2 - \cdots\right) \qquad (2.210)$$

The first two terms in Eq. (2.210) are in quadrature with I and on the average do not do any work; but the third term is 180° out of phase with I, and work is done against the field by the impressed electromotive force. Then, the in-phase component of this force is

$$Re(V^i) = -lRe(E_z) = \frac{\eta}{6\pi}\beta^2 l^2 I = \frac{2\pi\eta l^2 I}{3\lambda^2} \qquad (2.211)$$

The work done by this force per second is $2\pi\eta l^2 I^2/(2 \times 3\lambda^2)$ which is equal to the total power radiated, W.

The ratio,

$$R = \frac{Re(V^i)}{I} = \frac{2\pi\eta l^2}{3\lambda^2} = 80\pi^2(l/\lambda_0)^2 \qquad (2.212)$$

for free space is called the radiation resistance of the current element.

2.16   Electromagnetic Field Produced by Given Distribution of Applied Electric and Magnetic Currents

Any given distribution of applied electric or magnetic currents may be subdivided into elements, and the resulting field can be obtained by the superposition of the fields of individual elements. If an infinitesimal volume bounded by the lines of flow and two surfaces normal to them are considered, then the current I in this element is $\mathbf{J}^i dS$, where dS is the cross-section of the tube of current flow. If dl is the length of this tube, then the moment Idl of the current

Fig. 2.9   Infinitesimal current element.

element is $\mathbf{J}^i \cdot dS \cdot dl = \mathbf{J}^i \cdot dV$, where dV is the volume of the element (see Fig. 2.9).

The magnetic vector potential due to the electric current distribution is

$$\mathbf{A} = \int_v \frac{\mathbf{J}^i e^{-\gamma r}}{4\pi r}\, dV \tag{2.213}$$

where r is the distance between a typical current element and a typical point in space, and is taken over the volume over which the current is distributed. Similarly, for a magnetic current distribution, the electric vector potential is

$$\mathbf{F} = \int_v \frac{\mathbf{M}^i e^{-\gamma r}}{4\pi r}\, dV \tag{2.214}$$

The scalar potentials V and U may then be evaluated by using Eqs. 2.168 and 2.169, and hence $\mathbf{E}$ and $\mathbf{H}$ can be calculated from Eqs. 2.172 and 2.173. If, however, $\mathbf{J}^i$ and $\mathbf{M}^i$ are differentiable, then V and U can be computed from the equations

$$V = -\int_v \frac{\operatorname{div}\mathbf{J}^i dV}{4\pi(\sigma + j\omega\epsilon)r}, \quad U = \int_v \frac{\operatorname{div}\mathbf{M}^i dV}{4\pi j\omega\mu r} \tag{2.215}$$

In nondissipative media

$$V = \int_v \frac{\rho e^{-j\beta r}}{4\pi\epsilon_r}\, dV, \quad U = \int_v \frac{m e^{-j\beta r}}{4\pi\epsilon_r}\, dV \tag{2.216}$$

2.17   Electromagnetic Field Produced by Impressed Currents Varying Arbitrarily with Time

In nondissipative media, the magnetic vector potential of a given electric current distribution varying harmonically with time (that is, varying as $e^{j\omega t}$) is

$$\mathbf{A} = \int_v \frac{\mathbf{J}^i e^{-j\beta r} e^{j\omega t}}{4\pi r}\, dV \tag{2.217}$$

The phase of $\mathbf{J}^i$ is ωt, and the phase of the corresponding component of the vector potential is $\omega t - \beta r = \omega(t - r/v)$, where $v = \omega/\beta =$ the characteristic velocity of the medium. The time delay r/v is independent of the frequency; hence all frequency components of a general function $\mathbf{J}^i(x, y, z; t)$ are shifted equally on the time scale, and $\mathbf{A}$ will depend on $\mathbf{J}^i(x, y, z; t - r/v)$. Thus we have

$$A(x, y, z; t) = \int_v \frac{\mathbf{J}^i(x, y, z; t - r/v)}{4\pi r} \, dV \qquad (2.218)$$

In a dissipative medium, no simple formula such as the one given by Eq. 2.218 is possible.

2.18 Field of Electric Current Element whose Current Varies Arbitrarily with Time

Consider an electric current element of length l along the z-axis at the origin, and suppose that current $I(t)$ is given by

$$
\begin{aligned}
I(t) &= 0 \qquad \text{for } t \leqslant 0 \\
&= I(t) \quad \text{for } t > 0
\end{aligned}
\qquad (2.219)
$$

Let dI/dt be finite. The charge $q(t)$ at the upper end is zero when $t \leqslant 0$, and is given by

$$q(t) = \int_0^t I(t) \, dt \quad \text{for } t > 0. \qquad (2.220)$$

At the lower end the charge is $-q(t)$.

The field of the current element may now be evaluated. Both $\mathbf{E}$ and $\mathbf{H}$ consist of three parts: $(\mathbf{E}', \mathbf{H}')$ depending only on the time derivative of the current, $(\mathbf{E}'', \mathbf{H}'')$ depending on the current alone, and $(\mathbf{E}''', \mathbf{H}''')$ depending on the charge. Thus

$$E = E' + E'' + E''' \qquad (2.221)$$

$$H = H' + H'' + H''' \qquad (2.222)$$

$$E'_\theta = \eta H'_\phi, \; H_\phi = \frac{lI'(t - r/v)}{4\pi vr} \sin \theta \qquad (2.223)$$

$$E'_r = 0, \; E''_r = 2E''_\theta \cot \theta \qquad (2.224)$$

$$E''_\theta = \eta H''_\phi, \; H''_\phi = \frac{lI(t - r/v)}{4\pi r^2} \sin \theta \qquad (2.225)$$

$$E'''_0 = \frac{lq(t - r/v)}{4\pi \epsilon r^3} \sin \theta \qquad (2.226)$$

$$E'''_r = \frac{lq(t - r/v)}{2\pi \epsilon r^3} \cos \theta \qquad (2.227)$$

At sufficiently great distances, only the radiation field $(\mathbf{E}', \mathbf{H}')$ is significant. The entire field is zero outside the spherical surface of radius vt with its centre at the element. This sphere is the wavefront of the wave emitted by the element, and on it both $(\mathbf{E}'', \mathbf{H}'')$ and $(\mathbf{E}''', \mathbf{H}''')$ vanish. At the wavefront, $\mathbf{E}$ and $\mathbf{H}$ are perpendicular to the radius or are tangential to the wavefront.

2.19 Reflection of Electromagnetic Field at Boundary Surface

Let an infinite homogeneous medium be divided into two homogeneous parts by a surface S, so that one of these regions, say, region 2, is source-free (see Fig. 2.10). If the sources are distributed in both the regions, then

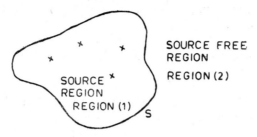

Fig. 2.10 Surface enclosing sources of field.

the total field may be regarded as the superposition of the two fields, each produced by sources located in one region only. If the two regions have the same electromagnetic properties, the field of these sources can be found by using the results given in Section 2.13. But if the electromagnetic properties of the two regions are different, the field (E^i, H^i) thus obtained is not the actual field. In region 1 which has the sources, this field represents the primary field of the sources and is called the impressed field. To obtain the actual field in region 1, a field (E^r, H^r) has to be added to (E^i, H^i). The field (E^r, H^r) is then called the reflected field because it is reflected from the boundary surface S. This field has no sources in region 1, and hence it satisfies the homogeneous form of Maxwell's equations

$$\text{curl } E^r = -j\omega\mu_1 H^r, \tag{2.228}$$

$$\text{curl } H^r = (\sigma_1 + j\omega\epsilon_1)E^r, \tag{2.229}$$

where (μ_1, ϵ_1, σ_1) are the primary electromagnetic constants of region 1.

Let the actual field in region 2 be (E^t, H^t). This field is called the transmitted (or refracted) field. It too should satisfy the homogeneous Maxwell's equations

$$\text{curl } E^t = -j\omega\mu_2 H^t, \tag{2.230}$$

$$\text{curl } H^t = (\sigma_2 + j\omega\epsilon_2)E^t, \tag{2.231}$$

where (μ_2, σ_2, ϵ_2) are the primary electromagnetic constants of region 2.

At the boundary surface S of the two media, the tangential components of E and H are continuous. Thus

$$E_t^i + E_t^r = E_t^t, \tag{2.232}$$

$$H_t^i + H_t^r = H_t^t. \tag{2.233}$$

The set of Eqs. 2.232 and 2.233 constitutes one formulation of the problem for determining the field of a given system of sources when the medium consists of two homogeneous regions. The method can be extended to any number of homogeneous regions.

If the boundary S is a perfectly conducting sheet, then the tangential component of E should vanish on S, or

$$E_t^i + E_t^r = 0$$

or

$$E_t^r = -E_t^i.$$

(2.234)

Such a sheet can support a finite electric current, and the tangential component of **H** is no longer continuous across S. In fact, the field in region 2 due to sources in region 1 is equal to zero. The component of **H** tangential to the surface S in region 1 represents the surface current density $\mathbf{J}_s$ on S. By the second law of electromagnetic induction, $\mathbf{J}_s$ is normal to the tangential component of **H**. Therefore, if **n** is a unit normal to S, regarded as positive as it points into the source-free region 2, then

$$\mathbf{J}_s = (\mathbf{H}_t^i + \mathbf{H}_t^r) \times \mathbf{n}.$$

(2.235)

Since the vector product of **n** and the normal component of **H** is zero, the subscript 't' for tangential component may be dropped, and we may write

$$\mathbf{J}_s = (\mathbf{H}^i + \mathbf{H}^r) \times \mathbf{n}$$

(2.236)

If the impedance Z_n normal to the boundary S is prescribed, then the relation between the tangential components in region 1 is,

$$\mathbf{E}_t^i + \mathbf{E}_t^r = Z_n(\mathbf{H}_t^i + \mathbf{H}_t^r) \times \mathbf{n}$$

(2.237)

The fields in waveguides and resonators are examples of reflection, and the total field is the sum of the impressed and reflected fields.

2.20   Induction Theorem

If we formulate the set of Eqs. 2.232 and 2.233 as

$$E_t^t - E_t^r = E_t^i,$$

(2.238)

$$H_t^t - H_t^r = H_t^i,$$

(2.239)

we can define the induced field as the field that comprises the reflected field (E^r, H^r) in region 1 and the transmitted field (E^t, H^t) in region 2. The induced field satisfies the homogeneous Maxwell's equations everywhere except on the surface S. It can be shown that this field may be obtained from a distribution of sources on S as well as from the original sources. Therefore

$$\mathbf{M}_s = (\mathbf{E}_t^t - \mathbf{E}_t^r) \times \mathbf{n} = \mathbf{E}_t^i \times \mathbf{n}$$

(2.240)

$$\mathbf{J}_s = \mathbf{n} \times (\mathbf{H}_t^t - \mathbf{H}_t^r) = \mathbf{n} \times \mathbf{H}_t^i$$

(2.241)

Since the vector product of **n** and the normal component of the field is zero, we may write

$$\mathbf{M}_s = \mathbf{E}^i \times \mathbf{n} \tag{2.242}$$

$$\mathbf{J}_s = \mathbf{n} \times \mathbf{H}^i \tag{2.243}$$

These two equations show that the induced field, which is a combination of the reflected fields ($\mathbf{E}^r$, $\mathbf{H}^r$) and the transmitted field ($\mathbf{E}^t$, $\mathbf{H}^t$) could be produced by the electric and magnetic current sheets $\mathbf{J}_s$ and $\mathbf{M}_s$ on the surface S given by Eqs. 2.240 and 2.241.

2.21   Equivalence Theorem

If regions 1 and 2 separated by the surface S have the same electromagnetic constants, then the entire region is homogeneous. Hence, the reflected field is zero, and the transmitted field is the actual field in the homogeneous region due to the sources in region 1. But this transmitted field can also be calculated from a suitable distribution of electric and magnetic current sheets over the surface S. The required surface current densities of these sheets are

$$\mathbf{J}_s = \mathbf{n} \times \mathbf{H}^t = \mathbf{n} \times \mathbf{H}^i \tag{2.244}$$

$$\mathbf{M}_s = \mathbf{E}^t \times \mathbf{n} = \mathbf{E}^i \times \mathbf{n} \tag{2.245}$$

Thus, in order to compute the electromagnetic field in a source-free region 1 bounded by a surface S, the source distribution outside S (in region 2) can be replaced by a distribution of surface electric and surface magnetic currents $\mathbf{J}_s$ and $\mathbf{M}_s$ over the surface S given by Eqs. (2.244) and (2.245). This is called the equivalence theorem. It is also known as Love's equivalence principle or Schelkunoff's equivalence principle.

The equivalence theorem is a very powerful tool in solving many electromagnetic problems involving radiation from aperture-type antennas.

2.22   Hertz Potentials

Hertz[1] has shown that, under ordinary conditions, an electromagnetic field can be derived by any one of the two vector functions, π_h and π_e, which are called Hertz vector potentials.

In a homogeneous isotropic source-free region, $\nabla \cdot \mathbf{E} = 0$. $\mathbf{E}$ may be derived from the curl of an auxiliary vector potential function which we shall call a magnetic type of Hertz vector potential. (The reason for using the adjective 'magnetic' will be explained in Section 2.23.) Let

$$\mathbf{E} = -j\omega\mu\nabla \times \pi_h \tag{2.246}$$

where π_h is the magnetic Hertz vector potential. Substituting Eq. 2.246 in Maxwell's equations, we obtain

$$\nabla \times \mathbf{H} = j\omega\epsilon\mathbf{E} = +\omega^2\mu\epsilon\nabla \times \pi_h \tag{2.247}$$

---

[1] H. R. Hertz, *Ann. Physik*, 36, 1, 1888.

so that

$$H = k^2\pi_h + \nabla\phi \qquad (2.248)$$

where $\omega^2\mu\epsilon = k^2$ and ϕ is an arbitrary scalar function. Substituting H in the other Maxwell's equations, we have

$$\nabla\times E = -j\omega\mu\nabla\times\nabla\times\pi_h = -j\omega\mu H = -j\omega\mu(k^2\pi_h + \nabla\phi) \quad (2.249)$$

or

$$\nabla\times\nabla\times\pi_h = \nabla\nabla\cdot\pi_h - \nabla^2\pi_h = k^2\pi_h + \nabla\phi \qquad (2.250)$$

Since both ϕ and $\nabla\cdot\pi_h$ are still arbitrary, we may make them satisfy a Lorentz type of condition, i.e., put $\nabla\cdot\pi_h = \phi$, so that the equation for π_h becomes

$$\nabla^2\pi_h + k^2\pi_h = 0. \qquad (2.251)$$

Since $\nabla\cdot H = 0$, from Eq. (2.248),

$$\text{div }(H) = k^2 \text{ div } \pi_h + \nabla\cdot\nabla\phi = 0$$

or

$$\nabla^2\phi + k^2\phi = 0 \qquad (2.252)$$

The fields E and H are now given by

$$E = -j\omega\nabla\times\pi_h \qquad (2.253)$$

$$H = k^2\pi_h + \nabla\nabla\cdot\pi_h = \nabla\times\nabla\times\pi_h \qquad (2.254)$$

where π_h is a solution of the Helmholtz Eq. 2.251. In practice, we need not determine the scalar function ϕ explicitly since it is equal to $\nabla\cdot\pi_h$ and has been eliminated from the equation determining the magnetic field H.

We now introduce the electric Hertz vector potential π_e in the same way as we introduced the magnetic Hertz vector potential π_h. Let

$$H = j\omega\epsilon\nabla\times\pi_e \qquad (2.255)$$

Thus, it can be shown that π_e is a solution of

$$\nabla^2\pi_e + k^2\pi_e = 0 \qquad (2.256)$$

and that the electric field E is given by

$$E = k^2\pi_e + \nabla\nabla\cdot\pi_e = \nabla\times\nabla\times\pi_e \qquad (2.257)$$

(in a lossy dielectric medium, ϵ is replaced by $\epsilon - (j\sigma/\omega)$.)

2.23 Electric and Magnetic Polarization Sources for Hertz Vector Potentials

When an electric field is applied to a dielectric material, the electron orbits of the various atoms and molecules involved are perturbed, resulting in a dipole polarization P per unit volume. Some materials, called electrets, have a permanent dipole polarization. But since these elementary dipoles are randomly oriented, there is no net resultant field as there is no applied external field. The application of an external field therefore tends to align the dipoles with the field, resulting in a decrease in the electric field intensity in the material. Thus, the relation between the electric displacement density $\mathbf{D}$, the electric field intensity $\mathbf{E}$, and the electric polarization vector $\mathbf{P}$, in the interior of a dielectric material, is

$$\mathbf{D} = \epsilon_0 \mathbf{E} + \mathbf{P} \qquad (2.258)$$

Here, $\mathbf{E}$ is the net field intensity in the dielectric, i.e., the vector sum of the applied field and the field arising from the dipole polarization. For many materials, $\mathbf{P}$ is collinear with and proportional to the applied field and hence proportional to $\mathbf{E}$. Therefore

$$\mathbf{P} = \chi_e \epsilon_0 \mathbf{E} \qquad (2.259)$$

where χ_e is called the electric susceptibility of the material, and is a dimensionless quantity. Consequently,

$$\mathbf{D} = \epsilon_0 \mathbf{E} + \mathbf{P} = \epsilon_0 (1 + \chi_e) \mathbf{E} = \epsilon \mathbf{E} \qquad (2.260)$$

so that

$$\epsilon = (1 + \chi_e)\epsilon_0 \qquad (2.261)$$

The relative permittivity or dielectric constant is

$$\epsilon_r = \frac{\epsilon}{\epsilon_0} = 1 + \chi_e \qquad (2.262)$$

Equation 2.261 is true only for materials having a high degree of symmetry in their crystal structure. Such materials are called isotropic materials.

In general, $\bar{\chi}_e$ is a dyadic or tensor quantity, and hence

$$\bar{\epsilon} = (\bar{I} + \bar{\chi}_e)\epsilon_0 \qquad (2.263)$$

where $\bar{I}$ is the unit dyadic. This is somewhat analogous to the situation for materials having a permeability different from μ_0, $\mathbf{H}$ being defined by the relation

$$\mathbf{H} = \frac{\mathbf{B}}{\mu_0} - \mathbf{M}, \qquad (2.264)$$

where $\mathbf{M}$ is the magnetic dipole polarization per unit volume. $\mathbf{M}$ is related to $\mathbf{H}$ as

$$\mathbf{M} = \chi_m \mathbf{H}, \qquad (2.265)$$

where χ_m is the magnetic susceptibility of the medium. Equation 2.265 leads to

$$\mu = (1 + \chi_m)\mu_0 \qquad (2.266)$$

As $\mathbf{P}$ is not always in the direction of $\mathbf{E}$ in the dielectric case, $\mathbf{M}$ is not always in the direction of $\mathbf{B}$ for materials having a permeability different from μ_0, and hence χ_m and μ are, in general, dyadic quantities. Materials whose μ or ϵ are dyadics are called anisotropic materials.

If μ and ϵ are not functions of position, materials are said to be homogeneous; otherwise, they are called non-homogeneous.

In materials with μ and ϵ as scalar functions of position, Maxwell's equations become

$$\nabla \times \mathbf{E} = -j\omega\mu_0\mathbf{H} - j\omega\mu_0 X_m\mathbf{H} \qquad (2.267)$$

$$\nabla \times \mathbf{H} = j\omega\epsilon_0\mathbf{E} + j\omega\epsilon_0 X_e\mathbf{E} + \mathbf{J}^i \qquad (2.268)$$

The term $j\omega\epsilon_0 X_e\mathbf{E}$ may be regarded as an equivalent electric polarization current $\mathbf{J}_e$, and $j\omega\mu_0 X_m\mathbf{H}$ may be considered as an equivalent magnetic polarization current $\mathbf{J}_m$.

Introducing the polarization vectors $\mathbf{P}$ and $\mathbf{M}$ in the divergence equations, we obtain

$$\nabla \cdot \mathbf{D} = \epsilon_0\nabla \cdot \mathbf{E} + \nabla \cdot \mathbf{P} = \rho \qquad (2.269)$$

$$\nabla \cdot \mathbf{B} = \mu_0\nabla \cdot \mathbf{H} + \mu_0\nabla \cdot \mathbf{M} = 0 \qquad (2.270)$$

Here, $\nabla \cdot \mathbf{P}$ and $\nabla \cdot \mathbf{M}$ may be interpreted as defining an electric polarization charge density $-\rho_e$ and a magnetic polarization charge density $-\rho_m$, respectively; or

$$-\nabla \cdot \mathbf{P} = \rho_e - \epsilon_0\nabla \cdot (X_e\mathbf{E}) = -\epsilon_0 X_e\nabla \cdot \mathbf{E} - \epsilon_0\mathbf{E} \cdot \nabla X_e \qquad (2.271)$$

$$-\nabla \cdot \mathbf{M} = \rho_m = -\nabla \cdot X_m\mathbf{H} = -X_m\nabla \cdot \mathbf{H} - \mathbf{H} \cdot \nabla X_m \qquad (2.272)$$

Now, we shall show that the source functions for the Hertz vector potentials $\boldsymbol{\pi}_h$ and $\boldsymbol{\pi}_e$ are the magnetic and electric polarizations $\mathbf{M}$ and $\mathbf{P}$, respectively. If we put

$$\mathbf{E} = -j\omega\mu_0\nabla \times \boldsymbol{\pi}_h \qquad (2.273)$$

then

$$\mathbf{H} = k_0^2\boldsymbol{\pi}_h + \nabla\phi \qquad (2.274)$$

where $k_0^2 = \omega^2\mu_0\epsilon_0$. Hence, from

$$\nabla \times \mathbf{E} = -j\omega\mu_0\mathbf{H} = -j\omega\mu_0\nabla \times \nabla \times \boldsymbol{\pi}_h$$

it follows that

$$\nabla \times \nabla \times \boldsymbol{\pi}_h = \frac{\mu_0\mathbf{H}}{\mu_0} = \frac{\mathbf{B}}{\mu_0} = \mathbf{H} + \mathbf{M}$$

or

$$\nabla\nabla \cdot \boldsymbol{\pi}_h - \nabla^2\boldsymbol{\pi}_h = k_0^2\boldsymbol{\pi}_h + \nabla\phi + \mathbf{M} \qquad (2.275)$$

since $B = \mu_0(\mathbf{H} + \mathbf{M})$. Putting

$$\phi = \nabla \cdot \boldsymbol{\pi}_h, \qquad (2.276)$$

we obtain

$$\nabla^2\boldsymbol{\pi}_h + k_0^2\boldsymbol{\pi}_h = -\mathbf{M}. \qquad (2.277)$$

Similarly, if we put

$$\mathbf{H} = j\omega\epsilon_0\nabla \times \boldsymbol{\pi}_e, \qquad (2.278)$$

then

$$\nabla^2\boldsymbol{\pi}_e + k_0^2\boldsymbol{\pi}_e = -\frac{\mathbf{P}}{\epsilon_0}, \qquad (2.279)$$

$$\mathbf{E} = k_0^2\boldsymbol{\pi}_e + \nabla\nabla \cdot \boldsymbol{\pi}_e = \nabla \times \nabla \times \boldsymbol{\pi}_e - \frac{\mathbf{P}}{\epsilon_0}. \qquad (2.280)$$

Usually, it is more convenient to absorb **M** into the parameter μ and **P** into the parameter ϵ. Then,

$$\nabla^2 \pi_h + k^2 \pi_h = 0, \; \nabla^2 \pi_e + k^2 \pi_e = 0,$$

where $k^2 = \omega^2 \mu \epsilon$.

2.24 Babinet's Principle

Maxwell's equations in a source-free region are

$$\nabla \times \mathbf{E} = -j\omega\mu\mathbf{H}, \tag{2.281}$$

$$\nabla \times \mathbf{H} = j\omega\epsilon\mathbf{E}. \tag{2.282}$$

They exhibit a high degree of symmetry in the expressions for **E** and **H**. If a field $(\mathbf{E}_1, \mathbf{H}_1)$ satisfies these equations, then a second field $(\mathbf{E}_2, \mathbf{H}_2)$ obtained from it by the transformations

$$\mathbf{E}_2 = \mp(\mu/\epsilon)^{1/2}\mathbf{H}_1, \tag{2.283}$$

$$\mathbf{H}_2 = \pm(\epsilon/\mu)^{1/2}\mathbf{E}_1 \tag{2.284}$$

is also a solution of the source-free field equations.

Equations 2.281–2.284 are a statement of the duality property of the electromagnetic field and constitute the basis of Babinet's principle. We must consider also the effect of the transformations 2.283 and 2.284 on the boundary conditions to be satisfied by the new field. Clearly, since the roles of **E** and **H** are interchanged, we must interchange all electric walls $E_t = 0$ for magnetic walls $H_t = 0$ and magnetic walls for electric walls. At a boundary where the conditions on $(\mathbf{E}_1, \mathbf{H}_1)$ specify that the tangential components be continuous across a surface with ϵ and μ equal to ϵ_1, μ_1 on one side and ϵ_2, μ_2 on the other side, the new boundary conditions on $(\mathbf{E}_2, \mathbf{H}_2)$ are still the continuity of the tangential components, and this changes the relative amplitudes of the fields on the two sides of the discontinuity surface.

2.25 General Solutions of the Electromagnetic Equations in Terms of the Sources: The Vector Kirchhoff Formula

In Section 2.13, the field produced by a given distribution of currents in an infinite homogeneous medium has been derived using the concepts of vector and scalar potentials. Stratton and Chu[1] and Silver[2] have expressed the field resulting from given sources in a manner that differs from the vector Kirchhoff formula.

[1]J. A. Stratton and L. J. Chu, 'Diffraction Theory of Electromagnetic Waves', *Phys. Rev.*, vol. 56, p. 99, 1939.

[2]S. Silver, *Microwave Antenna Theory and Design*, McGraw-Hill Book Co., New York, 1949, Chap. 3.

Maxwell's equations for time-periodic fields can be expressed as:

$$\nabla \times \mathbf{E} + j\omega\mu\mathbf{H} = -\mathbf{M}^i, \tag{2.285}$$

$$\nabla \times \mathbf{H} - j\omega\epsilon\mathbf{E} = \mathbf{J}^i, \tag{2.286}$$

$$\nabla \cdot \mathbf{H} = \frac{\rho_m}{\mu}, \tag{2.287}$$

$$\nabla \cdot \mathbf{E} = \frac{\rho}{\epsilon}, \tag{2.288}$$

$$\nabla \cdot \mathbf{J}^i + j\omega\rho = 0, \tag{2.289}$$

$$\nabla \cdot \mathbf{M}^i + j\omega\rho_m = 0, \tag{2.290}$$

where ρ and ρ_m are the electric and magnetic charge densities respectively. $\mathbf{E}$ and $\mathbf{H}$ then satisfy the wave equations

$$\nabla \times \nabla \times \mathbf{E} - k^2\mathbf{E} = -j\omega\mu\mathbf{J}^i - \nabla \times \mathbf{M}^i, \tag{2.291}$$

$$\nabla \times \nabla \times \mathbf{H} - k^2\mathbf{H} = -j\omega\epsilon\mathbf{M}^i + \nabla \times \mathbf{J}^i. \tag{2.292}$$

Let V be a region shown in Fig. 2.11 bounded by the surfaces S_1, $S_2,\ldots$, S_n. If $\mathbf{F}$ and $\mathbf{G}$ are two vector functions of position in this region, both of them being continuous and having continuous first and second derivatives everywhere within V and on the boundary surfaces. If $\mathbf{n}$ is the unit vector normal to a boundary surface directed into the region V, the vector Green's Theorem states that

$$\int_V (\mathbf{F} \cdot \nabla \times \nabla \times \mathbf{G} - \mathbf{G} \cdot \nabla \times \nabla \times \mathbf{F})\, dv$$

$$= -\int_{S_1+S_2+\ldots S_n} (\mathbf{G} \times \nabla \times \mathbf{F} - \mathbf{F} \times \nabla \times \mathbf{G}) \cdot \mathbf{n}\, dS. \tag{2.293}$$

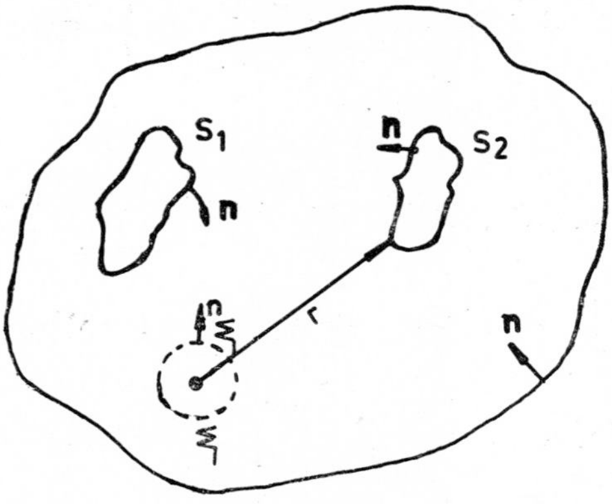

Fig. 2.11   Notation for vector Green's Theorem.

Let there be an electromagnetic field $(\mathbf{E}, \mathbf{H})$ in region V which satisfy the continuity conditions of Green's Theorem.

A vector function **G** of position is defined as

$$\mathbf{G} = \frac{\exp(-jkr)}{r}\,\mathbf{a} = \psi\mathbf{a}, \tag{2.294}$$

where r is the distance from point P to any other point in the region, and **a** is an arbitrary but constant vector. Then **G** also satisfies the continuity conditions of Green's Theorem. The point P is surrounded by a sphere Σ of radius r_0. Considering the portion V' of V which is bounded by the surfaces $S_1, S_2, \ldots, S_n$ and Σ, and putting $\mathbf{F} = \mathbf{E}$ and using **G** as defined by (2.294), Green's Theorem is applied to **E** and **G**. We then have

$$\int_{V'} (\psi\mathbf{a}\cdot\nabla\times\nabla\times\mathbf{E} - \mathbf{E}\cdot\nabla\times\nabla\times\psi\mathbf{a})\,dv$$

$$= -\int_{S_1+S_2+\ldots+S_n+\Sigma} (\mathbf{E}\times\nabla\times\psi\mathbf{a} - \psi\mathbf{a}\times\nabla\times\mathbf{E})\cdot\mathbf{n}\,dS. \tag{2.295}$$

Using the vector identity given by

$$\nabla\times\nabla\times\mathbf{P} = \nabla(\nabla\cdot\mathbf{P}) - \nabla^2 P \tag{2.296}$$

we have

$$\nabla\times\nabla\times\psi\mathbf{a} = \nabla(\mathbf{a}\cdot\nabla\psi) + k^2\psi\mathbf{a} \tag{2.297}$$

Using Eq. 2.297 with Eq. 2.291, we obtain

$$\psi\mathbf{a}\cdot\nabla\times\nabla\times\mathbf{E} - \mathbf{E}\cdot\nabla\times\nabla\times\psi\mathbf{a}$$

$$= \mathbf{a}\cdot(-j\omega\mu\mathbf{J}^i\psi - \psi\nabla\times\mathbf{M}^i) - \mathbf{E}\cdot\nabla(\mathbf{a}\cdot\nabla\psi) \tag{2.298}$$

Using the transformations,

$$\mathbf{E}\cdot\nabla(\mathbf{a}\cdot\psi) = \nabla\cdot[\mathbf{E}(\mathbf{a}\cdot\nabla\psi)] - (\mathbf{a}\cdot\nabla\psi)\nabla\cdot\mathbf{E}$$

$$= \nabla\cdot[\mathbf{E}(\mathbf{a}\cdot\nabla\psi)] - \frac{\rho}{\epsilon}\,\mathbf{a}\cdot\nabla\psi \tag{2.299}$$

and

$$\psi\nabla\times\mathbf{M}^i = \nabla\times\mathbf{M}^i\psi + \mathbf{M}^i\times\nabla\psi \tag{2.300}$$

Equation 2.295 becomes

$$\mathbf{a}\cdot\int_{V'} \left(j\omega\mu\mathbf{J}^i\psi + \mathbf{M}^i\times\nabla\psi - \frac{\rho}{\epsilon}\nabla\psi\right) + \mathbf{a}\cdot\int_{V'} \nabla\times\psi\mathbf{M}^i\,dv$$

$$+ \int_{V'} \nabla\cdot[\mathbf{E}(\mathbf{a}\cdot\nabla\psi)]\,dv$$

$$= \int_{S_1+S_2+\ldots S_n+\Sigma} [(\mathbf{E}\times\nabla\times\nabla\psi\mathbf{a})\cdot\mathbf{n} - (\psi\mathbf{a}\times\nabla\times\mathbf{E})\cdot\mathbf{n}\,dS \tag{2.301}$$

Putting

$$\mathbf{a}\cdot\int_{V'} \nabla\times\psi\mathbf{M}^i\,dv = -\mathbf{a}\int_{S_1+\ldots+\Sigma} \mathbf{n}\times\mathbf{M}^i\,dS \tag{2.302}$$

and

$$\int_{V'} \nabla\cdot[\mathbf{E}(\mathbf{a}\cdot\nabla\psi)]\,dv = -\int_{S_1+\ldots+\Sigma} (\mathbf{n}\cdot\mathbf{E})(\mathbf{a}\cdot\nabla\psi)\,dS$$

$$= -\mathbf{a}\cdot\int_{S_1+\ldots+\Sigma} (\mathbf{n}\cdot\mathbf{E})\nabla\psi\,dS \tag{2.303}$$

Now using the transformations

$$[\mathbf{E}\times\nabla\times\psi\mathbf{a}]\cdot\mathbf{n} = [\mathbf{E}\times(\nabla\psi\times\mathbf{a})]\cdot\mathbf{n} = [(\mathbf{n}\times\mathbf{E})\times\nabla\psi]\cdot\mathbf{a} \tag{2.304}$$

$$\psi(\mathbf{a} \times \nabla \times \mathbf{E}) \cdot \mathbf{n} = -j\omega\mu(\mathbf{a} \times \mathbf{H}) \cdot \mathbf{n} - \psi(\mathbf{a} \times \mathbf{M}^i) \cdot \mathbf{n}$$

$$= j\omega\mu\mathbf{a} \cdot (\mathbf{n} \times \mathbf{H}) + \psi\mathbf{a} \cdot (\mathbf{n} \times \mathbf{M}^i) \qquad (2.305)$$

we obtain

$$\mathbf{a} \cdot \int_{V'} \left(j\omega\mu\mathbf{J}^i + \mathbf{M}^i \times \nabla\psi - \frac{\rho}{\epsilon}\nabla\psi \right) dv = \mathbf{a} \cdot \int_{S_1+\ldots+\Sigma} [-j\omega\mu\psi(\mathbf{n} \times \mathbf{H})$$
$$+ (\mathbf{n} \times \mathbf{E}) \times \nabla\psi + (\mathbf{n} \cdot \mathbf{E})\nabla\psi]\, dS \qquad (2.306)$$

Since Eq. 2.306 must hold for every vector $\mathbf{a}$, the factor $\mathbf{a}$ can be removed from both sides of Eq. 2.306. Then,

$$\int_{\Sigma} [-j\omega\mu\psi(\mathbf{n} \times \mathbf{H}) + (\mathbf{n} \times \mathbf{E}) \times \nabla\psi + (\mathbf{n} \cdot \mathbf{E})\nabla\psi]\, dS$$

$$= \int_{V'} \left(j\omega\mu\psi\mathbf{J}^i + \mathbf{M}^i \times \nabla\psi - \frac{\rho}{\epsilon}\nabla\psi \right) dv$$

$$- \int_{S_1+S_2+\ldots+S_n} [-j\omega\mu\psi)(\mathbf{n} \times \mathbf{H}) + (\mathbf{n} \times \mathbf{E}) \times \nabla\psi + (\mathbf{n} \cdot \mathbf{E})\nabla\psi]\, dS$$
$$(2.307)$$

In the limit as $\Sigma \to 0$ and shrinks to the point P, the integral depends only on the field at P.

Consider the integral over Σ. On the surface of this sphere, we have

$$(\nabla\psi) = \left[\frac{d}{dr} \left(\frac{\exp(-jkr)}{r} \right) \right] \mathbf{n} = -\left(jk + \frac{1}{r_0} \right) \frac{\exp(-jkr_0)}{r_0} \mathbf{n} \quad (2.308)$$

The normal $\mathbf{n}$ is directed along the radius out from P. If $d\Omega$ is the solid angle subtended at P by an element of surface dS on Σ, then the surface integral on Σ can be written as

$$\int_{\Sigma} [\ ]\, dS = -jr_0 \exp(-jkr_0) \int_{\Sigma} \{\omega\mu(\mathbf{n} \times \mathbf{H}) + k(\mathbf{n} \times \mathbf{E}) \times \mathbf{n} + (\mathbf{n} \cdot \mathbf{E})\mathbf{n}\}\, d\Omega$$

$$- \exp(-jkr_0) \int_{\Sigma} [(\mathbf{n} \times \mathbf{E}) \times \mathbf{n} + (\mathbf{n} \cdot \mathbf{E})\mathbf{n}]\, d\Omega$$

$$= -4\pi jr_0 \exp(-jkr_0)(\omega\mu\overline{\mathbf{n} \times \mathbf{H}} + k\mathbf{E}) - 4\pi \exp(-jkr_0)\, \mathbf{E}$$
$$(2.309)$$

where the overline denotes the mean value of the function over the surface of the sphere. If now the sphere Σ shrinks to zero, the term containing r_0 vanishes because the field vectors are finite in the neighbourhood of P. Also $\mathbf{E}$ approaches E_P, the value of $\mathbf{E}$ and P. Therefore,

$$\lim_{r_0 \to 0} \int_{\Sigma} [\ ]\, dS = -4\pi E_P \qquad (2.310)$$

Then V' becomes V, and Eq. 2.307 becomes

$$\mathbf{E}_P = -\frac{1}{4\pi} \int_V \left(j\omega\mu\mathbf{J}^i + \mathbf{M}^i \times \nabla\psi - \frac{\rho}{\epsilon}\nabla\psi \right) dv$$

$$+ \frac{1}{4\pi} \int_{S_1+S_2+\ldots+S_n} [-j\omega\mu\psi(\mathbf{n} \times \mathbf{H}) + (\mathbf{n} \times \mathbf{E}) \times \nabla\psi + (\mathbf{n} \cdot \mathbf{E})\nabla\psi]\, dS$$
$$(2.311)$$

Similarly,

$$\mathbf{H}_P = -\frac{1}{4\pi} \int_V \left[(j\omega\epsilon\mathbf{M}^i - \mathbf{J}^i \times \nabla\psi) - \frac{\rho_m}{\mu} \nabla\psi \right] dv$$

$$+ \frac{1}{4\pi} \int_{S_1 + \ldots + S_n} [j\omega\epsilon(\mathbf{n} \times \mathbf{E})\psi + (\mathbf{n} \times \mathbf{H}) \times \nabla\psi + (\mathbf{n} \cdot \mathbf{H})\nabla\psi] \, dS$$

$$(2.312)$$

Equations 2.311 and 2.312 express the fields at the point P as the sum of contributions from the sources distributed in region V and from fields existing on the boundary surfaces. These surface integrals may be thought of as the contributions to the field from sources lying outside the region V. For example, if S_i enclosed an exterior volume V_i, then the surface integral over S_i represents the effects of sources within V_i $(\mathbf{n} \times \mathbf{H})$, $(\mathbf{n} \times \mathbf{E})$, $(\mathbf{n} \cdot \mathbf{E})$ and $(\mathbf{n} \cdot \mathbf{H})$ on the surfaces $S_1, S_2, \ldots, S_n$ may be replaced by $\mathbf{J}_S^i$, $+\mathbf{M}_S^i$, ρ_S/ϵ and ρ_{mS}/μ, where $\mathbf{J}_S^i$, $\mathbf{M}_S^i$, ρ_S and ρ_m may be considered as the surface electric current density, surface magnetic current density, surface electric charge density and surface magnetic charge density respectively on these surfaces.

In Schelkunoff's Equivalence Principle discussed in Section 2.20, only surface current $\mathbf{J}_S$ and $\mathbf{M}_S$ are considered and these surface currents are produced by the fields. $\mathbf{H}^i$ and $\mathbf{E}^i$ on the surfaces which are in their turn produced by the sources distributed within the volume. Hence, essentially Schelkunoff's Equivalence Principle is equivalent to the vector Kirchhoff integral. In fact the equivalence principle holds good for open or closed surfaces, but the vector Kirchhoff integral is based on Green's theorem which holds good for closed surfaces only.

In applying either of the methods to a practical antenna, certain approximations have to be made in the assumptions used for either theory. The accuracy of the theory depends on the accuracy of the assumptions used.

2.26   The Huyghens-Green Formula for the Electromagnetic Field

Given the values of the electric and magnetic field vectors $\mathbf{E}$ and $\mathbf{H}$ over an equiphase surface, S shown in Fig. 2.12 how can the field vectors at a specified field point be determined? Consider the region bounded by S and the sphere at infinity. Since the sources lie outside this region, in Eqs. 2.311 and 2.312, the volume integrals vanish, and they become

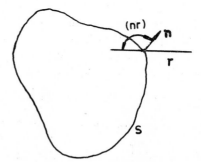

Fig. 2.12   Equiphase surface.

$$\mathbf{E}_P = \frac{1}{4\pi} \int_S [-j\omega\mu(\mathbf{n} \times \mathbf{H})\psi + (\mathbf{n} \times \mathbf{E}) \times \nabla\psi + (\mathbf{n} \cdot \mathbf{E})\nabla\psi] \, dS \quad (2.313)$$

and

$$\mathbf{H}_P = \frac{1}{4\pi} \int_S [j\omega\mu(\mathbf{n}\times\mathbf{E})\psi + (\mathbf{n}\times\mathbf{H})\times\nabla\psi + (\mathbf{n}\cdot\mathbf{H})\nabla\psi]\, dS \qquad (2.314)$$

where $\psi = \dfrac{\exp(-jkr)}{r}$ and $\mathbf{n}$ is the unit normal vector to S. Equations 2.313 and 2.314 may be called the vector form of the Huyghens-Fresnel principle. The scalar Huyghens-Fresnel principle states that each point on a given wavefront can be regarded as a secondary source which gives rise to a spherical wavelet, and the wave at a field point is obtained by the superposition of all these spherical wavelets with due regard to their phase differences. In the vector form, the sources of the wavelets are the surface electric and magnetic currents and charges.

2.27 Geometrical Optics

It is sometimes convenient to approach the subject of wave propagation from the simpler principles of geometrical optics, in which the successive positions of equiphase surfaces, and the associated system of rays are considered.

Let $L(x, y, z) = L_0$ be the wavefront at time t_0 and let $L(x, y, z) = L_0 + \delta L$ be the wavefront at time $t_0 + \delta t$ as shown in Fig. 2.13. Geometrical optics is concerned with the form of the wavefront surface and also with the point-to-point transformation from one wavefront into the succeeding one. The point-to-point transformation of the wavefronts is given by the rays which are a family of curves having at each point the direction of the energy flow in the field. In the electromagnetic field a ray can be traced by proceeding in the direction of the Poynting vector at each point. The rays are nearly normal to the wavefronts.

The field of the wave is characterized by two velocities, namely, the wave velocity and the ray velocity. The wave velocity is the rate of displacement of the wavefront in the direction normal to the wavefront surface, while the ray velocity is the velocity of energy propagation. In an isotropic medium the two velocities are identical, while in an anisotropic medium they may differ in general.

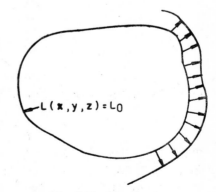

Fig. 2.13   Wavefront

The wavefronts are chosen as the surfaces of constant phase. The phase increment between two successive wavefront surfaces is $(\omega/c)\delta L$, where c is the phase velocity, and is also equal to $\omega\delta t$. If δs_n is the distance between the two surfaces and if v is the wave velocity, then

$$\frac{\omega}{c}\,\delta L = \omega\,\delta t = \frac{\omega \delta s_n}{v} \tag{2.315}$$

If

$$\delta L = |\nabla L|\delta s_n \tag{2.316}$$

then,

$$|\nabla L| = \frac{c}{v} = n \tag{2.317}$$

where n is the refractive index of the medium, which may be a function of position. Then,

$$|\nabla L|^2 = \left(\frac{\partial L}{\partial x}\right)^2 + \left(\frac{\partial L}{\partial y}\right)^2 + \left(\frac{\partial L}{\partial z}\right)^2 = n^2 \tag{2.318}$$

In a homogeneous medium, the rays are straight lines, while in an inhomogeneous medium they are curved. If $\mathbf{s}$ is a unit vector in the direction of the ray at any point, it is normal to the wavefront and has the direction ∇L. Then from Eq. 2.317,

$$\mathbf{s} = \frac{\nabla L}{n} \tag{2.319}$$

If $\mathbf{N}$ is a unit vector in the direction of the radius of curvature of the ray at the same point and if ρ is the radius of curvature, then the vector curvature of the ray is $\mathbf{N}/\rho$ and is also given by $d\mathbf{s}/ds$. Therefore

$$\frac{d\mathbf{s}}{ds} = (\mathbf{s}\cdot\nabla)\mathbf{s} = -\mathbf{s}\times(\nabla\times\mathbf{s}) \tag{2.320}$$

Then,

$$\frac{\mathbf{N}}{\rho} = -\mathbf{s}\times(\nabla\times\mathbf{s}) \tag{2.321}$$

Then,

$$\frac{1}{\rho} = -\mathbf{N}\cdot(\mathbf{s}\times\nabla\times\mathbf{s}) = -(\mathbf{N}\times\mathbf{s})\cdot(\nabla\times\mathbf{s}) \tag{2.322}$$

Substituting Eq. 2.319 for Eq. 2.322, we have

$$\frac{1}{\rho} = \mathbf{N}\cdot\nabla(\ln n). \tag{2.323}$$

Since the radius of curvature is always considered positive, it can be seen from Eq. 2.323 that the ray bends toward the region of higher index of refraction n. In a homogeneous medium, $1/\rho = 0$ or the radius of curvature is infinite, and the ray is a straight line.

From Eq. 2.321, it follows that in a homogeneous medium

$$\nabla\times\mathbf{s} = 0 \tag{2.324}$$

which means that there is a family of surfaces orthogonal to the field of the vectors **s**, and these surfaces are the equiphase surfaces or wave-fronts. Let there be two wavefronts L_1 and L_2 and a tube of rays that cut them in elements of areas dA_1 and dA_2 as shown in Fig. 2.14. No power will flow across the sides of the tube, and if S_1 and S_2 are the rates of power flow per unit area at the two wavefronts, then,

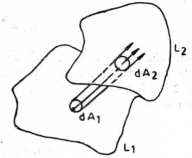

Fig. 2.14   Two wavefronts and tube of rays.

$$S_1 \, dA_1 = S_2 \, dA_2 \qquad (2.325)$$

S can be taken as the magnitude of the Poynting vector and put equal to $\frac{1}{2}(\epsilon/\mu) |E|^2$, so that Eq. 2.325 becomes

$$\epsilon_1^{1/2} |E_1|^2 \, dA_1 = \epsilon_2^{1/2} |E_2|^2 \, dA_2 \qquad (2.326)$$

or

$$n_1 |E_1|^2 \, dA_1 = n_2 |E_2|^2 \, dA_2. \qquad (2.327)$$

Equation 2.327 gives a one-to-one correspondence between the field E_1 at one point and the field E_2 at another point.

The optical path length ΔL along a curve C between two points P_1 and P_2 is defined as the line integral

$$\Delta L = \int_C n|ds| \qquad (2.328)$$

where n is the refractive index at the line element ds. Between two adjacent equiphase surfaces or wavefronts $L(x, y, z) = L_0$ and $L(x, y, z) = L_0 + \Delta L$, the increment ΔL is given by

$$\Delta L = \frac{c}{v} \, \delta s = n \delta s. \qquad (2.329)$$

However, an optical path length need not be along a ray, and it can be determined along any curve.

Fermat's Principle states that the optical ray or rays from a source at a point P_1 to a point of observation P_2 is the curve along which the optical path length is stationary with respect to infinitesimal variations in path.

2.28   Radiation Condition

While solving electromagnetic radiation problems uniquely, a specific boundary condition is to be imposed on the fields at infinity. Consider the solution of the scalar wave equation

$$\nabla^2 \psi(r) + k^2 \psi(r) = -\delta(r) \qquad (2.330)$$

in the absence of boundary conditions, the two solutions of Eq. 2.330 are

$$\psi_1 = \frac{\exp(-jkr)}{4\pi r} \qquad (2.331)$$

and

$$\psi_2 = \frac{\exp(+jkr)}{4\pi r}. \tag{2.332}$$

Both solutions are mathematically valid. However, only Eq. 2.331 is physically possible, because it represents a radially outward wave, while Eq. 2.332 is a radially inward wave. For a scalar field, Sommerfeld stated the radiation condition as

$$|r\psi| < K \tag{2.333}$$

$$\lim_{r \to \infty} r \left(\frac{\partial \psi}{\partial r} + jk\psi \right) = 0 \tag{2.334}$$

for a time dependence of $\exp(j\omega t)$ and where K is a finite constant. It can be seen that ψ_1 given by Eq. 2.331 satisfies Eqs. 2.333 and 2.334 while ψ_2 does not.

Equation 2.333 is satisfied whenever Eq. 2.334 is satisfied and, therefore, it can be dropped.

PROBLEMS

1. Show that the average Poynting vector of a circularly-polarized wave is twice that of a linearly-polarized wave if the maximum electric field E is the same for both waves.

2. A wave travelling in the z-direction is the resultant of two elliptically polarized waves, one with components $E_x = 4 \cos \omega t$ and $E_y = 3 \sin \omega t$ and the other with components $E_x = 3 \cos \omega t$ and $E_y = 5 \sin \omega t$. For the resultant wave, calculate the axial ratio, and the angle between the major axis of the polarization ellipse and the positive axis. Does E rotate clockwise or counter-clockwise?

3. Show that the instantaneous Poynting vector of a plane circularly-polarized travelling wave is a constant.

4. A half-space of air (medium 1) and a half-space of dielectric (medium 2) are separated by a 10 μm thick copper sheet which has constants $\sigma = 58$ mmhos and $\mu_r = \epsilon_r = 1$. A plane linearly polarized 3 GHz travelling wave in medium 1 (air) with electric field $E = 10$ Vm^{-1} rms is incident normally on the copper sheet. The constants for medium 2 are $\sigma = 0$, $\mu_r = 1$, $\epsilon_r = 5$. Determine the rms value of E (i) inside the copper sheet adjacent to medium 1 (air), (ii) inside the copper sheet adjacent to medium 2 (dielectric) and (iii) at a distance of 2 m from the copper sheet in medium 2. Also, determine H at these points.

5. Show that, if the boundary conditions are satisfied for the electric intensity and the field satisfies Maxwell's equations, the boundary conditions for the magnetic intensity are automatically satisfied, and vice versa.

6. A sheet of glass having a relative dielectric constant of 8 and negligible conductivity is coated with a silver plate. Show that at a frequency of 100 MHz, the surface impedance will be less for a 0.001 cm coating than for a 0.002 cm coating and explain why this is so.

7. A particle with a negative charge of 10^{-17} coulombs and a mass of 10^{-26} kg is at rest in a field-free space. If a uniform electric field $E = 1$ kVm^{-1} is applied for 1 μs, find the velocity of the particle and the radius of curvature of the particle path if

the particle enters a magnetic field $B = 2$ milliweber, per square meter, with the velocity moving normal to B.

8. Wherever the displacement current is negligible with respect to the conduction current $(\sigma/j\omega\epsilon \gg 1)$, the field approximately satisfies the equations

$$\nabla \times \mathbf{E} + \mu \frac{\partial \mathbf{H}}{\partial t} = 0, \quad \nabla \times \mathbf{H} = \mathbf{i}.$$

Show that to the same approximation

$$\mathbf{E} = \mu \frac{\partial \mathbf{A}}{\partial t}, \quad \mathbf{H} = \nabla \times \mathbf{A}$$

$$\nabla^2 \mathbf{A} - \mu \frac{\partial \mathbf{A}}{\partial t} = 0, \quad \nabla \cdot \mathbf{A} = 0,$$

and that the current density satisfies

$$\nabla^2 \mathbf{i} - \mu\sigma \frac{\partial \mathbf{i}}{\partial t} = 0, \quad \nabla \cdot \mathbf{i} = 0.$$

The foregoing equations govern the distribution of current in metallic conductors at all frequencies in the radio spectrum and below.

9. A circular cylindrical wire of radius a, which is much greater than the skin depth δ, has a uniform electric field applied in the axial direction at its surface. Use the surface-impedance concept to find the total current on the wire. Show that the ratio of the ac impedance of the wire to the dc resistance is

$$\frac{Z_{ac}}{R_{dc}} = \frac{a\sigma}{2} Z_m,$$

where Z_m is the surface impedance. Evaluate this ratio for copper at $f = 10^6$ cycles for $\sigma = 5.8 \times 10^7$ mhos/m, $a = 0.05$ cm, and $\mu = \mu_0$.

10. Show that, when the relaxation time for a material is little as compared to the period of the time-harmonic field, the displacement current may be neglected in comparison with the conduction current.

11. Show that, when a uniform plane wave is incident on an interface between two dielectric media, with its electric vector parallel to the plane of incidence (vertical or parallel polarization), the ratio of the reflected electric field E_r to the incident electric field E_i is given by

$$\frac{E_r}{E_i} = \frac{(\epsilon_2/\epsilon_1) \cos \theta_1 - [(\epsilon_2/\epsilon_1) - \sin^2 \theta_1]^{1/2}}{(\epsilon_2/\epsilon_1) \cos \theta_1 + [(\epsilon_2/\epsilon_1) - \sin^2 \theta_1]^{1/2}}$$

where ϵ_1 and ϵ_2 are the permittivities of the two media and θ_1 is the angle of incidence. Discuss the possibility of not obtaining any reflection. Show that, when Brewster's angle $\tan \theta_1 = \sqrt{\epsilon_2/\epsilon_1}$ there is no reflection. Find Brewster's angle when $\epsilon_1 = \epsilon_0$, $\epsilon_2 = 5\epsilon_0$.

12. Obtain expressions for the standing-wave patterns of $\mathbf{E}$ and $\mathbf{H}$ when a uniform plane wave is incident normally at an interface between a perfect dielectric and a perfect conductor.

SUGGESTED READING

1. Collin, R. E. and Plonsey, R., *Principles and Applications of Electromagnetic Fields*, McGraw-Hill, New York, 1961.

2. Fano, R. M., Chu, L. J. and Adler, R. B., *Electromagnetic Fields, Energy and Forces*, Wiley, New York, 1960.

3. Harrington, R. F., *Time-harmonic Electromagnetic Fields*, McGraw-Hill, New York, 1961.

4. Hayt, W. H., *Engineering Electromagnetics*, 2nd edn., McGraw-Hill, New York, 1967.

5. Johnk, C. T. A., *Engineering Electromagnetic Fields and Waves*, Wiley, New York, 1975.

6. Jones, D. S., *The Theory of Electromagnetism*, Pergamon Press, Oxford, 1964.

7. Jordan, E. C. and Balamain, K. G., *Electromagnetic Waves and Radiating Systems*, 2nd edn., Prentice-Hall, Englewood Cliffs, New Jersey, 1968.

8. Kong, J. A., *Theory of Electromagnetic Waves*, Wiley-Interscience, New York, 1975.

9. Kraus, J. D., *Electromagnetics*, McGraw-Hill, New York, 1953.

10. Mittra, R., *Computer Techniques for Electromagnetics*, Pergamon Press, Oxford, 1973.

11. Parton, J. E. and Owen, S. J. T., *Applied Electromagnetics*, Macmillan, London, 1975.

12. Portis, A. M., *Electromagnetic Fields: Sources and Media*, Wiley, New York, 1978.

13. Schelkunoff, S. A., *Electromagnetic Waves*, D. Van Nostrand, New York, 1943.

14. Silver, S., *Microwave Antenna Theory and Design*, McGraw-Hill, New York, 1949.

15. Stratton, J. A., *Electromagnetic Theory*, McGraw-Hill, New York, 1941.

16. Van Bladel, J., *Electromagnetic Fields*, McGraw-Hill, New York, 1964.

3

THIN LINEAR ANTENNA

3.1 Short Current Element or Short Electric Dipole

Any linear antenna may be considered as made up of a large number of very short conductors connected in series. Hence, the short current element or short electric dipole will be considered first as an elementary antenna or radiator.

Figure 3.1(a) represents such a short dipole. Plates are provided at the ends to give capacitive loading. The plates and the fact that the length L of the dipole $\ll \lambda$, ensure uniform current I along the entire length of the antenna. The dipole is energized by a balanced transmission line and it is assumed that the radiation from the end plates is negligible. The dipole is thin so that its diameter $d \ll L$. Hence, a simple equivalent of the short dipole is shown in Fig. 3.1(b) which consists of a thin conductor of length L carrying a uniform current I and with point charges $+q$ and $-q$ at the ends. The current I and charge q are related by

$$\frac{dq}{dt} = I \tag{3.1}$$

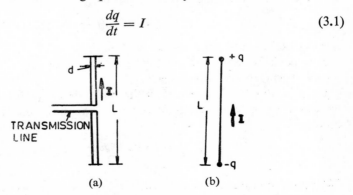

(a)   (b)

Fig. 3.1   Short electric dipole: (a) Short dipole, (b) Its equivalent.

Assuming that the medium surrounding the dipole is air or vacuum, we proceed to find the electromagnetic field produced by the dipole. Figure 3.2

represents the coordinate systems used for this purpose. The centre of the dipole is located at the origin of the coordinates as shown in Figs. 3.2(a) and (b).

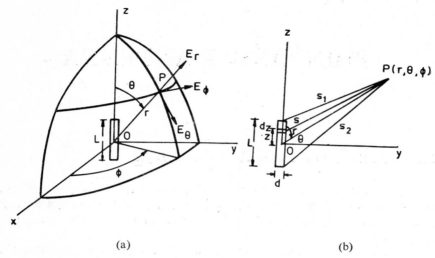

(a)   (b)

Fig. 3.2   Geometry of short dipole. (a) Vector components of field in spherical polar coordinates, (b) Geometry for use in analysis.

Since the current I is in the z-direction, the retarded magnetic vector potential has only the z component, namely A_z, which is given by

$$A_z = \frac{\mu_0}{4\pi} \int_{-L/2}^{L/2} \frac{[I]}{s}\, dz \tag{3.2}$$

where $[I]$ is the retarded current given by

$$[I] = I_0 \exp\left[j\omega\{t - (s/c)\} \right] \tag{3.3}$$

z = distance to a point on the conductor;
I_0 = peak value in time of current;
μ_0 = permeability of free space.
If $r \gg L$ and $\lambda \gg L$, $s \simeq r$, then the phase difference of the field contributions from different parts of the wire can be neglected. Then Eq. 3.3 becomes

$$A_z = \frac{\mu_0 L I_0 \exp\left[j\omega\{t - (r/c)\} \right]}{4\pi r} \tag{3.4}$$

The retarded scalar electric potential V due to the two charges $+q$ and $-q$ at the ends of the dipole is given by

$$V = \frac{1}{4\pi\epsilon_0} \left\{ \frac{[q]}{s_1} - \frac{[q]}{s_2} \right\} \tag{3.5}$$

(Figs. 3.1(b) and 3.2(b)).

From Eqs. 3.1 and 3.3 $[q]$ is given by

$$[q] = \int [I] \, dt = I_0 \int \exp\left[j\omega\{t - (s/c)\}\right] dt = \frac{[I]}{j\omega} \tag{3.6}$$

Using Eq. 3.6 in Eq. 3.5, we have

$$V = \frac{I_0}{4\pi\epsilon_0 j\omega}\left[\frac{\exp\left[j\omega\{t - (s_1/c)\}\right]}{s_1} - \frac{\exp\left[j\omega\{t - (s_2/c)\}\right]}{s_2}\right] \tag{3.7}$$

When $r \gg L$, the lines connecting the ends of the dipole and point P may be considered parallel so that

$$s_1 \doteq r - \frac{L}{2}\cos\theta \tag{3.8}$$

and

$$s_2 = r + \frac{L}{2}\cos\theta \tag{3.9}$$

Using Eqs. 3.8 and 3.9 in Eq. 3.7, we obtain

$$V = \frac{I_0 \exp\left[j\omega\{t - (r/c)\}\right]}{4\pi\epsilon_0 j\omega}$$

$$\times \left[\frac{\exp\left(j\,\dfrac{\omega L\cos\theta}{2c}\right)\left(r + \dfrac{L}{2}\cos\theta\right) - \exp\left(-\dfrac{j\omega L\cos\theta}{2c}\right)\left(r - \dfrac{L}{2}\cos\theta\right)}{r^2}\right] \tag{3.10}$$

(by neglecting the term $L^2/4\cos^2\theta$ in the denominator in comparison with r^2, because $r \gg L$).

$$V = \frac{I_0 \exp\left[j\omega\{t - (r/c)\}\right]}{4\pi\epsilon_0 j\omega r^2}\left[\left(\cos\frac{\omega L\cos\theta}{2c} + j\sin\frac{\omega L\cos\theta}{2c}\right)\left(r + \frac{L}{2}\cos\theta\right)\right.$$

$$\left. - \left(\cos\frac{\omega L\cos\theta}{2c} - j\sin\frac{\omega L\cos\theta}{2c}\right)\left(r - \frac{L}{2}\cos\theta\right)\right] \tag{3.11}$$

If $\lambda \gg L$, then

$$\cos\frac{\omega L\cos\theta}{2c} = \cos\frac{\pi L\cos\theta}{\lambda} \simeq 1 \tag{3.12}$$

and

$$\sin\frac{\omega L\cos\theta}{2c} \simeq \frac{\omega L\cos\theta}{2c} \tag{3.13}$$

Using Eqs. 3.12 and 3.13 in Eq. 3.11, we have

$$V = \frac{I_0 L\cos\theta \exp\left[j\omega\{t - (r/c)\}\right]}{4\pi\epsilon_0 c}\left(\frac{1}{r} + \frac{c}{j\omega}\frac{1}{r^2}\right) \tag{3.14}$$

Knowing the magnetic vector potential **A** and the electric scalar potential

V, the electric and magnetic fields may be obtained from the relations

$$\mathbf{E} = -j\omega\mathbf{A} - \nabla V \tag{3.15}$$

$$\mathbf{H} = \frac{1}{\mu_0} \nabla \times \mathbf{A} \tag{3.16}$$

$\mathbf{H} = \nabla \times \mathbf{A}$ or $\mathbf{H} = \dfrac{1}{\mu_0} \nabla \times \mathbf{A}$ can be used (see section 2.13). $\mathbf{A}$ has only a z component A_z given by Eq. 3.4 and V is given by Eq. 3.14. Using spherical polar coordinates, the components A_r and A_θ of $\mathbf{A}$ are given by

$$A_r = A_z \cos \theta, \tag{3.17}$$

$$A_\theta = -A_z \sin \theta \tag{3.18}$$

and

$$\nabla V = \mathbf{u}_r \frac{\partial V}{\partial r} + \mathbf{u}_\theta \frac{1}{r} \frac{\partial V}{\partial \theta} + \mathbf{u}_\phi + \frac{1}{r \sin \theta} \frac{\partial V}{\partial \phi}. \tag{3.19}$$

Using Eqs. 3.4 and 3.14 in Eqs. 3.15 and 3.16, and using Eqs. 3.17 to 3.19, the r, θ and ϕ components of $\mathbf{E}$ are given by

$$E_r = \frac{I_0 L \cos \theta \, \exp\left[\,j\omega\{t - (r/c)\}\right]}{2\pi\epsilon_0} \left(\frac{1}{cr^2} + \frac{1}{j\omega r^3}\right), \tag{3.20}$$

$$E_\theta = \frac{I_0 L \sin \theta \, \exp\left[\,j\omega\{t - (r/c)\}\right]}{4\pi\epsilon_0} \left(\frac{j\omega}{c^2 r} + \frac{1}{cr^2} + \frac{1}{j\omega r^3}\right) \tag{3.21}$$

and

$$E_\phi = 0 \tag{3.22}$$

$\left(\text{remembering that } \dfrac{1}{c^2} = \mu_0\epsilon_0\right)$.

Similarly, the r, θ and ϕ components of H are given by

$$H_r = H_\theta = 0 \tag{3.23}$$

and

$$H_\phi = \frac{I_0 L \sin \theta \, \exp\left[\,j\omega\{t - (r/c)\}\right]}{4\pi} \left(\frac{j\omega}{cr} + \frac{1}{r^2}\right) \tag{3.24}$$

Therefore, the field of the short dipole has only the three components E_r, E_θ and H_ϕ given by Eqs. 3.20, 3.21 and 3.24 respectively. Examining the expressions for these three components, it can be noticed that E_r has a $1/r^2$ term and a $1/r^3$ term; E_θ has a $1/r$ term, a $1/r^2$ term and a $1/r^3$ term; and H_ϕ has a $1/r$ term and a $1/r^2$ term. The $1/r^2$ term is called the 'induction field' and the $1/r^3$ term is called the 'electrostatic field'. These two terms are significant only very close to the dipole and therefore are important in the 'near field' of the dipole. For very large r, the $1/r^2$ and $1/r^3$ terms may be neglected and only the $1/r$ term need be considered. This $1/r$ term is called the 'far field'. Therefore in the 'far field', E_r is negligible and only E_θ and H_θ are important. They are given by

$$E_\theta = \frac{j\omega I_0 L \sin \theta \, \exp\left[\,j\omega\{t - (r/c)\}\right]}{4\pi\epsilon_0 c^2 r} \tag{3.25}$$

and

$$H_\phi = \frac{j\omega I_0 L \sin\theta \exp\left[\, j\omega\{t - (\,r/c)\}\right]}{4\pi cr} \tag{3.26}$$

and also

$$\frac{E_\theta}{H_\phi} = \frac{1}{\epsilon_0 c} = \sqrt{\frac{\mu_0}{\epsilon_0}} = 377 \text{ ohms} \tag{3.27}$$

which is the intrinsic or characteristic impedance of free space.

Examining Eqs. 3.25 and 3.26, it can be seen that E_θ and H_ϕ are in time phase in the far field, and that the field patterns of both are proportional to $\sin\theta$ but independent of ϕ. Therefore, the space pattern is a figure of revolution and is doughnut-shaped (Fig. 3.3).

Referring to Eqs. 3.20, 3.21 and 3.24, it can be noted that for all small r the electric field has two components E_r and E_θ which are both in time phase quadrature with the magnetic field, as in a resonator. At intermediate distances, E_θ and E_ϕ can approach time quadrature so that the total electric field vector rotates in a plane parallel to the direction of propagation, thus exhibiting the phenomenon of cross-field. The near-field pattern for E_r is proportional to $\cos\theta$ as shown in Fig. 3.4.

Fig. 3.3   Near- and far-field patterns of E_θ and H_ϕ components of short dipole.

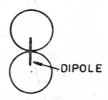

Fig. 3.4   Near-field pattern of E_r component short dipole.

At very low frequencies, it is called the quasi-stationary case. Since from Eq. 3.6,

$$[I] = j\omega[q]. \tag{3.28}$$

Equations 3.20, 3.21 and 3.24 can be rewritten as

$$E_r = \frac{[q]L\cos\theta}{2\pi\epsilon_0}\left(\frac{j\omega}{cr^2} + \frac{1}{r^3}\right), \tag{3.29}$$

$$E_\theta = \frac{[q]L\sin\theta}{2\pi\epsilon_0}\left(-\frac{\omega^2}{c^2 r} + \frac{j}{cr^2} + \frac{1}{r^3}\right) \tag{3.30}$$

and

$$H_\phi = \frac{[I]L\sin\theta}{4\pi}\left(\frac{j\omega}{cr} + \frac{1}{r^2}\right). \tag{3.31}$$

As $\omega \to 0$ at low frequencies, the terms with ω in the numerator can be neglected, and

$$[q] = q_0 \exp\left[j\omega\left(t - \frac{r}{c}\right)\right] = q_0 \tag{3.32}$$

and

$$[I] = I_0. \tag{3.33}$$

Hence, for the quasi-stationary case, or dc case, the field components of the short dipole become

$$E_r = \frac{q_0 L \cos \theta}{2\pi\epsilon_0 r^3}, \tag{3.34}$$

$$E_\theta = \frac{q_0 L \sin \theta}{4\pi\epsilon_0 r^3}, \tag{3.35}$$

$$H_\phi = \frac{I_0 L \sin \theta}{4\pi r^2}. \tag{3.36}$$

The expressions for E_r and E_θ as given by Eqs. 3.34 and 3.35 are identical to the electrostatic field of two point charges $+q_0$ and $-q_0$ separated by a distance L. The expression for H_ϕ as given by Eqs. 3.36 is the Biot-Savart relation for the magnetic field of a short element carrying a steady or slowly varying electric current.

The far-field components are E_θ and H_ϕ, so that the Poynting vector has a radial component P_r given by

$$P_r = \tfrac{1}{2} \operatorname{Re} (E_\theta H_\phi^*) = \tfrac{1}{2} \operatorname{Re} Z_0 H_\phi H_\phi^*$$

$$= \tfrac{1}{2} |H_\phi|^2 \operatorname{Re} Z_0 = \tfrac{1}{2} |H_\phi|^2 \sqrt{\frac{\mu_0}{\epsilon_0}} \tag{3.37}$$

$$\left(\text{where } Z_0 = \sqrt{\frac{\mu_0}{\epsilon_0}}\right).$$

The total power radiated is given by

$$W = \iint_{\text{Large sphere}} P_r \, da = \frac{1}{2}\sqrt{\frac{\mu_0}{\epsilon_0}} \int_0^{2\pi} \int_0^{\pi} |H_\phi|^2 r^2 \sin \theta \, d\theta \, d\phi \tag{3.38}$$

and from Eq. 3.26,

$$|H_\phi| = \frac{\omega I_0 L \sin \theta}{4\pi c r} \tag{3.39}$$

Therefore,

$$W = \frac{1}{32}\sqrt{\frac{\mu_0}{\epsilon_0}} \frac{\beta^2 I_0^2 L^2}{\pi^2} \int_0^{2\pi} \int_0^{\pi} \sin^3 \theta \, d\theta \, d\phi = \sqrt{\frac{\mu_0}{\epsilon_0}} \frac{\beta^2 I_0^2 L^2}{12\pi} \tag{3.40}$$

This power is equated to $I^2 R$, where I is the rms current on the dipole and R is a resistance, called the radiation resistance of the dipole. Therefore,

$$\sqrt{\frac{\mu_0}{\epsilon_0}} \frac{\beta^2 I_0^2 L^2}{12\pi} = \frac{I_0^2}{2} R \tag{3.41}$$

and hence

$$R = \sqrt{\frac{\mu_0}{\epsilon_0}} \frac{\beta^2 L^2}{6\pi} = 377 \frac{\beta^2 L^2}{6\pi} = 80\pi^2 (L/\lambda)^2 = 80\pi^2 L_\lambda^2 \tag{3.42}$$

where $L_\lambda = L/\lambda$.

3.2 Thin Dipole Antenna

The antennas are symmetrically fed at the centre by a balanced transmission line as shown in Fig. 3.5, and the current distribution is assumed to be sinusoidal. When the diameter of the conductor $< \lambda/100$, this sinusoidal distribution of current is approximately correct (this will be proved in Chapter 4).

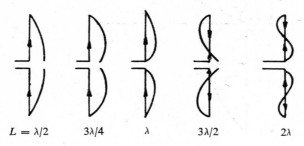

$$L = \lambda/2 \qquad 3\lambda/4 \qquad \lambda \qquad 3\lambda/2 \qquad 2\lambda$$

Fig. 3.5   Current distribution on thin centre-fed dipoles.

In order to study the properties of such an antenna as a radiator, the far-field of the antenna is derived. The retarded value of the current at any point z on the antenna referred to a point $P(r, \theta, \phi)$ which is at radial distance s from the point z is

$$I = I_0 \sin \left[\frac{2\pi}{\lambda}\left(\frac{L}{2} \pm z\right)\right] \exp \left[j\omega\left(t - \frac{s}{c}\right)\right] \qquad (3.43)$$

In Eq. 3.43, $(L/2 + z)$ is used when $z < 0$ and $(L/2 - z)$ is used when $z > 0$.

The antenna is considered as made up of a series of infinitesimally short dipoles of length dz. The far field components of one of these short dipoles dz at P (in Fig. 3.6) are

$$dE_\theta = \frac{j60\pi[I] \sin \theta \, dz}{s\lambda} \qquad (3.44)$$

$$dH_\phi = \frac{j(I) \sin \theta \, dz}{2s\lambda} \qquad (3.45)$$

Since $E_\theta/H_\phi = (\mu_0/\epsilon_0)^{1/2} = Z_0$, it is enough if H_ϕ is evaluated by integrating dH_ϕ over the length of the antenna. Thus, H_ϕ is given by

$$H_\phi = \int_{-L/2}^{L/2} dH_\phi \qquad (3.46)$$

Using Eq. 3.43 in Eq. 3.46, we obtain

$$H_\phi = \frac{jI_0 \sin \theta \exp (j\omega t)}{2\lambda} \left\{ \int_{-L/2}^{0} \frac{1}{s} \sin \left[\frac{2\pi}{\lambda}\left(\frac{L}{2} + z\right)\right] \exp \left(\frac{-j\omega s}{c}\right) dz \right.$$

$$\left. + \int_{0}^{L/2} \frac{1}{s} \sin \left[\frac{2\pi}{\lambda}\left(\frac{L}{2} - z\right)\right] \exp \left(-j\frac{\omega s}{c}\right) dz \right\} \qquad (3.47)$$

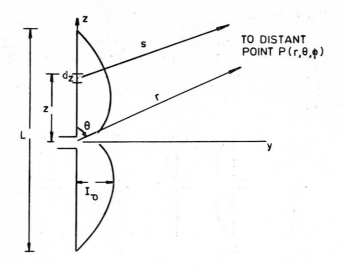

Fig. 3.6   Thin linear centre-fed dipole.

From Fig. 3.6

$$s = r - z \cos \theta \qquad (3.48)$$

And at a large distance $s \simeq r$. Then Eq. 3.47 becomes

$$H_\phi = \frac{jI_0 \sin \theta \exp\left[j\omega\{t-(r/c)\}\right]}{2\lambda r} \left\{ \int_{-L/2}^{0} \sin\left[\frac{2\pi}{\lambda}\left(\frac{L}{2}+z\right)\right] \exp\left(\frac{j\omega \cos \theta}{c}z\right) dz \right.$$

$$\left. + \int_{0}^{L/2} \sin\left[\frac{2\pi}{\lambda}\left(\frac{L}{2}-z\right)\right] \exp\left(j\frac{\omega \cos \theta}{c}z\right) dz \right\}$$

$$= \frac{j\beta I_0 \sin \theta \exp\left[j\omega\{t-(r/c)\}\right]}{4\pi\lambda} \left\{ \int_{-L/2}^{0} \exp\left(j\beta z \cos \theta\right) \sin\left[\beta\left(\frac{L}{2}+z\right)\right] dz \right.$$

$$\left. + \int_{0}^{L/2} \exp\left(j\beta z \cos \theta\right) \sin\left[\beta\left(\frac{L}{2}-z\right)\right] dz \right\} \qquad (3.49)$$

(since $\beta = \omega/c = 2\pi/\lambda$ and $\beta/4\pi = \lambda/2$). Since the two integrals in Eq. 3.49 are of the form

$$\int \exp\left(a_1 x\right) \sin\left(a_3 + a_2 x\right) dx = \frac{\exp\left(a_1 x\right)}{a_{12}^{22} + a} \left[a_1 \sin\left(a_3 + a_2 x\right)\right.$$

$$\left. - a_2 \cos\left(a_3 + a_2 x\right)\right]. \qquad (3.50)$$

Equation 3.49 becomes

$$H_\phi = \frac{j[I_0]}{2\pi r} \frac{\left[\cos\{(\beta L \cos \theta)/2\} - \cos\left(\beta L/2\right)\right]}{\sin \theta}. \qquad (3.51)$$

The E_θ is given by

$$E_\theta = 120 H_\phi = \frac{j60[I_0][\cos\{(\beta L \cos \theta)/2\} - \cos\left(\beta L/2\right)]}{r \sin \theta}, \qquad (3.52)$$

where $[I_0] = I_0 \exp\left[j\omega\left(t - \frac{r}{c}\right)\right]$.

Equations 3.51 and 3.52 give the expressions for the far-field components of the centre-fed thin dipole antenna. The factor in the brackets in Eqs. 3.51 and 3.52 determines the shape of the far-field pattern. Figure 3.7 shows the far-field pattern of linear antennas of lengths $\lambda/2$ and $3\lambda/2$.

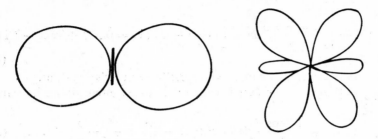

Fig. 3.7   Far-field patterns of thin centre-fed dipoles: (a) $L=2$, (b) $L=3/2$.

To find the radiation resistance of a thin linear antenna, the Poynting vector is integrated over a large sphere to obtain the total power radiated W, and this total power is equated to $I_0^2 R/2$, where R is the radiation resistance at a current maximum point. W is given by

$$W = \frac{15I_0^2}{\pi} \int_{\phi=0}^{2\pi} \int_{\theta=0}^{\pi} \frac{[\cos\{(\beta L \cos\theta)/2\} - \cos(\beta L/2)]^2}{\sin\theta}\, d\theta\, d\phi$$

$$= 30I_0^2 \int_0^{\pi} \frac{[\cos\{(\beta L \cos\theta)/2\} - \cos(\beta L/2)]^2}{\sin\theta}\, d\theta \qquad (3.53)$$

Equating W to $I_0^2 R/2$, the radiation resistance R is given by

$$R = 60 \int_0^{\pi} \frac{[\cos\{(\beta L \cos\theta)/2\} - \cos(\beta L/2)]^2}{\sin\theta}\, d\theta \qquad (3.54)$$

This radiation resistance given by Eq. 3.54 is referred to the current maximum. For a $\lambda/2$ dipole, the current maximum is at the centre of the antenna.

To evaluate Eq. 3.54, put $u = \cos\theta$, then

$$R = 60 \int_{-1}^{+1} \frac{\left(\cos\dfrac{\beta L}{2} u - \cos\dfrac{\beta L}{2}\right)^2}{(1 - u^2)}\, du$$

$$= 30 \int_{-1}^{+1} \left[\frac{(\cos ku - \cos k)^2}{(1 + u)} + \frac{(\cos ku - \cos k)^2}{(1 - u)}\right] du \qquad (3.55)$$

For $L = \lambda/2$, Eq. 3.55 becomes

$$R = 30 \int_{-1}^{+1} \left[\frac{\cos^2(\pi u/2)}{(1 + u)} + \frac{\cos^2(\pi u/2)}{(1 - u)}\right] du \qquad (3.56)$$

In Eq. 3.56, putting $1 + u = v/\pi$ in the first term and $1 - u = v'/\pi$ in the second term, we obtain,

$$R = 60 \int_0^{2\pi} \frac{\cos^2 (v - \pi)/2}{v}\, dv = 30 \int_0^{2\pi} \frac{1 + \cos (v - \pi)}{v}\, dv$$

$$= 30 \int_0^{2\pi} \frac{1 - \cos v}{v}\, dv = 30\, \text{Cin}\, (2\pi) \tag{3.57}$$

where

$$\text{Cin}\, (x) = \int_0^x \frac{1 - \cos v}{v}\, dv = \ln\, (\gamma x) - \text{Ci}\, (x) = 0.577 + \ln x - \text{Ci}\, (x) \tag{3.58}$$

where $\gamma = e^c = 1.781$, or $\ln \gamma = c = 0.577 = $ Euler's constant.

$\text{Ci}\, (x) = [\ln (x) - \text{Cin}\, (x)]$ is called the cosine integral and its value is given by

$$\text{Ci}\, (x) = \int_\infty^x \frac{\cos v}{v}\, dv = \ln\, (\gamma x) - \frac{x^2}{2!2} + \frac{x^4}{4!4} - \frac{x^6}{6!6} + \ldots \tag{3.59}$$

When x is small (< 0.2),

$$\text{Ci}\, (x) \simeq \ln\, (\gamma x) = 0.577 + \ln x \tag{3.60}$$

and when x is large $(x \gg 1)$,

$$\text{Ci}\, (x) = \frac{\sin x}{x} \tag{3.61}$$

Figure 3.8 shows Ci (x) as a function of x. Hence Eq. 3.57 becomes

$$R = 30\, \text{Cin}\, (2\pi) = 30 \times 2.44 = 73 \text{ ohms} \tag{3.62}$$

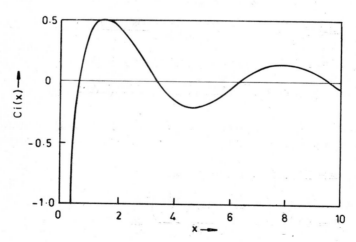

Fig. 3.8   Cosine integral Ci (x).

Equation 3.62 gives the value of the radiation resistance of a half-wave dipole. The terminal impedance also includes some inductive reactance as will be shown in Chap. 4. To make the antenna resonant, this reactance should be zero, and this requires that the antenna be a little less than $\lambda/2$ long, which also makes the radiation resistance a little less than 73 ohm.

It can be shown that the general expression for the radiation resistance of a centre-fed dipole of length L is given by

$$R = 30\{S_1(b) - [S_1(2b) - {}_,\text{Si}(b)]\cos b + [\text{Si}(2b) - \text{Si}(b)]\sin b$$
$$+ (1 + \cos b)S_1(b) - \sin b\,\text{Si}(b)\} \qquad (3.63)$$

where

$$S_1(b) = \int_0^b \frac{(1 - \cos v)}{v}\,dv = \text{Cin}(b) \qquad (3.64)$$

$$b = \beta L$$

and

$$\text{Si}(b) = \int_0^b \frac{\sin v}{v}\,dv = x + \frac{x^3}{3!3} + \frac{x^5}{5!5} + \ldots = \text{sine integral} \quad (3.65)$$

Figure 3.9 shows Si (x) as a function of x.

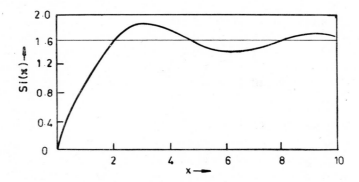

Fig. 3.9   Sine integral Si (x).

3.3   Radiation Resistance of a Dipole at a Point which is Not a Current Maximum

For a dipole of length L, the current maximum may not be at the feed point as shown in Fig. 3.10. The expression given by Eq. 3.63 is the radiation resistance at the current maximum. To obtain the radiation resistance at the feed point, R can be transformed to the value that would appear across the terminals of the transmission line connected at the centre of the dipole. This can be done by making use of the fact that the power supplied by the transmission line, given by $I_1^2 R_1/2$ is equal to the total radiated power $I_0^2 R/2$, where I_1 is the current amplitude at the feed terminals, I_0 is the current

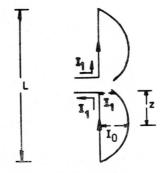

Fig. 3.10   Current distribution on dipole of length L.

maximum and R_1 is the radiation resistance at the feed terminals. Thus,

$$\frac{I_1^2 R_1}{2} = \frac{I_0^2}{2} R \tag{3.66}$$

Therefore the radiation resistance R_1 at the feed terminals is given by

$$R_1 = \frac{I_0^2}{I_1^2} R_0 \tag{3.67}$$

Since the current distribution on the dipole is sinusoidal,

$$I_1 = I_0 \cos(\beta z) \tag{3.68}$$

Hence

$$R_1 = \frac{R}{\cos^2(\beta z)} \tag{3.69}$$

when $z = 0$, $R_1 = R$; however when $z = \lambda/4$, $R_1 = \infty$ if $R \neq 0$. But the radiation resistance at a current minimum, as at the feed points of dipole of length λ, is not infinite as can be calculated from Eq. 3.69, since an actual antenna is not infinitesimally thin and the current at a minimum point is not zero (as will be shown in Chap. 4). Nevertheless, the radiation resistance at a current minimum is very large in practice (may be thousands of ohms).

3.4   Linear Travelling Wave Antenna

The linear antennas considered so far have sinusoidal current distribution which may be considered as a standing wave produced by two uniform travelling waves of equal amplitude travelling in opposite directions along the antenna. However, there are antennas like Beverage, rhombic, helical, thick conducting or dielectric cylindrical etc. (Fig. 3.11) which carry a single travelling wave approximately. On such antennas, the

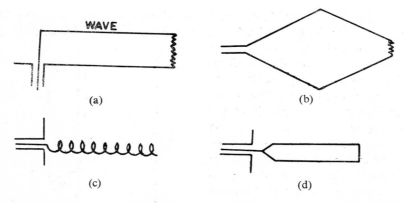

Fig. 3.11   Travelling wave antennas. (a) Beverage antenna, (b) Rhombic antenna, (c) Helical antenna, (d) Thick conducting or dielectric cylindrical antenna.

current amplitude distribution is uniform, but the phase changes linearly as shown in Fig. 3.12.

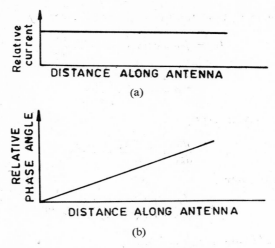

Fig. 3.12 Current amplitude and phase distribution along linear travelling wave antenna. (a) Amplitude, (b) Phase.

Figure 3.13 shows a linear travelling wave antenna of length b along the z-axis. A single travelling wave is assumed to be travelling along the antenna in the z-direction and the retarded current distribution is given by

$$[I] = I_0 \sin \omega \left(t - \frac{r}{c} - \frac{z_1}{v} \right) \qquad (3.70)$$

where z_1 is a point on the conductor. The phase velocity of the travelling wave is v and let $p = v/c$. The relative phase velocity is p.

Since the current is in the z-direction, the Hertz vector π has only a z

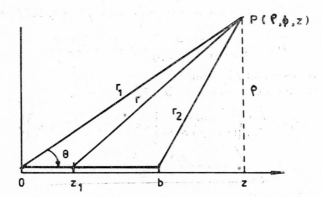

Fig. 3.13 Cylindrical coordinate system used to study the linear travelling wave antenna.

component given by

$$\pi_z = \frac{1}{4\pi j\omega\epsilon_0} \int_0^b \frac{[I]}{r} dz_1 \tag{3.71}$$

The ϕ component of $\mathbf{H}$ at the distant point is given by

$$H_\phi = j\omega\epsilon_0(\nabla \times \boldsymbol{\pi})_\phi = -j\omega\epsilon \frac{\partial \pi_z}{\partial \rho} \tag{3.72}$$

Put

$$u = t - \frac{r}{c} - \frac{z_1}{v} \tag{3.73}$$

Since $r = [(z - z_1)^2 + \rho^2]^{1/2}$,

$$\frac{du}{dz_1} = \frac{z - z_1}{rc} - \frac{1}{pc} \tag{3.74}$$

Then Eq. 3.71 becomes,

$$\pi_z = \frac{I_0 c}{4\pi j\omega_0\epsilon_0} \int_{u_1}^{u_2} \frac{\sin \omega u}{(z - z_1 - r/p)} du \tag{3.75}$$

where $u_1 = t - \frac{r_1}{c}$ and $u_2 = t - \frac{r_2}{c} - \frac{b}{v}$.

Using Eq. 3.75 in Eq. 3.72, we have

$$H_\phi = -\frac{I_0 c}{4\pi} \frac{\partial}{\partial \rho} \int_{u_1}^{u_2} \frac{\sin \omega u}{(z - z_1 - r/p)} du \tag{3.76}$$

In the far field, $r \gg b$ and z_1 can be neglected in the denominator of the integrand in Eq. 3.76. Then this becomes

$$H_\phi = -\frac{I_0 c}{4\pi\omega} \frac{\partial}{\partial \rho} \left(\frac{-\cos \omega u_2 + \cos \omega u_1}{(z - r/p)} \right)$$

$$= \frac{I_0 c}{4\pi r} \left[\frac{(z - r/p)(\sin \omega u_2 - \sin \omega u_1) + (\lambda/2\pi)(\cos \omega u_2 - \cos \omega u_1)}{(z - r/p)^2} \right] \tag{3.77}$$

For $|(z - r/p)| \gg \frac{\lambda}{2\pi}$ and for $(\sin \omega u_2 - \sin \omega u_1) \neq 0$, Eq. 3.77 becomes,

$$H_\phi = \frac{I_0}{4\pi r} \frac{\sin \theta}{(\cos \theta - 1/p)} (\sin \omega u_2 - \sin \omega u_1) \tag{3.78}$$

where $z/r = \cos \theta$ and $\rho/r = \sin \theta$.

Using the values of u_1 and u_2 in Eq. 3.78, we have

$$H_\phi = \frac{I_0 p}{2\pi r_1} \left\{ \frac{\sin \theta}{(1 - p \cos \theta)} \left[\sin \frac{\omega b}{2pc} (1 - p \cos \theta) \right] \right\}$$

$$\times \angle \omega\left(t - \frac{r_1}{c} \right) + \frac{\omega b}{2pc} (1 - p \cos \theta) \tag{3.79}$$

$$E_\theta = Z_0 H_\phi \tag{3.80}$$

where $Z_0 = (\mu_0/\epsilon_0)^{1/2} = 377$ ohms.

The amplitude of H_ϕ gives the shape of the far-field pattern, and the angle gives the phase of the far-field referred to the origin of coordinates as phase centre.

Figure 3.14 shows the far-field patterns of $a\lambda/2$ linear travelling wave antenna for two values of p. Figure 3.15 shows the far-field pattern of a 5λ linear travelling-wave antenna. It can be noticed that as the phase velocity v of the travelling wave is reduced, the tilt angle is increased and the beam width reduced. Even for $v = c$ ($p = 1$), the lobes are sharper and tilted forward as compared to the standing wave antenna. As the length of the antenna is increased, the tilt angle increases further, reaching a value of $78°$ when the length is 20λ (for $p = 1$).

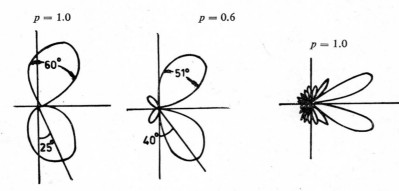

Fig. 3.14 Far-field patterns of $\lambda/2$ linear travelling wave antenna. (a) $p = 1.0$, (b) $p = 0.6$.

Fig. 3.15 Far-field pattern of 5λ linear travelling wave antenna.

3.5 Total Field of a Linear Travelling Wave Antenna

Only the far-field of a linear travelling-wave antenna was considered in the previous section. In this section the total field including the far-field and near-field will be considered.

Figure 3.16 shows a linear travelling wave antenna along z-axis, extending from $z = z_1$ to $z = z_2$. The current $I(z)$ on the antenna is of the form

$$I(z', t) = |I(z', t)| \, e^{j\psi(z', t)} \qquad (3.81)$$

where both amplitude $|I(z', t)|$ and phase $\psi(z', t)$ vary with time t and distance z'. If the source is monochromatic (single frequency ω), the time variation is of the form $e^{j\omega t}$, and Eq. 3.81 can be written as

$$I(z') = |I(z')| \, e^{j\psi(z')} \qquad (3.82)$$

The magnetic and electric fields of this source given by Eq. 3.82 are:

$$\mathbf{H} = \frac{1}{\mu} \, \nabla \times \mathbf{A} \qquad (3.83)$$

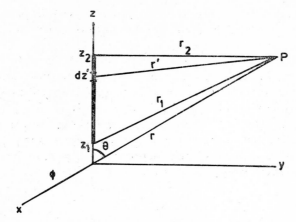

Fig. 3.16   Electric current travelling-wave linear antenna.

$$E = \frac{-j}{\omega\mu_0\epsilon_0} \nabla \times \nabla \times A = \frac{-j}{\omega\epsilon_0} \nabla \times H \qquad (3.84)$$

There are no variations in the ϕ direction because of symmetry, and therefore, the point P can be located in the y-z plane (Fig. 3.16). Then

$$r = (y^2 + z^2)^{1/2} \qquad (3.85)$$

$$r_1 = [y^2 + (z - z_1)^2]^{1/2} \qquad (3.86)$$

$$r' = [y^2 + (z - z')^2]^{1/2} \qquad (3.87)$$

$$r_2 = [y^2 + (z - z_2)^2]^{1/2} \qquad (3.88)$$

Since the current is in the z-direction, A has only a z-component given by

$$A_z = \frac{\mu_0}{4\pi} \int_{z_1}^{z_2} I(z') \frac{e^{-jkr'}}{r} \, dz' \qquad (3.89)$$

Therefore from Eq. 3.83, H has only a ϕ component H_ϕ given by

$$H_\phi = -\frac{1}{\mu_0} \frac{\partial A_z}{\partial \rho} \qquad (3.90a)$$

Since P is in the y-z plane, $\rho = y$, and therefore,

$$H_\phi = -\frac{1}{\mu_0} \frac{\partial A_z}{\partial y} = -\frac{1}{4\pi} \int_{z_1}^{z_2} \frac{\partial}{\partial y} \left[\frac{I(z')e^{-jkr'}}{r'} \right] dz'$$

Using Eq. 3.89

$$H_\phi = \frac{1}{4\pi} \int_{z_1}^{z_2} \left[\frac{jkyI(z')e^{-jkr'}}{r'^2} + \frac{yI(z')e^{-jkr'}}{r'^3} \right] dz' \qquad (3.90b)$$

For an arbitrary source, an exact solution for Eq. 3.90 may not be possible; however, numerical integration methods can be used to obtain approximate results.

For the special case $I(z') = I_0 e^{-jkz'}$, where I_0 is a constant and k is the phase constant of the travelling wave, Eq. 3.90 can be integrated. Then,

$$H_\phi = \frac{I_0}{4\pi} \int_{z_1}^{z_2} \left\{ \frac{jkye^{-jk(z'+r')}}{r'^2} + \frac{ye^{-jk(z'+r')}}{r'^3} \right\} dz \qquad (3.91)$$

It can be shown that

$$\frac{d}{dz'} \left[\frac{e^{-jk(z'+r')}}{r'(r'+z'-z)} \right] = \frac{-jke^{-jk(z'+r')}}{r'^2} - \frac{e^{-jk(z'+r')}}{r'^3} \qquad (3.92)$$

Using Eq. 3.92 in Eq. 3.91, we obtain

$$H_\phi = -\frac{I_0 y}{4\pi} \left[\frac{e^{-jk(z_2+r_2)}}{r_2(r_2+z_2-z)} - \frac{e^{-jk(z_1+r_1)}}{r_1(r_1+z_1-z)} \right] \qquad (3.93)$$

Using Eqs. 3.86 and 3.88 in Eq. 3.93, we have

$$H_\phi = -\frac{I_0}{4\pi y} \left[\left(1 + \frac{z-z_2}{r_2} \right) e^{-jk(z_2+r_2)} - \left(1 + \frac{z-z_1}{r_1} \right) e^{-jk(z_1-r_1)} \right] \quad (3.94)$$

Converting to spherical coordinates, Eq. 3.94 becomes

$$H_\phi = -\frac{I_0}{4\pi r \sin \theta} \left[\left(1 + \frac{r \cos \theta - z_2}{r_2} \right) e^{-jk(z_2+r_2)} \right.$$
$$\left. - \left(1 + \frac{r \cos \theta - z_1}{r_1} \right) e^{-jk(z_1+r_1)} \right] \qquad (3.95)$$

where

$$r_2 = (r^2 + z_2^2 - 2rz_2 \cos \theta)^{1/2}$$
$$r_1 = (r^2 + z_1^2 - 2rz_1 \cos \theta)^{1/2}$$

From Eq. 3.84, the electric field components E_r and E_θ can be obtained as

$$E_r = \frac{1}{j\omega\epsilon_0 r \sin \theta} \frac{\partial}{\partial \theta} (H_\phi \sin \theta)$$

$$= \frac{I_0}{j\omega\epsilon_0 4\pi r} \left\{ \left[\frac{z_2(r \cos \theta - z_2)}{r_2^3} + \frac{jkz_2(r \cos \theta - z_2)}{r_2^2} \right. \right.$$
$$\left. + \frac{(1+jkz_2)}{r_2} \right] e^{-jk(z_2+r_2)} - \left[\frac{z_1(r \cos \theta - z_1)}{r_1^3} \right.$$
$$\left. \left. + \frac{jkz_1(r \cos \theta - z_1)}{r_1^2} + \frac{(1+jkz_1)}{r_1} \right] e^{-jk(z_1+r_1)} \right\} \qquad (3.96)$$

and

$$E_\theta = \frac{1}{j\omega\epsilon_0 r} \frac{\partial}{\partial r} (rH_\phi)$$

$$= \frac{-I_0}{j\omega\epsilon_0 4\pi r \sin \theta} \left(\left\{ \left[\frac{(r \cos \theta - z_2)}{r_2^3} + \frac{jk(r \cos \theta - z_2)}{r_2^2} + \frac{jk}{r_2} \right] \right. \right.$$
$$\left. \times (r - z_2 \cos \theta) - \frac{\cos \theta}{r_2} \right\} e^{-jk(z_2+r_2)} - \left\{ \left[\frac{(r \cos \theta - z_1)}{r_1^3} \right. \right.$$
$$\left. \left. + \frac{jk(r \cos \theta - z_1)}{r_1^2} + \frac{jk}{r_1} \right] (r - z_1 \cos \theta) - \frac{\cos \theta}{r_1} \right\} e^{-jk(z_1+r_1)} \right)$$
$$(3.97)$$

PROBLEMS

1. Assume that the current distribution on a $3\lambda/4$ thin linear antenna is sinusoidal. Calculate and plot the radiation pattern in the plane of the antenna.

2. Prove the formula given in Eq. 3.63 for the radiation resistance of a centre-fed dipole of length L.

3. Calculate the relative amplitudes of the radiation, induction and electrostatic fields at a distance of 2 wavelengths from a short current element.

4. If a half-wave dipole is located parallel to and at $\lambda/2$ from a plane conducting sheet, sketch the lines of current flow in the sheet. Use the image principle and the relation $J = \mathbf{n} \times \mathbf{H}$.

SUGGESTED READING

1. Collin, R. E. and Zucker, F. J., *Antenna Theory*, Parts 1 and 2, McGraw-Hill, New York, 1969.

2. Jasik, H., *Antenna Engineering Handbook*, McGraw-Hill, 1961.

3. Jordan, E. C. and Balmain, K. G., *Electromagnetic Waves and Radiating Systems*, Prentice-Hall of India, 1969.

4. Kraus, J. D., *Antennas*, McGraw-Hill, New York, 1950.

5. Schelkunoff, S. A. and Friis, H. T., *Antennas, Theory and Practice*, Chapman and Hall, London, 1952.

4

CYLINDRICAL ANTENNA

In Chapter 3, it has been assumed that the current distribution of an infinitesimally thin linear conducting antenna is sinusoidal. However, in practice the thin linear antenna is a cylinder of a very small diameter. In this chapter the current distribution on a cylindrical antenna is determined by using Hallen's integral-equation method. Hallen[1] and King[2] have considered this problem as a boundary-value problem.

4.1 Current Distribution on the Cylindrical Conducting Antenna

Figure 4.1 shows a centre-fed cylindrical antenna of length $2l$ and diameter $2a$. The boundary conditions to be satisfied are:

$$E'_z = E_z \tag{4.1}$$

along the cylindrical surface,

$$E'_\rho = E_\rho \tag{4.2}$$

at the two ends of the antenna, where E'_z and E_z are the z-components of the electric field at $\rho = a - da$ and $\rho = a + da$ respectively, and E'_ρ and E_ρ are the ρ-components of the electric field at $z = l - dl$ and $z = l + dl$ respectively, as shown in Fig. 4.2.

It is assumed that $l \gg a$ and that $\beta a \ll 1$. Then the effect of the end-faces can be neglected and the current I_z at $z = \pm l$ can be assumed to be zero. Then

$$E'_z = ZI_z \tag{4.3}$$

where Z is the impedance of the conductor in ohms per unit length of the conductor due to skin-effect, and I_z is the current in the cylinder.

---

[1] Erik Hallen, 'Theoretical Investigations into the Transmitting and Receiving Qualities of Antennae', *Nova Acta Regiae Soc. Sci. Upsaliensis*, Ser IV, 11, No. 4, 1–44, 1938.
[2] L. V. King, 'On the Radiation Field of a Perfectly Conducting Base-insulated Cylindrical Antenna over a Perfectly conducting Plane Earth, and the Calculation of the Radiation Resistance and Reactance', *Phil. Trans. Roy. Soc.* (London), 236, 381–422, 1937.

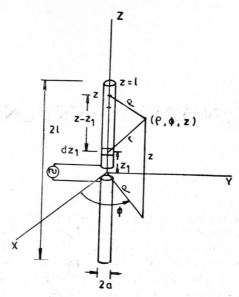

Fig. 4.1   Centre-fed cylindrical antenna.

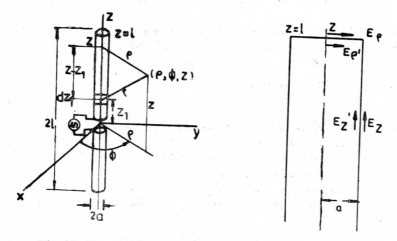

Fig. 4.2   Tangential components at the surfaces of the antenna.

If $\mathbf{A}$ is the magnetic vector potential due to the electric current I_z, then

$$\mathbf{E} = -j\frac{c^2}{\omega}\nabla(\nabla\cdot\mathbf{A}) - j\omega\mathbf{A} \tag{4.4}$$

Since the current is entirely in the z-direction, $\mathbf{A}$ has only a z-component A_z, and then Eq. 4.4 becomes,

$$E_z = -j\frac{\omega}{\beta^2}\left(\frac{\partial^2 A_z}{\partial z^2} + \beta^2 A_z\right) \tag{4.5}$$

From Eqs. 4.3 and 4.5, we have

$$\frac{\partial^2 A_z}{\partial z^2} + \beta^2 A_z = \frac{j\beta^2}{\omega} ZI_z \tag{4.6}$$

Equation 4.6 is a one-dimensional wave equation of the second order and first degree whose solution for A_z has to be determined. If $Z = 0$, then Eq. 4.6 becomes a homogeneous equation, and its solution A_z will be the sum of a complementary function A_c and a particular integral A_r. Thus,

$$A_z = A_c + A_r$$

$$= -\frac{j}{c} [C_1 \cos (\beta z) + C_2 \sin (\beta z)] + \frac{jZ}{c} \int_0^z I(s) \sin (z - s) \, ds \tag{4.7}$$

If the antenna is excited symmetrically by a pair of closely spaced terminals, then

$$I_z(z) = I_z(-z) \text{ and } A_z(z) = A_z(-z) \tag{4.8}$$

By differentiating Eq. 4.7 with respect to z and letting $z \to 0$,

$$\lim_{z \to 0} \left(\frac{\partial A_z}{\partial z} \right) = -\frac{j\beta}{c} C_2 \tag{4.9}$$

Using,

$$\nabla \cdot \mathbf{A} = -j \frac{\omega}{c^2} V \tag{4.10}$$

$$\frac{\partial A_z}{\partial z} = -j \frac{\omega}{c^2} V$$

so that

$$\frac{\partial A_z(+z)}{\partial z} = -j \frac{\omega}{c^2} V(+z) \text{ and } \frac{\partial A_z(-z)}{\partial z} = -j \frac{\omega}{c^2} V(-z) \tag{4.11}$$

Since

$$V(+z) = V(-z)$$

$$V_T = 2 \lim_{z \to 0} V(+z) = \frac{2jc^2}{\omega} \lim_{z \to 0} \frac{\partial A_z(+z)}{\partial z} \tag{4.12}$$

where V_T is the applied terminal voltage. Using Eq. 4.12 in Eq. 4.9, we obtain,

$$C_2 = \tfrac{1}{2} V_T \tag{4.13}$$

A_z can be expressed in terms of I_z as follows:

$$A_z = \frac{\mu_0}{4\pi} \int_{-l}^{+l} \frac{|I_{z1}|}{r} \, dz_1 = \frac{\mu_0 e^{j\omega t}}{4\pi} \int_{-l}^{+l} \frac{I_{z1} e^{-j\beta r}}{r} \, dz_1 \tag{4.14}$$

where $r = [(\rho^2 + (z - z_1)^2]^{1/2}$.

Using Eqs. 4.13 and 4.14 in Eq. 4.7, we obtain Hallen's integral equation

$$\frac{jc\mu_0 e^{j\omega t}}{4\pi} \int_{-l}^{+l} \frac{I_{z1} e^{-j\beta r}}{r} \, dz_1 = C_1 \cos(\beta z) + \frac{V_T}{2} \sin(\beta|z|)$$

$$- Z \int_0^z I(s) \sin(z - s) \, ds \qquad (4.15)$$

$|z|$ is used in the second term on the right side of Eq. 4.15 because of the symmetry condition given by Eq. 4.8. Equation 4.15 is to be solved for I_{z1}, which gives the current distribution on the antenna as a function of the antenna dimensions and the conductor impedance.

If the cylinder is a very good conductor, $Z \simeq 0$, so that Eq. 4.15 can be written as

$$30j \int_{-l}^{+l} \frac{I_{z1} e^{-j\beta r}}{r} \, dz_1 = C_1 \cos(\beta z) + \frac{V_T}{2} \sin(\beta|z|) \qquad (4.16)$$

In order to solve Eq. 4.16, add and subtract I_z to the integral in Eq. 4.16 as follows:

$$\int_{-l}^{+l} \frac{I_{z1} e^{-j\beta r}}{r} \, dz_1 = \int_{-l}^{+l} \frac{(I_z + I_{z1} e^{-j\beta r} - I_z)}{r} \, dz_1$$

$$= I_z \int_{-l}^{l} \frac{dz_1}{r} = \int_{-l}^{+l} \frac{(I_{z1} e^{-j\beta r} - I_z)}{r} \, dz_1 \qquad (4.17)$$

At $\rho = a$, i.e., on the surface of the cylinder,

$$\int_{-l}^{+l} \frac{dz_1}{r} = \Omega + \ln[1 - (z/l)^2] + \delta \qquad (4.18)$$

where

$$\Omega = 2 \ln(2l/a) \qquad (4.19)$$

and

$$\delta = \ln\{[\tfrac{1}{4}\sqrt{(a/(l-z))^2} + 1][\sqrt{1 + (a/(l+z))^2} + 1]\} \qquad (4.20)$$

Substituting Eq. 4.18 in Eq. 4.17, and then in Eq. 4.16, we obtain

$$I_z = -\frac{j}{30\Omega} \{C_1 \cos(\beta z) + \tfrac{1}{2}V_T \sin(\beta|z|)\}$$

$$- \frac{1}{\Omega} \left\{ I_z \ln[1 - (z/l)^2] + I_z\delta + \int_{-l}^{l} \frac{I_{z1} e^{-j\beta r} - I_z}{r} \right\} dz_1 \qquad (4.21)$$

when $z = l$, $I_z = 0$, and hence Eq. 4.21 becomes

$$0 = -\frac{j}{30\Omega} \{C_1 (\cos \beta l) + \tfrac{1}{2}V_T \sin(\beta l)\} + \frac{1}{\Omega} \int_{-l}^{+l} \frac{I_{z1} e^{-j\beta r_1}}{r_1} \, dz_1 \qquad (4.22)$$

where $r_1 = \sqrt{(l - z_1)^2 + a^2}$.

Subtracting Eq. 4.22 from Eq. 4.21, we have

$$I_z = -\frac{j}{30\Omega} [C_1 \{\cos(\beta z) - \cos(\beta l)\} + \tfrac{1}{2}V_T \{\sin(\beta|z|) - \sin(\beta l)\}]$$

$$- \frac{1}{\Omega} \left\{ I_z \ln[1 - (z/l)^2] + I_z\delta + \int_{-l}^{+l} \frac{(I_{z1} e^{-j\beta r} - I_z)}{r} \, dz_1 \right.$$

$$\left. - \int_{-l}^{+l} \frac{I_{z1} e^{-j\beta r_1}}{r_1} \, dz_1 \right\} \qquad (4.23)$$

$$\frac{1}{\Omega}\left\{I_z \ln [1 - (z/l)^2] + I_z\delta + \int_{-l}^{+l} \frac{(I_{z1}e^{-j\beta r} - I_z)}{r}\, dz_1 - \int_{-l}^{+l} \frac{I_{z1}e^{-j\beta r_1}}{r}\, dz_1\right\}$$

is put equal to zero, so that the zero-order approximation for I_z is

$$I_{z0} = -\frac{1}{30\Omega}(C_1 F_{0z} + \tfrac{1}{2}V_T G_{0z}) \tag{4.24}$$

where

$$F_{0z} = \cos (\beta z) - \cos (\beta l)$$
$$G_{0z} = \sin (\beta z) - \sin (\beta l) \tag{4.25}$$

Substituting Eq. 4.24 in Eq. 4.23, we obtain the first-order approximation I_{z1} as follows:

$$I_{z1} = -\frac{j}{30\Omega}\left[C_1\left(F_{0z} + \frac{F_{1z}}{\Omega}\right) + \tfrac{1}{2}V_T\left(G_{0z} + \frac{G_{1z}}{\Omega}\right)\right] \tag{4.26}$$

where

$$F_{1z} = F_1(z) - F_1(l) \tag{4.27}$$

$$F_1(z) = -F_{0z} \ln [1 - (z/l)^2] + F_{0z}\delta - \int_{-l}^{+l} \frac{(F_{0z1}e^{-j\beta r_1} - F_{0z})}{r}\, dz_1 \tag{4.28}$$

$$F_1(l) = -\int_{-l}^{+l} \frac{F_{0z1}e^{-j\beta r_1}}{r_1}\, dz_1 \tag{4.29}$$

$$G_1(z) = G_1(z) - G(l) \tag{4.30}$$

$G_1(z)$ is the same as $F_1(z)$ and $G_1(l)$ is the same as $F_1(l)$ with G substituted for F. If Eq. 4.26 is used in Eq. 4.21, the second-order approximation for the current is obtained. This process can be continued in order to obtain higher-order approximations, and I_z takes the form

$$I_z = -\frac{j}{30\Omega}\left[C_1\left(F_{0z} + \frac{F_{1z}}{\Omega} + \frac{F_{2z}}{\Omega} + \dots\right) + \tfrac{1}{2}V_T\left(G_{0z} + \frac{G_{1z}}{\Omega} + \frac{G_{2z}}{\Omega} + \dots\right)\right] \tag{4.31}$$

At $z = l$, $I_z = 0$ in Eq. 4.31. Hence,

$$C_1 = -\tfrac{1}{2}V_T\left[\frac{G_0(l) + (1/\Omega)G_1(l) + \dots}{F_0(l) + (1/\Omega)F_1(l) + \dots}\right] \tag{4.32}$$

Using Eq. 4.32 in Eq. 4.31, we obtain

$$I_z = j\frac{V_T}{60\Omega}\left[\frac{\sin \beta(l - |z|) + (b_1/\Omega) + (b_2/\Omega^2) + \dots}{\cos (\beta l) + d_1/\Omega + d_2/\Omega^2 + \dots}\right] \tag{4.33}$$

where

$$b_1 = F_1(z) \sin (\beta l) - F_1(l) \sin \beta|z| + G_1(l) \cos (\beta z) - G_1(z) \cos (\beta l) \tag{4.34}$$
$$d_1 = F_1(l) \dots \text{etc.} \tag{4.35}$$

Neglecting b_2, d_2 and higher order terms, the first-order solution for I_z is

$$I_z = \frac{jV_T}{60\Omega}\left[\frac{\sin \beta(l - |z|) + (b_1/\Omega)}{\cos (\beta l) + (d_1/\Omega)}\right] \tag{4.36}$$

King and Harrison[1] have calculated b_1 and d_1 for several values of l.

[1]Ronald King and C. W. Harrison, Jr., 'The Distribution of Current Along a Symmetrical Centre-Driven Antenna', Proc. I.R.E., 31, 548–567, Oct. 1943.

We have assumed that $l \gg a$ and $\beta a \ll 1$ in the foregoing analysis. Usually $l/a \geqslant 60$. When $l/a = 60$, $\Omega = 2 \ln 2l/a = 2 \ln 120 \simeq 9.6$. Unless l/a is large enough, Ω will not be large enough, and Eq. 4.33 will not asymptotically converge.

For antennas of non-circular cross-section, this analysis can be made to apply by considering an equivalent radius. For example, for a square cross-section of side length b, the equivalent radius is $a = 0.59b$; while for a thin strip of width w, the equivalent radius is $a = 0.25w$.

The amplitude and phase angle of the current distribution I_z as given by Eq. 4.36 can be calculated. Figures 4.3 and 4.4 show the amplitude and phase angle for a $\lambda/2$ long antenna and $1\frac{1}{4}\lambda$ long antenna respectively, as calculated by King and Harrison. For each length, two values of l/a, namely 75 and ∞ are considered. When the antenna is infinitesimally thin, the current amplitude distribution is almost sinusoidal but for thicker

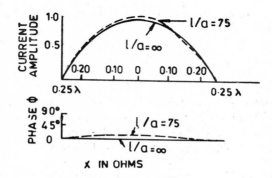

Fig. 4.3   Relative current amplitude and phase along a centre-fed 1/2 wavelength cylindrical antenna for $1/a = 75$ and $1/a = \infty$ (after King and Harrison).

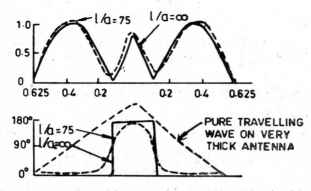

Fig. 4.4   Relative current amplitude and phase distribution along a centre-fed 5/4 wavelength cylindrical antenna for $1/a = 75$ and $1/a = \infty$ (after King and Harrison).

antennas, the current minimum has a value greater than zero. The phase variation is constant over 1/2 wavelength and changes by 180° for infinitesimally thin antennas, but for thicker antennas the change is less abrupt. For a very thick antenna the phase would tend to approach that of a pure travelling wave as shown by the straight line dashed curves in Fig. 4.4.

4.2   Input Impedance of Cylindrical Conducting Antenna

The input impedance Z_i of a centre-fed cylindrical antenna is given by the ratio of the input voltage V_T to the input current I_T. Therefore,

$$Z_i = V_T/I_T = R_i + jX_i \tag{4.37}$$

where $I_T = I_z(0)$.

Therefore, by putting $z = 0$ in Eq. 4.36, we obtain

$$Z_i = -j60\Omega \left[\frac{\cos (\beta l) + (d_1/\Omega)}{\sin (\beta l) + (b_1/\Omega)} \right] \tag{4.38}$$

which is a first-order approximation. If second-order terms are included, then

$$Z_i = -j60\Omega \left[\frac{\cos (\beta l) + d_1/\Omega + d_2/\Omega^2}{\sin (\beta l) + b_1/\Omega + b_2/\Omega^2} \right] \tag{4.39}$$

Equation 4.39 has been evaluated and presented in the form of impedance spirals in Fig. 4.5 for cylindrical centre-fed antennas. It can be seen from this figure that the variation of the input impedance with the frequency is much less for a thicker cylindrical antenna than for a thinner antenna.

These cylindrical antennas are usually operated at or near resonance,

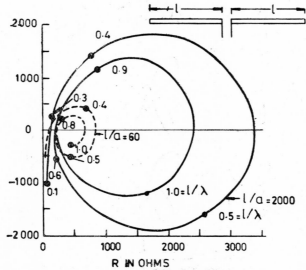

Fig. 4.5   Calculated input impedance $(R + jX)$ in ohms for cylindrical centre-fed antenna for (i) $1/a = 60$ and (ii) $1/a = 2000$.

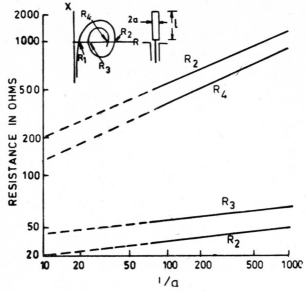

Fig. 4.6   Resonance resistance of cylindrical stub antenna
with ground plane as a function of $1/a$.

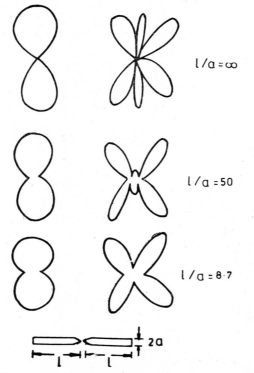

Fig. 4.7   Field patterns of cylindrical antennas for
different values of $1/a$.

where the input impedance is pure resistive. The resonant resistances calculated by Hallen are shown in Fig. 4.6 for the first four resonances of a cylindrical stub antenna with large ground plane as a function of l/a. The input impedance of a stub antenna of length l is $1/2$ the input impedance of a centre-fed antenna of length $2l$.

4.3 Field Patterns of Cylindrical Antennas

Figure 4.7 depicts some field patterns of cylindrical antennas for several values of l/a. It can be noticed that for thicker antennas the nulls are not zero and some of the minor lobes are missing.

4.4 Thin Cylindrical Antenna

If it is assumed that $l/a \rightarrow \infty$ ($\Omega \rightarrow \infty$), the current distribution given by Eq. 4.33 or Eq. 4.36 reduces to

$$I_z = \frac{jV_T}{60\Omega} \frac{\sin \beta(l - |z|)}{\cos \beta l} \tag{4.40}$$

As $\Omega \rightarrow \infty$, V_T/Ω may be kept constant by making $V_T \rightarrow \infty$.
Then

$$I_z = k \sin \beta(l - |z|) \tag{4.41}$$

where $k =$ constant.

The input impedance then becomes,

$$Z_T = \frac{V_T}{I_T} = -j60 \cot (\beta l) \tag{4.42}$$

In Eq. 4.42, Ω is large but finite. Z_T is a reactance as given by Eq. 4.42 and is the same as the input impedance of an open-circuited lossless transmission line of length l, if 60Ω is taken to be the characteristic impedance of the line. If 60Ω is taken as the average impedance Z_k of the cylindrical antenna, then

$$Z_k = 60\Omega = 120 \ln (2l/a)$$

We shall see in Chap. 5 that Eq. 4.42 is of the same form as the characteristic impedance Z_k of a thin biconical antenna.

SUGGESTED READING

1. Collin, R. E. and Zucker, F. J., *Antenna Theory, Parts 1 and 2*, McGraw-Hill, New York, 1969.

2. Jasik, H., *Antenna Engineering Handbook*, McGraw-Hill, 1961.

3. Jordan, E. C. and Balmain, K. G., *Electromagnetic Waves and Radiating Systems*, Prentice-Hall of India, 1969.

4. Kraus, J. D., *Antennas*, McGraw-Hill, New York, 1950.

5. Schelkunoff, S. A. and Friis, H. T., *Antennas, Theory and Practice*, Chapman and Hall, London, 1952.

5

BICONICAL ANTENNA

5.1 The Infinitely Long Biconical Antenna

The infinitely long biconical antenna as shown in Fig. 5.1 acts as a guide for spherical waves just as a uniform transmission line acts as a guide for plane waves. The generator connected to the terminals of the infinitely long biconical antenna causes spherical phase fronts to travel radially. These waves produce currents on the cones. If V is the voltage between points on the upper and lower cones a distance r from the terminals, and I is the total current on the surface of one of the cones at the distance r, then V/I may be called the characteristic impedance of the antenna. If it can be shown that V/I is independent of r, then the characteristic impedance is uniform. To determine V and I, the electromagnetic field supported by the bicone should be determined.

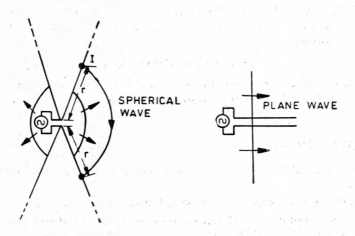

(a) Infinite biconical antenna   (b) Uniform transmission line

Fig. 5.1 Analogy between (a) infinite biconical antenna and (b) uniform transmission line.

Though the biconical transmission line can support an infinite number of modes, let it be assumed that it supports only the fundamental TEM mode. For this mode, both **E** and **H** are transverse to r, **E** being equal to $\mathbf{u}_\theta E_\theta$ and **H** being equal to $\mathbf{u}_\phi H_\phi$, both E_θ and H_ϕ being independent of ϕ.

Then the two Maxwell's equations

$$\nabla \times \mathbf{E} = -j\omega\mu_0\mathbf{H} \tag{5.1}$$

and

$$\nabla \times \mathbf{H} = j\omega\epsilon_0\mathbf{E} \tag{5.2}$$

reduce to

$$\frac{1}{r}\frac{\partial(rE_\theta)}{\partial r} = -j\omega\mu_0 H_\phi \tag{5.3}$$

and

$$\frac{\partial}{\partial r}(rH_\phi) = -j\omega\epsilon_0(rE_\theta) \tag{5.4}$$

Since $E_r = 0$, from Eq. 5.2

$$\frac{\partial(\sin\theta H_\phi)}{\partial\theta} = 0 \tag{5.5}$$

From Eq. 5.5,

$$H_\phi \propto \frac{1}{\sin\theta} \tag{5.6}$$

From Eqs. 5.3 and 5.4, it can be shown that

$$\frac{\partial^2}{\partial r^2}(rH_\phi) = -\omega^2\mu_0\epsilon_0(rH_\phi) \tag{5.7}$$

A solution of Eq. 5.7, using Eq. 5.6, can be written as

$$H_\phi = \frac{1}{r\sin\theta}H_0 e^{-j\beta r} \tag{5.8}$$

where

$$\beta = \omega(\mu_0\epsilon_0)^{1/2} = 2\pi/\lambda \tag{5.9}$$

Equation 5.9 represents a travelling wave travelling in the $+r$ direction. It can be shown that

$$E_\theta = Z_0 H_\phi = \frac{Z_0}{r\sin\theta}H_0 e^{-j\beta r} \tag{5.10}$$

where $Z_0 = (\mu_0/\epsilon_0)^{1/2}$.

The voltage $V(r)$ between points 1 and 2 on the cones at a distance r from the terminals (Fig. 5.2) is given by

$$V(r) = \int_{\theta_c}^{\pi-\theta_c} E_\theta r\, d\theta \tag{5.11}$$

where θ_c is the half angle of the cone.

Using Eq. 5.10 in Eq. 5.11, we obtain,

$$V(r) = Z_0 H_0 e^{-j\beta r}\int_{\theta_c}^{\pi-\theta_c}\frac{d\theta}{\sin\theta} = Z_0 H_0 e^{-j\beta r}\ln\frac{\cot(\theta_c/2)}{\tan(\theta_c/2)} \tag{5.12}$$

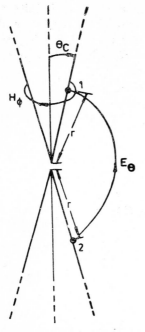

Fig. 5.2   The electromagnetic field
of the bicone.

or
$$V(r) = 2Z_0 H_0 e^{-j\beta r} \ln \cot (\theta_c/2)$$
$$(5.13)$$

The total current $I(r)$ on the cone at a distance r from the terminals is given by

$$I(r) = \int_0^{2\pi} H_\phi r \sin \theta \; d\theta$$

$$= 2\pi r H_\phi \sin \phi$$

$$= 2\pi H_0 e^{-j\beta r}$$

Therefore the characteristic impedance Z_k of the biconical antenna is given by

$$Z_k = \frac{V(r)}{I(r)} = \frac{Z_0}{\pi} \ln \cot (\theta_c/2)$$
$$(5.14)$$

For free space $= Z_0 = (\mu_0/\epsilon_0)^{1/2} = 120\pi$.

Hence
$$Z_k = 120 \ln \cot (\theta_c/2) \quad (5.15)$$

If θ_c is small ($<20°$), $\cot (\theta_c/2) \simeq 2/\theta_c$) so that

$$Z_k = 120 \ln (2/\theta_c) \text{ ohms} \qquad (5.16)$$

which is a pure resistance. Comparing Eq. 5.16 with Eq. 4.42 for a thin cylinder, if $a/l = \theta_c$, the two expressions are identical. Because Eqs. 5.14 and 5.16 are independent of r, it can be said that the biconial transmission line or antenna has a uniform characteristic impedance. The input impedance Z_i

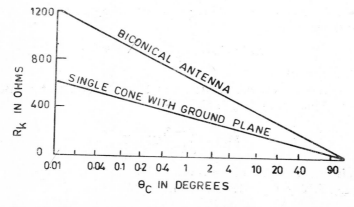

Fig. 5.3   Characteristic resistance of infinite biconial antenna and
single cone with ground plane.

of the biconial antenna is $V(r)/I(r)$ as $r \to 0$, and is equal to Z_k. Therefore,

$$Z_i = Z_k \tag{5.17}$$

Figure 5.3 shows the characteristic impedance Z_k (or R_k) as a function of the biconical antenna. For the single cone with a ground plane, this characteristic resistance is R_k.

5.2 Finite Length Biconical Antenna

For a finite length biconical antenna, there will be higher-order modes in addition to the TEM mode. All these modes will be reflected at the sphere of radius l, where l is the length of the cones. This is equivalent to a load impedance Z_L connected across the open end of the cones as shown in Fig. 5.4. If the effect of the end caps of the cone is neglected, the finite length biconical antenna can be treated as a transmission line, having a characteristic impedance Z_k, terminated in a load impedance Z_L. Then the input impedance Z_i at the input terminals of the biconical antenna is given by

$$Z_i = Z_k \frac{Z_L + jZ_k \tan (\beta l)}{Z_k + jZ_L \tan (\beta l)} \tag{5.18}$$

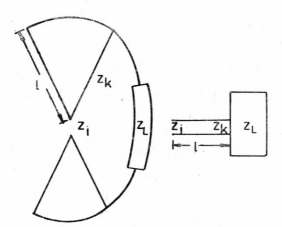

Fig. 5.4   Finite length biconical antenna.

Schelkunoff[1] has calculated Z_L by first calculating Z_m at a current maximum on a very thin biconical antenna as shown in Fig. 5.5, by assuming a sinusoidal current distribution. Z_m is the impedance which appears between the current maximum on one cone and the current maximum on the other cone. This impedance occurs at $\lambda/4$ distance from the open end of the antenna, and therefore, Z_L can be transformed over a $\lambda/4$ line.

---

[1] S. A. Schelkunoff, *Electromagnetic Waves*, D. Van Nostrand Company, Inc., New York, 1943, Chap. 11, p. 441.

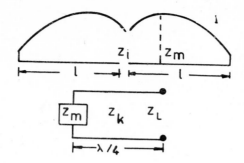

Fig. 5.5   Thin finite-length biconical antenna and
its transmission-line equivalent.

Finally Z_i is Z_L transformed over a line of characteristic impedance Z_k
and length l. Hence

$$Z_L = Z_k \frac{Z_m + jZ_k (\tan \beta x)}{Z_k + jZ_m (\tan \beta x)} \tag{5.19}$$

where $\beta x = \pi/2$ and hence

$$Z_L = Z_k^2/Z_m \tag{5.20}$$

Z_m is of the form $R_m + jX_m$. Schelkunoff has given the following expressions for R_m and X_m for a thin biconical antenna.

$$R_m = 60 \operatorname{Cin} (2\beta l) + 30[0.577 + \ln (\beta l) - 2 \operatorname{Ci} (2\beta l) + \operatorname{Ci} (4\beta l)]$$

$$\times \cos (2\beta l) + 30[\operatorname{Si} (4\beta l) - 2\operatorname{Si} (2\beta l)] \sin (2\beta l) \tag{5.21}$$

and

$$X_m = 60 \operatorname{Si} (2\beta l) + 30[\operatorname{Ci} (4\beta l) - \ln (\beta l) - 0.577] \sin (2\beta l)$$

$$- 30[\operatorname{Si} (4\beta l)] \cos (2\beta l) \tag{5.22}$$

Then

$$Z_i = Z_k \frac{Z_k + jZ_m \tan (\beta l)}{Z_m + jZ_k \tan (\beta l)} \tag{5.23}$$

where $Z_m = R_m + jX_m$ and Z_k is given by Eq. 5.16.

Z_m is independent of the cone angle for thin cones. However Z_k depends on the cone angle θ_c. Therefore Z_i is a function of θ_c.

Figure 5.6 shows the calculated input impedance of two biconical antennas having θ_c equal to 2.7° and 0.027°. It can be seen that for the larger cone angle, the impedance is less frequency dependent, and hence the biconical antenna can be considered to be a broad-band antenna for slightly larger cone angles.

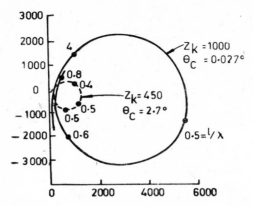

Fig. 5.6   Calculated input impedance of two biconical antennas
with $\theta_e = 2.7°$ and 0.027° as function of $1/\lambda$.

PROBLEMS

1. Prove that the characteristic impedance Z_k for a single cone and ground plane is half
    of Z_k for a biconical antenna.

2. If the length of the cone is $3\lambda/4$ and the biconical antenna has a half angle of 1°,
    calculate the terminal impedance.

SUGGESTED READING

1. Collin, R. E. and Zucker, F. J., *Antenna Theory*, Parts 1 and 2, McGraw-Hill, New
    York, 1969.

2. Jasik, H., *Antenna Engineering Handbook*, McGraw-Hill, 1961.

3. Jordan, E. C. and Balmain, K. G., *Electromagnetic Waves and Radiating Systems*,
    Prentice-Hall of India, 1969.

4. Kraus, J. D., *Antennas*, McGraw-Hill, New York, 1950.

5. Schelkunoff, S. A. and Friis, H. T., *Antennas, Theory and Practice*, Chapman and
    Hall, London, 1952.

6

ANTENNA ARRAYS

One of the usual methods of obtaining higher directive gain is an arrangement of several individual antennas so spaced and phased that their individual contributions add in one preferred direction and cancel in other directions. Such an arrangement is known as an array of antennas.

6.1 Linear Arrays of n Isotropic Point Sources of Equal Amplitude and Spacing

Figure 6.1 shows a linear array of n isotropic point source of equal amplitude and spacing. Though a point source is fictitious for electromagnetic waves, it is a convenient concept. This is because if we confine our attention only to the far-field of the antenna where the electric and magnetic fields vary as $1/r$, then we may assume, by extrapolating inward along the radii of a sphere, that the waves originate at a fictitious point source at the centre of the observation circle.

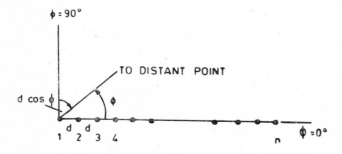

Fig. 6.1   Linear array of n point sources.

The total field E at a large distance in the direction ϕ due to the linear array shown in Fig. 6.1 is given by

$$E = 1 + e^{j\psi} + e^{2j\psi} + e^{3j\psi} + \ldots + e^{j(n-1)\psi} \tag{6.1}$$

where

$$\psi = d_r \cos \phi + \delta \qquad (6.1a)$$

$$d_r = \frac{2\pi d}{\lambda}$$

and δ is the phase difference of adjacent sources.

Let the amplitudes of the fields of all the sources be taken as equal to unity. Source 1 is considered as the phase centre.

Multiplying Eq. 6.1 by $e^{j\psi}$, we obtain

$$Ee^{j\psi} = e^{j\psi} + e^{j2\psi} + e^{j3\psi} + \dots e^{jn\psi} \qquad (6.2)$$

Subtracting Eq. 6.2 from Eq. 6.1 and dividing by $(1 - e^{j\psi})$, we obtain

$$E = \frac{1 - e^{jn\psi}}{1 - e^{j\psi}} = e^{j\xi} \frac{\sin (n\psi/2)}{\sin (\psi/2)} = \frac{\sin (n\psi/2)}{\sin (\psi/2)} \angle \xi \qquad (6.3)$$

where

$$\xi = \left(\frac{n-1}{2} \right) \psi \qquad (6.4)$$

Considering the centre point of the array instead of source 1 as the phase centre

$$E = \frac{\sin (n\psi/2)}{\sin (\psi/2)} \qquad (6.5)$$

Equation 6.1 may be considered as the addition of several vectors, and the total field E may be obtained by a graphical vector addition as shown in Fig. 6.2. The phase of the field is constant wherever E has a value but

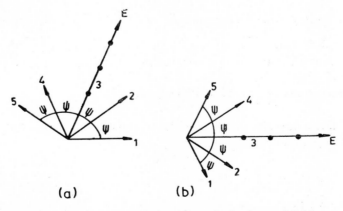

Fig. 6.2   Vector addition of fields to obtain the total field at a large distance from a linear array of 5 point sources of equal amplitude and equal spacing. (a) Source 1 as phase centre, (b) Source 3 (mid-point) as phase centre.

changes by π in directions for which $E = 0$ (null directions) and Eq. 6.5 changes sign.

When $\psi = 0$, Eqs. 6.4 and 6.5 are indeterminate, and hence E must be obtained as the limit of Eq. 6.5 as $\psi \to 0$. Thus for $\psi = 0$, we have

$$E = n \qquad (6.6)$$

Equation 6.6 gives the maximum value of E, and putting $E_{max} = n$, we obtain the normalized value of E as

$$E = \frac{1}{n} \frac{\sin (n\psi/2)}{\sin (\psi/2)} \qquad (6.7)$$

Equation 6.7 is known as the array factor. Figure 6.3 shows the array factor as a function of $\pm\psi$ for various values of n.

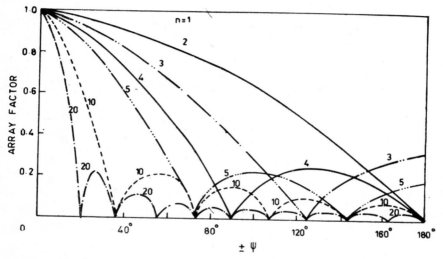

Fig. 6.3   Array factor as a function of ψ for various values of n.

The field is maximum when $\psi = 0$. But for some cases, ψ may not become zero. Then the field is maximum at the minimum value of ψ.

6.2   Linear Broadside Array of Point Sources

In a linear array of n isotropic point sources of the same amplitude and phase, $\delta = 0$, and therefore

$$\psi = d_r \cos \phi \qquad (6.8)$$

For $\psi = 0$, $\phi = (2k + 1) \dfrac{\pi}{2}$, where $k = 0, 1, 2, 3, \ldots$. Therefore the field is maximum when

$$\phi = \frac{\pi}{2} \quad \text{and} \quad \frac{3\pi}{2}$$

so that the maximum field is in a direction normal to the array. Such an array is called a broadside array. For example, Fig. 6.4 shows the field and phase patterns of a broadside array of four point sources in phase, spaced $\lambda/2$ from each other.

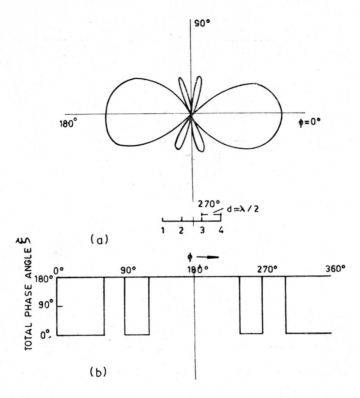

Fig. 6.4 Field and phase patterns of broadside array of 4 point sources $(d = \lambda)$: (a) Field pattern, (b) Phase pattern with phase centre at mid-point of array.

6.3 Ordinary End-Fire Array of Point Sources

For an end-fire array, the maximum occurs at $\phi = 0$, i.e., in the direction of the array. Hence $\psi = 0$ when $\phi = 0$, and therefore from Eq. 6.1a,

$$\delta = -d_r \qquad (6.9)$$

which means that the phase between adjacent sources is retarded progressively by the same amount as the spacing between sources in radians.

For example, for an end-fire array of 4-point sources with $d = \lambda/2$,

$\delta = -d_r = -\dfrac{2\pi d}{\lambda} = -\dfrac{2\pi}{2} = -\pi$. Figure 6.5 shows the field and phase patterns of such an end-fire array.

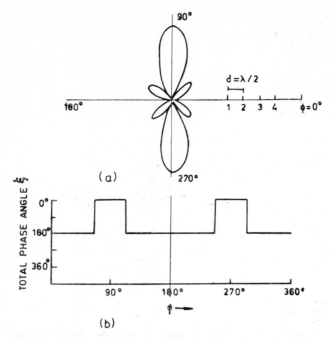

Fig. 6.5   Field and phase patterns of end-fire array of 4-point
sources $(d = \lambda/2)$. (a) Field pattern, (b) phase-
pattern with mid-point of array as phase centre.

6.4   End-fire Arrary with Increased Directivity

Hansen and Woodyard[1] have shown that a higher directivity is obtained
by making

$$\delta = -\left(d_r + \frac{\pi}{n}\right) \tag{6.10}$$

This condition is called 'the Hansen and Woodyard's condition of increased directivity'. For example, for a linear array of 4 point sources with
$d = \lambda/2$, $\delta = -(\pi + \pi/4)$. $= -5\pi/4$. The field pattern of such an array is
shown in Fig. 6.6. In this pattern the main lobe is considerably sharper
but the back lobe is much larger.

To realize a better pattern with increased directivity, the spacing d
should be reduced and n should be increased so that $|\psi|$ should be about
π/n at $\phi = 0$ and about π at $\phi = 180°$. For example, if $n = 10$ and $d = \lambda/4$,
then $\delta = -0.6\pi$. Figure 6.7 shows the field pattern of such an array which
definitely has increased directivity.

[1]W. W. Hansen and J. R. Woodyard, A New Principle in Directional Antenna Design',
Proc. I.R.E., 26, March 1938, 333–345.

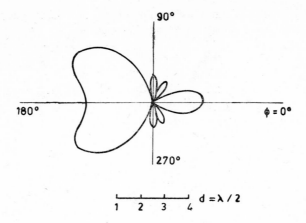

Fig. 6.6 Field pattern of end-fire array of 4 point sources
($d = \lambda/2$) and $\delta = -5\pi/4$.

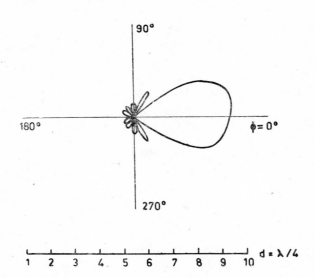

Fig. 6.7 Field pattern of end-fire array of 10 point sources
($d = \lambda/4$) with increased directivity.

6.5 Array with Maximum Field in an Arbitrary Direction

If the linear array of n point sources is to have a maximum in some
arbitrary direction ϕ_1 not equal to $k\pi/2$, where $k = 0$, 1, 2, or 3, then

$$0 = d_r \cos \phi_1 + \delta \qquad (6.11)$$

By specifying the value of d_r and knowing ϕ_1, δ can be found out.

6.6 Direction of Nulls and Maxima for Arrays of n Isotropic Point Sources of Equal Amplitude and Spacing

6.6.1 Nulls

From Eq. 6.3 it can be seen that the null directions for an array of n isotropic point sources of equal amplitude and spacing occur when $E = 0$, i.e., when

$$e^{jn\psi} = 1 \tag{6.12}$$

provided

$$e^{j\psi} \neq 1 \tag{6.13}$$

Equation 6.12 requires that

$$n\psi = \pm 2K \tag{6.14}$$

where $K = 1, 2, 3, \ldots$.

This makes

$$\psi = d_r \cos \phi_0 + \delta = \pm \frac{2K\pi}{n} \tag{6.15}$$

or

$$\phi_0 = \cos^{-1} \left[\left(\pm \frac{2K\pi}{n} - \delta \right) \frac{1}{d_r} \right] \tag{6.16}$$

where ϕ_0 gives the direction of the pattern nulls. The condition (Eq. 6.13) requires that the values of $K = mn$ should be excluded, where $m = 1, 2, 3, \ldots$.

In a *broadside array*, $\delta = 0$, so that

$$\phi_0 = \cos^{-1} \left(\pm \frac{2K}{nd_r} \right) = \cos^{-1} \left(\pm \frac{K}{nd} \right) \tag{6.17}$$

If γ_0 is the complementary angle of ϕ_0, then $\gamma_0 = \pi/2 - \phi_0$, so that

$$\gamma_0 = \sin^{-1} \left(\pm \frac{K\lambda}{nd} \right) \tag{6.18}$$

For a long array for which $nd \gg k\lambda$,

$$\gamma_0 \simeq \pm k\lambda/nd \tag{6.19}$$

For an ordinary end-fire array,

$$\delta = -d_r,$$

so that from Eq. 6.16

$$\cos \phi_0 - 1 = \pm \frac{2k\pi}{nd_r} \tag{6.20}$$

or

$$\frac{\phi_0}{2} = \sin^{-1} \left\{ \pm \sqrt{\frac{K\pi}{nd_r}} \right\} \tag{6.21}$$

for a long array for which $nd \gg K\lambda$,

$$\phi_0 \simeq \pm \sqrt{\frac{2K\lambda}{nd}} \tag{6.22}$$

Hence the total beamwidth of the main lobe between first nulls for a long ordinary end-fire array is

$$2\phi_{01} \simeq \pm \sqrt{\frac{2\lambda}{nd}} \qquad (6.23)$$

For end-fire arrays with increased directivity as proposed by Hansen and Woodyard, Eq. 6.16 becomes

$$d_r (\cos \phi_0 - 1) - \frac{\pi}{n} = \pm 2 \frac{K\pi}{n} \qquad (6.24)$$

or

$$\frac{\phi_0}{2} = \sin^{-1} \left\{ \pm \sqrt{\frac{\pi}{2nd_r} (2K - 1)} \right\} \qquad (6.25)$$

or

$$\phi_0 = 2 \sin^{-1} \sqrt{\frac{\lambda}{4nd} (2K - 1)} \qquad (6.26)$$

For a long array for which $nd \gg K\lambda$,

$$\phi_0 \simeq \pm \sqrt{\frac{\lambda}{nd} (2K - 1)} \qquad (6.27)$$

Hence the total beamwidth of the main lobe between first nulls for a long end-fire array with increased directivity is given by

$$2\phi_{01} \simeq 2\sqrt{\frac{\lambda}{nd}} \qquad (6.28)$$

which is $1/\sqrt{2}$ or 71 per cent of the beamwidth of the ordinary end-fire array. Figure 6.8 shows the beamwidth between first nulls as a function of nd_λ for array of n isotropic point sources of equal amplitude, for ordinary end-fire, end-fire with increased directivity and for broadside.

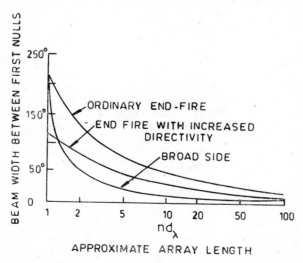

Fig. 6.8   Beamwidth between first nulls as a function of nd_λ for arrays of n isotropic point sources of equal amplitude.

6.6.2 Maxima

The major lobe maximum usually occurs when $\psi = 0$, for broadside or ordinary end-fire arrays. For the broadside array this occurs at $\phi = 90°$ and $\phi = 270°$, while for an ordinary end-fire array it occurs at $\phi = 0°$ and $\phi = 180°$. For the end-fire array with increased directivity, the main lobe maximum occurs at $\psi = \pm\pi/n$ with the main lobe at $0°$ or $180°$.

The maxima of the minor lobes are situated between the first and higher order nulls. These maxima occur approximately whenever the numerator of Eq. 6.7 is a maximum, i.e., when

$$\sin (n\psi/2) = 1 \qquad (6.29)$$

The numerator $\sin (n\psi/2)$ of Eq. 6.7 varies as a function of ψ more rapidly than the denominator $\sin (\psi/2)$, especially true when n is large. This can be seen in Fig. 6.9. Therefore, the maxima occur approximately when $\sin (n\psi/2) = 1$, which requires that

$$\frac{n\psi}{2} = \pm(2K + 1)\,\frac{\pi}{2} \qquad (6.30)$$

where $K = 1, 2, 3, \ldots$.

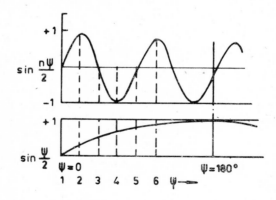

Fig. 6.9 $\sin (n\psi/2)$ and $\sin (\psi/2)$ versus ψ for $n = 0$. Points 1, 2, 3, 4, 5, 6 represent the first maximum for broadside or ordinary end-fire array, first maximum for end-fire with increased directivity, first null, approximate maximum of first minor lobe, second null, and approximate maximum of second minor lobe respectively.

Using Eq. 6.30 in Eq. 6.1a, we have

$$d_r \cos \phi_m + \delta = \frac{\pm(2K + 1)\pi}{n} \qquad (6.31)$$

or

$$\phi_m \simeq \cos^{-1}\left\{\left[\frac{\pm(2K + 1)\pi}{n} - \delta\right]\frac{1}{d_r}\right\} \qquad (6.32)$$

where ϕ_m = direction of the minor lobe maxima.

For a broadside array, $\delta = 0$, so that Eq. 6.32 becomes

$$\phi_m \simeq \cos^{-1}\left\{\frac{\pm(2K+1)\lambda}{2nd}\right\} \tag{6.33}$$

For an ordinary end-fire array, $\delta = -d_r$ so that

$$\phi_m \simeq \cos^{-1}\left[\frac{\pm(2K+1)\lambda}{2nd} + 1\right] \tag{6.34}$$

and for an end-fire array with increased directivity $\delta = -(d_r + \pi/n)$ so that

$$\phi_m \simeq \cos^{-1}\left\{\frac{\lambda}{2nd}[1 \pm (2K+1)] + 1\right\} \tag{6.35}$$

Since $\sin(n\psi/2)$ is approximately equal to unity at the maximum of a minor lobe, the relative amplitude of a minor lobe maximum E_{mL} is given by

$$E_{mL} \simeq \frac{1}{n\sin(\psi/2)} \tag{6.36}$$

or

$$E_{mL} \simeq \frac{1}{n\sin[(2K+1)\pi/2n]} \tag{6.37}$$

When $n \gg K$, i.e., for the first few minor lobes of an array of a large number of sources,

$$E_{mL} \simeq \frac{2}{(2K+1)\pi} \tag{6.38}$$

In a broadside ordinary end-fire array, the major-lobe maximum is unity so that the relative amplitudes of the maximum and first five minor lobes are 1, 0.22, 0.13, 0.09, 0.07 and 0.06 respectively. For an end-fire array with increased directivity the maximum for $\phi = 0$ and $n = 20$ occurs at $\psi = \pi/20 = 9°$, and at this value the array factor is 0.63. Hence the relative amplitudes for such an array become 1, 0.35, 0.21, 0.14, 0.11 and 0.09.

The maximum value of the smallest minor lobe occurs for $2K + 1 = n$. Then,

$$\sin\left[\frac{(2K+1)\pi}{2n}\right] \simeq 1 \tag{6.39}$$

and

$$E_{mL} \simeq \frac{1}{n} \tag{6.40}$$

6.7 Two Isotropic Point Sources of Unequal Amplitude and Any Phase Difference

Let the two sources 1 and 2 be situated on the x-axis at a distance d as shown in Fig. 6.10(a). Let source 1 have a field E_0 at a large distance r, and let the field from source 2 at the same distance be aE_0. Then from

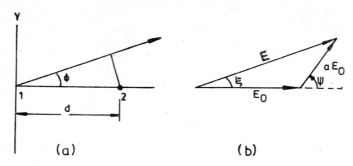

Fig. 6.10   Two isotropic sources of unequal amplitude and arbitrary phase: (a) Geometry, (b) Vector addition of fields $\xi = \underline{/\text{arc tan}\ (a\ \sin/1 + a\ \cos)}$.

Fig. 6.10(b), the magnitude and phase angle of the total field E is given by

$$E = E_0 \sqrt{(1 + a \cos \psi)^2 + a^2 \sin^2 \psi} \Big/ \tan^{-1} \left(\frac{a \sin \psi}{1 + a \cos \psi} \right) \quad (6.41)$$

where $\psi = d_r \cos \phi + \delta$, and the phase angle of E is referred to source 1.

6.8 Nonisotropic Similar Point Sources and the Principle of Pattern Multiplication

An isotropic point source radiates equally in all directions, while a non-isotropic point source does not do so. Two nonisotropic point sources are said to be 'similar' if the variation of the amplitudes and phases of their radiation fields with absolute angle ϕ is the same. The patterns must not only be of the same shape but also must be oriented in the same direction. The maximum amplitudes of the individual sources may be unequal. If the maximum amplitudes are equal, then the sources are not only 'similar' but 'identical'. The principle of 'pattern multiplication' is expressed as follows:

'The total field pattern of an array of nonisotropic similar sources is the product of the individual source pattern of an array of isotropic point sources each located at the phase centre of the individual source and having the same relative amplitude and phase, and the total phase pattern referred to the phase centre of the array is the sum of the phase patterns of the individual source and the array of isotropic point sources'.

In symbols this principle may be written as

$$E = f(\theta, \phi) F(\theta, \phi) \underline{/f_p(\theta, \phi) + F_p(\theta, \phi)} \quad (6.42)$$

where

$f(\theta, \phi) =$ field pattern of individual source

$f_p(\theta, \phi) =$ phase pattern of individual source

$F(\theta, \phi) =$ field pattern of array of isotropic source

$F_p(\theta, \phi) =$ phase pattern of array of isotropic source

In Eq. 6.42, the patterns are expressed as functions of both polar angles θ and ϕ to indicate that the principle of pattern multiplication applies to space patterns as well as to two-dimensional cases.

For example, consider an array of two nonisotropic point sources as shown in Fig. 6.11. Let each individual source pattern be given by

$$E_0 = \cos \phi \qquad (6.43)$$

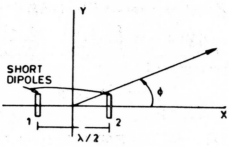

Fig. 6.11   Array of two non-isotropic point source.

which is produced by short dipoles oriented parallel to the y-axis and spaced $\lambda/2$ from each other as shown in Fig. 6.12(a). By the principle of pattern multiplication, the total normalized field is given by

$$E = \cos \phi \cos \left(\frac{\pi}{2} \cos \phi \right) \qquad (6.44)$$

Figure 6.12(b) shows the array pattern and Fig. 6.12(c) shows the total pattern.

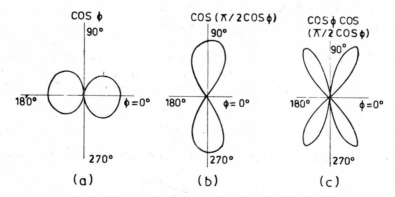

Fig. 6.12   Pattern multiplication for array of two non-isotropic similar point sources. (a) Individual pattern, (b) Array pattern, (c) Total pattern.

6.9   Array of Nonisotropic and Dissimilar Point Sources

If the sources are dissimilar, the principle of pattern multiplication cannot be applied. On the other hand, the fields of the sources must be added at each pair of angles (θ, ϕ) for which the total field is calculated. Thus, for two dissimilar sources 1 and 2 situated on the x-axis with source 1 at the origin, and the sources being separated by a distance d as shown in Fig. 6.10(a), the total field is given by

$$E = E_1 + E_2$$

$$= E_0\{[f(\phi) + aF(\phi) \cos \psi]^2 + [aF(\phi) \sin \psi]^2\}^{\mu/2}$$

$$\bigg/ f_p(\phi) + \tan^{-1}\left\{\frac{aF(\phi) \sin \psi}{f(\phi) + aF(\phi) \cos \psi}\right\} \tag{6.45}$$

where the field from source 1 is

$$E_1 = E_0 f(\phi) \diagup f(\phi) \tag{6.46}$$

and the field from source 2 is

$$E_2 = aE_0 F(\phi) \diagup F_p(\phi) + d_r \cos \phi + \delta \tag{6.47}$$

where

$E_0 = $ constant

$a = $ ratio of maximum amplitude of source 2 to source 1 $(0 \leqslant a \leqslant 1)$

$\psi = d_r \cos \phi + \delta - f_p(\phi) + F_p(\phi)$

$\delta = $ relative phase of source 2 with respect to source 1

$f(\phi) = $ relative field pattern of source 1

$f_p(\phi) = $ phase pattern of source 1

$F(\phi) = $ relative field pattern of source 2

$F_p(\phi) = $ phase pattern of source 2

The vector-addition of the fields is shown in Fig. 6.13. In this figure

$$\xi^1 = \tan^{-1}\left\{\frac{aF(\phi) \sin \psi}{f(\phi)aF(\phi) \cos \psi}\right\} \tag{6.48}$$

and

$$\xi = \xi^1 + f_p(\phi) \tag{6.49}$$

As an example, consider two nonisotropic dissimilar point sources 1

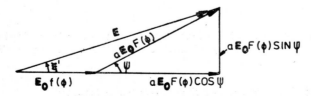

Fig. 6.13   Vector addition of the fields of two non-isotropic dissimilar point sources.

and 2 having field patterns E_1 and E_2 given by

$$E_1 = \cos \phi \angle 0 \qquad (6.50)$$

and

$$E_2 = \sin \phi \angle \psi \qquad (6.51)$$

where $\psi = d_r \cos \phi + \delta$. Let $d = \lambda/4$ and $\delta = \pi/2$. Then

$$\psi = \frac{\pi}{2} (\cos \phi + 1) \qquad (6.52)$$

Figure 6.14 shows the two individual patterns and the total pattern.

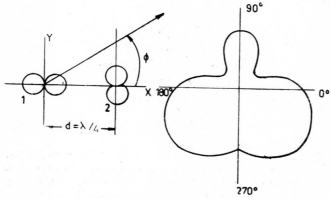

Fig. 6.14   Two non-isotropic dissimilar point sources (δ=90°).
(a) Individual pattern, (b) Total pattern.

6.10   Linear Broadside Arrays with Nonuniform Amplitude Distribution

There are four types of nonuniform amplitude distribution of broadside arrays, which are important. These are:

(i) Uniform

(ii) Binomial

(iii) Edge

(iv) Optimum

As an example, consider a broadside array of 5 isotropic point sources with $\lambda/2$ spacing. Figure 6.15 shows the normalized field patterns for the four types of nonuniform distribution. From this figure, it can be seen that though the binomial distribution gives no side lobes, the uniform distribution gives the minimum beamwidth of major lobe but large side lobes. However, the optimum distribution gives a compromise between the major lobe beamwidth and the side lobe levels. The edge distribution gives several large lobes. The optimum distribution is called the Dolph-Tchebyscheff distribution, which will be considered in the next section.

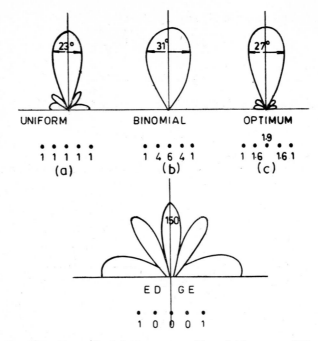

Fig. 6.15   Normalized field patterns of broadside arrays of 5 isotropic point sources spaced λ/2 apart. (a) Uniform distribution, (b) Bionomial distribution, (c) Optimum
distribution, (d) Edge distribution.

6.11   Linear Arrays with Optimum or Dolph-Tchebyscheff Distribution

Consider a linear array of even number n_e or odd number n_0 of isotropic point sources of uniform spacing d and of the same phase as shown in Fig. 6.16. The individual sources have amplitudes $A_0, A_1, A_2, \ldots$ etc., the amplitude distribution being symmetrical about the centre of the array.

For an even number of elements, the total field E_{ne} is given by

$$E_{ne} = 2A_0 \cos(\psi/2) + 2A_1 \cos(3\psi/2) + \ldots + 2A_k \cos\left(\frac{n_e - 1}{2}\psi\right)$$

(6.53)

where

$$\psi = \frac{2\pi d}{\lambda} \sin\theta = d_r \sin\theta$$

(6.54)

E_{ne} can be rewritten as

$$E_{ne} = 2 \sum_{k=0}^{k=N-1} A_k \cos\left(\frac{2k+1}{2}\psi\right)$$

(6.55)

where $N = n_e/2$.

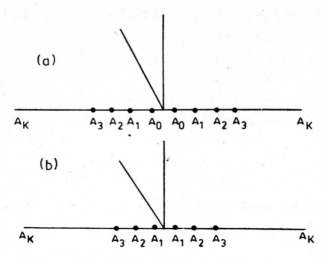

Fig. 6.16   Linear broadside arrays of n isotropic sources with uniform spacing: (a) n even, (b) n odd.

For an odd number of elements, the total field E_{n0} is given by

$$E_{n0} = 2A_0 + 2A_1 \cos \psi + 2A_2 \cos (2\psi) + \ldots + 2A_k \cos \left(\frac{n_0 - 1}{2} \psi\right)$$

$$= 2 \sum_{k=0}^{k=N} A_k \cos \left(2k \frac{\psi}{2}\right) \tag{6.56}$$

where

$$N = \left(\frac{n_0 - 1}{2}\right)$$

The series represented in Eqs. 6.55 and 6.56 are finite Fourier series of N terms.

In the Dolph-Tchebyscheff distribution, it can be shown that the coefficients of the pattern series can be uniquely determined so as to produce a pattern of minimum beamwidth for a specified side-lobe level. Both Eqs. 6.55 and 6.56 are polynomials of degree $(n_e - 1)$ and $(n_0 - 1)$ respectively.

A broadside array will be considered so that $\delta = 0$.

Consider the series

$$\cos (m\psi/2) = \cos^m \frac{\psi}{2} - \frac{m(m - 1)}{2!} \cos^{m-2} \frac{\psi}{2} \sin^2 \frac{\psi}{2}$$

$$+ \frac{m(m - 1)(m - 2)(m - 3)}{4!} \cos^{m-4} \frac{\psi}{2} \sin^4 \frac{\psi}{2} \ldots \tag{6.57}$$

[expanding $\cos (m\psi/2) = Re(\cos \psi/2 + j \sin \psi/2)^m$]. Putting $x = \cos \psi/2$, .

$\sin^2 \psi/2 = 1 - \cos^2 \psi/2 = (1 - x^2)$ and $T_m(x) = \cos (m\psi/2)$, we have

$$T_0(x) = 1$$
$$T_1(x) = x$$
$$T_2(x) = 2x^2 - 1$$
$$T_3(x) = 4x^3 - 3x$$
$$T_4(x) = 8x^4 - 8x^2 + 1 \qquad (6.58)$$
$$T_5(x) = 16x^5 - 20x^3 + 5x$$
$$T_6(x) = 32x^6 - 48x^4 + 18x^2 - 1$$
$$T_7(x) = 64x^7 - 112x^5 + 56x^3 - 7x$$

$T_m(x)$, $m = 0, 1, 2, \ldots$ are called the Tchebyscheff polynomials.

$$T_m(x) = 0 \text{ when } \cos (m\psi/2) = 0 \text{ or } m(\psi/2) = (2k - 1)\pi/2 \qquad (6.59)$$

or

$$x = x' = [\cos (2k - 1)\pi/2m] \qquad (6.60)$$

Figure 6.17 show the Tchebyscheff polynomials $T_m(x)$ as functions of x, for $m = 0, 1, 2, 3, 4, 5$. From this figure, it can be seen that:

(i) All the $T_m(x)$ polynomials pass through the point (1, 1).

(ii) For values of x in the range $-1 \leqslant x \leqslant +1$, all the polynomials lie between ordinate values of $+1$ and -1. All roots occur in this range, and all maximum or minimum values in this range are ± 1.

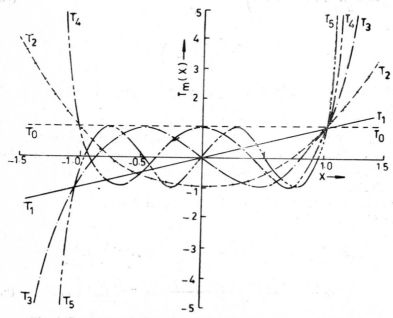

Fig. 6.17   Tchebyscheff polynomials of degree $m = 0, 1, 2, 3, 4, 5$.

Dolph-Tchebyscheff method of optimizing the linear broadside array is as follows. The field pattern of a linear broadside array of sources is a polynomial of degree equal to the number of sources less 1. This array polynomial is equated to the Tchebyscheff polynomial of the same degree. For example, for an array of 6 sources, the array polynomial which is of degree 5 is equated to the T-polynomial of degree 5 (Fig. 6.18). Let the ratio of the main lobe maximum to the minor lobe level be equal to R.

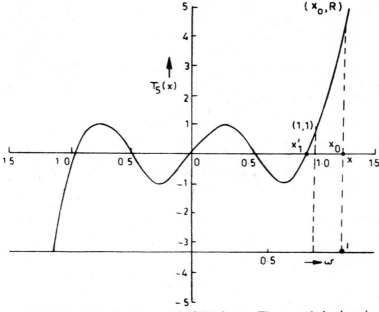

Fig. 6.18   Tchebyscheff polynomial of 5th degree. The w-scale is given in addition to the x-scale.

Then the point (x_0, R) on the $T_5(x)$ polynomial corresponds to the main lobe maximum, while all the minor lobes have a maximum value of 1. The roots of the polynomial correspond to the nulls of the field pattern. An important property of the T-polynomial is that if the ratio R is specified, the beamwidth to the first null ($x = x_1'$) is minimized, and if the beamwidth is specified then the ratio R is maximized.

The field polynomials (Eqs. 6.55 or 6.56) are equated to $T_{n-1}(x)$, where n is the number of sources n_e or n_0. Put $m = n - 1$ and let

$$T_m(x) = T_{n-1}(x_0) = R \qquad (6.61)$$

For $R > 1$, $x_0 > 1$ (from Figs. 6.17 and 6.18). But since $x = \cos \phi/2$, $|x| < 1$. To avoid this difficulty, a new variable w is introduced, where

$$w = x/x_0 \qquad (6.62)$$

and w is put equal to $\cos \psi/2$. Then $-1 \leqslant w \leqslant +1$. Then the pattern polynomial (Eq. 6.55 or 6.56) is expressed as a polynomial in w or in

(x/x_0). This array polynomial E_n is equated to the T-polynomial $T_{n-1}(x)$, i.e.

$$E_n(x/x_0) = T_{n-1}(x) \tag{6.63}$$

The coefficients of the array polynomial are obtained from Eq. 6.63, and hence the amplitude distribution is obtained.

To prove that the Tchebyscheff polynomial gives the optimum distribution, consider any other polynomial $P_5(x)$ of degree 5 which passes through (x_0, R) and through the highest root x_1' in Fig. 6.18, and lies between -1 and $+1$ for all smaller values of x. If $P_5(x)$ is less than ± 1 at its maxima and minima, then $P_5(x)$ would give a smaller side-lobe level for this beamwidth, and $T_5(x)$ will not be optimum. Since $P_5(x)$ lies between $+1$ and -1 in the range $-x_1' \leqslant x \leqslant +x_1'$, it must intersect the curve $T_5(x)$ in at least $5 + 1 = 6$ points, including (x_0, R). The polynomials of the same degree $m = 5$ which intersect in $m + 1 = 5 + 1 = 6$ points must be the same polynomial. Therefore,

$$P_5(x) = T_5(x) \tag{6.64}$$

and, therefore, $T_5(x)$ is the optimum polynomial.

6.12 Dolph-Tchebyscheff Distribution for a Linear Broadside Array of Eight Isotropic Point Sources

Let an array of 8 in-phase point sources spaced $\lambda/2$ apart, have a side-lobe level of 26 dB below the main-lobe level. It is required to find the amplitude distribution of such an array to produce minimum beamwidth between first nulls.

$$20 \log_{10} R = 26 \text{ dB} \tag{6.65}$$

and therefore,

$$R = 20 \tag{6.66}$$

The T-polynomial of degree $m - 1 = 8 - 1 = 7$ is $T_7(x)$.

Therefore,

$$T_7(x_0) = 20 \tag{6.67}$$

Hence

$$x_0 = \tfrac{1}{2}[R + \sqrt{R^2 - 1}]^{1/m} + [R - \sqrt{R^2 - 1}]^{1/m} \tag{6.68}$$

or

$$x_0 = \tfrac{1}{2}[20 + \sqrt{20^2 - 1}]^{1/7} + [20 - \sqrt{20^2 - 1}]^{1/7} \tag{6.68a}$$

Hence

$$x_0 = 1.15 \tag{6.69}$$

Putting $w = x/x_0 = \cos \psi/2$ in Eq. 6.53, we have,

$$
\begin{aligned}
E_8 &= A_0 w + A_1(4w^3 - 3w) + A_2(16w^5 - 20w^3 + 5w) \\
&\quad + A_3(64w^7 - 112w^5 + 56w^3 - 7w) \\
&= \frac{64A_3}{x_0^7} x^7 + \frac{(16A_2 - 112A_3)}{x_0^5} x^5 + \frac{(4A_1 - 20A_2 + 56A_3)}{x_0^3} x^3 \\
&\quad + \frac{(A_0 - 3A_1 + 5A_2 - 7A_3)}{x_0} x
\end{aligned} \tag{6.70}
$$

Equation 6.58 is equated to $T_7(x)$ given by

$$T_7(x) = 64x^7 - 112x^5 + 56x^3 = 7x \tag{6.71}$$

or

$$E_8 = T_7(x) \tag{6.72}$$

which requires

$$64A_3/x_0^7 = 64 \tag{6.73}$$

or

$$A_3 = x_0^7 = (1.15)^7 = 2.66 \tag{6.74}$$

Equating the other coefficients of the terms of like degree in Eq. 6.72, we have

$$A_2 = 4.56$$
$$A_1 = 6.82 \tag{6.75}$$
$$A_0 = 8.25$$

Therefore, the relative amplitudes of the 8 sources are

$$1, 1.7, 2.6, 3.1, 2.6, 1.7, 1$$

Using these amplitudes for the 8 sources in $E_8(x)$, and remembering that

$$\frac{\psi}{2} = \frac{dr \sin \theta}{2}$$

$$\cos \frac{\psi}{2} = w \tag{6.76}$$

and

$$x/x_0 = w$$

so that

$$x = x_0 \cos \frac{dr \sin \theta}{2} \tag{6.77}$$

$E_8(x)$ can be evaluated. Figure 6.19 shows the relative field pattern of a

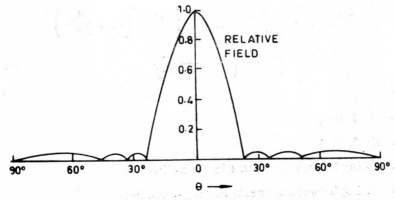

Fig. 6.19 Relative field pattern of broadside array of 8 isotropic point sources spaced λ/2 apart, with optimum Tchebyscheff distribution.

broadside array of 8 isotropic point sources spaced $\lambda/2$ apart having the optimum Tchebyscheff distribution.

6.13 Planar and Volume Arrays

While a linear array can steer the beam in only one dimension by either varying the phase difference δ between adjacent elements or changing the frequency, or varying the amplitudes of the elements, a planar array can steer the beam in two dimensions.

Let there be a *planar array* consisting of M elements along the x-direction and N elements along the y-direction in the z-plane as shown in Fig. 6.20. Let the mn-th element be excited by a source $V_{mn} \exp\{j(m\psi_x + n\psi_y)\}$;

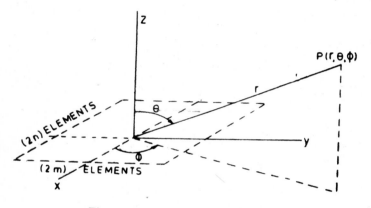

Fig. 6.20   Planar array in the z-plane.

where the steering phases ψ_x and ψ_y are the incremental phase shifts between elements in the x and y directions respectively. Then the array factor of the planar array of MN elements is

$$S(T_x, T_y) = \sum_{m=1}^{M-1} \sum_{n=1}^{N-1} V_{mn} \exp\left\{ j \frac{2\pi mb}{\lambda}\left(T_x + \frac{\psi_x}{2\pi b/\lambda}\right) + \frac{2\pi nd}{\lambda}\left(T_y + \frac{\psi_y}{2\pi d/\lambda}\right) \right\} \tag{6.78}$$

where

$T_x = \sin\theta \cos\phi$

$T_y = \sin\theta \sin\phi$

$b = $ spacing between elements in the x-direction

$d = $ spacing between elements in the y-direction

If the array elements are excited uniformly in amplitude, so that

$V_{mn} = V_{00} \exp \{j(m\psi_x + n\psi_y)\}$, then the array factor becomes

$$S_0(T_x, T_y) = V_0 \sum_{m=1}^{M-1} \exp \left\{ j \frac{2\pi mb}{\lambda} \left(T_x + \frac{\psi_x}{2\pi b/\lambda} \right) \right\}$$

$$\times \sum_{n=1}^{N-1} \exp \left\{ j \frac{2\pi nd}{\lambda} \left(T_y + \frac{\psi_y}{2\pi d/\lambda} \right) \right\}$$

$$= S_x S_y \tag{6.79}$$

where

$$S_x = \sum_{m=1}^{M-1} \exp \left\{ j \frac{2\pi mb}{\lambda} \left(T_x + \frac{\psi_x}{2\pi b/\lambda} \right) \right\} \tag{6.80}$$

= array factor of linear array of M elements in the x-direction and

$$S_y = \sum_{n=1}^{N-1} \exp \left\{ j \frac{2\pi nd}{\lambda} \left(T_y + \frac{\psi_y}{2\pi d/\lambda} \right) \right\} \tag{6.81}$$

= array factor of linear array of N elements in the y-direction.
The total field of the planar array is given by

$$E = E_e(T_x, T_y) S(T_x, T_y) \tag{6.82}$$

where

$$E_e(T_x, T_y) = E_e(\sin \theta \cos \phi, \sin \theta \sin \phi)$$

= far field pattern of the individual antenna element of the array.

From Eq. 6.82, it can be seen that by properly varying the steering phases ψ_x, ψ_y and also the amplitude V_{mn} of the elements, the main beam of the antenna can be made to scan the whole space.

A *volume array* or a *three-dimensional array* can also be used for this purpose (Fig. 6.21). The array factor of such an array is

$$S(\alpha, \beta, \theta) = \sum_{m=1}^{M-1} \sum_{n=1}^{N=1} \sum_{p=1}^{P-1} a_{mnp} \exp [jk(ps_x \cos \alpha + ms_y \cos \beta + ns_z \cos \theta)]$$

$$\tag{6.83}$$

where M, N and P are the number of elements and s_x, s_y and s_z are the spacings between consecutive elements in the x, y and z directions respectively, α, β and θ are the angles made by the radius vector to the distant point with the x, y and z axes and a_{mnp} is the current in the mnp-th element.

Let

$$a_{mnp} = A_{mnp} \exp [-jk(m\psi_x + n\psi_y + p\psi_z)] \tag{6.84}$$

where ψ_x, ψ_y and ψ_z are the progressive phase shifts in the x, y and z directions respectively. Then,

$$S(\alpha, \beta, \theta) = \sum_{m=1}^{M-1} \sum_{n=1}^{N-1} \sum_{p=1}^{P-1} A_{mnp} \exp [-jk(m\psi_x + n\psi_y + p\psi_z)$$

$$+ jk(ms_x \cos \alpha + ns_y \cos \beta + ps_z \cos \theta)] \tag{6.85}$$

The array factor $S(\alpha, \beta, \theta)$ given by Eq. 6.85 has to be multiplied by the element factor $E(\theta, \phi)$ to obtain the total pattern.

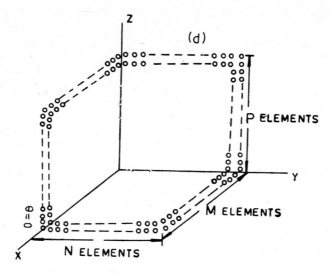

Fig. 6.21   Three-dimensional array.

If the array consists of a large number of elements, usually the array factor is more directional than the element pattern, and hence the main beam is determined by the array factor. For an equally spaced uniform cubic array, the array factor $S(\alpha, \beta, \theta)$ can be put in the form

$$S(\alpha, \beta, \theta) = S_x S_y S_z \qquad (6.86)$$

where S_x, S_y and S_z are the array factors of the linear arrays in the x, y and z directions respectively.

Examining the array factor given by Eq. 6.78 for the planar array, it can be seen that because of the periodicity of the array factor in T_x and T_y, the shape of the main beam and associated side-lobes will be repeated every λ/b and λ/d intervals in the T_x and T_y axes respectively. Such repetitions of the main lobe are called *grating lobes* which actually result in more than one principal beam and associated secondary maxima. Such grating lobes can exist for cubic arrays also.

Arrays can be designed on surfaces of other shapes which are non-planar. Such arrays are called *conformal arrays*, and the word 'conformal' signifies that the array conforms to the shape of the object on which the array is to be designed. Such conformal arrays are very useful on ships or flying vehicles. It is possible to study theoretically the behaviour of conformal arrays on simple surfaces like cylinders, spheres, etc.

PROBLEMS

1. Derive an expression for the array factor for an array of 4 isotropic point sources located at the positions (1, 0), (0, 1), (−1, 0) and (0, −1) in the x-y plane.

2. For a linear array of 6 equal isotropic point sources located along the x-axis, with spacing equal to $\lambda/3$, and with phase difference of 45° between adjacent sources, derive the array pattern.

3. Locate the maxima and minima for the array of Problem 2.

4. Derive the array factor of a bionomial array of 7 isotropic point sources.

5. Derive the Dolph-Tchebyscheff distribution for a broadside array of 10 point sources to give a minimum beamwidth for a side-lobe level of 26 dB. Let the spacing between sources be $3\lambda/2$.

SUGGESTED READING

1. Collin, R. E. and Zucker, F. J., *Antenna Theory*, Parts 1 and 2, McGraw-Hill, New York, 1969.

2. Jasik, H., *Antenna Engineering Handbook*, McGraw-Hill, 1961.

3. Jordan, E. C. and Balmain, K. G., *Electromagnetic Waves and Radiating Systems*, Prentice-Hall of India, 1969.

4. Kraus, J. D., *Antennas*, McGraw-Hill, New York, 1950.

5. Schelkunoff, S. A. and Friis, H. T., *Antennas Theory and Practice*, Chapman and Hall, London, 1952.

7

LOOP ANTENNA

7.1 Equivalence of Small Loop to Short Magnetic Dipole

A small loop of area A and carrying an electric current I can be replaced by an equivalent magnetic dipole of length l and carrying a fictitious magnetic current I_m, as shown in Fig. 7.1(a) and (b) respectively.

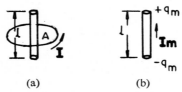

(a)   (b)

Fig. 7.1 Small loop antenna and its equivalent. (a) Small loop, (b) Equivalent magnetic dipole.

If q_m is the pole strength at each end as shown in Fig. 7.1(b), then $q_m l$ is the moment of the magnetic dipole. The magnetic current I_m is then given by

$$I_m = -\mu \frac{dq_m}{dt} \tag{7.1}$$

where $I_m = I_{m0} e^{j\omega t}$.

Integrating Eq. 7.1, we obtain

$$q_m = -I_m / j\omega\mu \tag{7.2}$$

Equating the magnetic moment $q_m l$ of the magnetic dipole to the magnetic moment IA of the small loop, we have

$$q_m l = IA \tag{7.3}$$

or

$$\frac{-I_m l}{j\omega\mu} = IA \tag{7.4}$$

by using Eq. 7.2 in Eq. 7.3.

Therefore,

$$I_m l = -j\omega\mu IA$$

$$= -j2\pi f \frac{\lambda}{\lambda} IA$$

$$= -j2\pi \frac{Z_0}{\lambda} \mu IA \tag{7.5}$$

or

$$I_m l = -j240\pi^2 I \frac{A}{\lambda} \tag{7.6}$$

(Substituting $Z_0 = 120\pi$ for free space). In retarded from Eq. 7.6 becomes

$$[I_m]l = -j240\pi^2[I] \frac{A}{\lambda} \tag{7.7}$$

where

$$[I_m] = I_{m0} \exp\left\{j\omega\left(t - \frac{r}{c}\right)\right\} \tag{7.8}$$

and

$$[I] = I_0 \exp\left\{j\omega\left(t - \frac{r}{c}\right)\right\} \tag{7.9}$$

7.2   Radiation or Far-Field of Short Magnetic Dipole and its Equivalent Small Loop

The method of finding the radiation field of a short-magnetic dipole is similar to that applied to a short electric dipole as in Section 3.1. The electric current I is replaced by a fictitious magnetic current I_m, E is replaced by H and the magnetic vector potential A is replaced by the electric vector potential F.

Let the magnetic dipole or magnetic current element be placed along the z-axis as shown in Fig. 7.2. The electric vector potential F of the magnetic current element is given by

$$\mathbf{F} = \mathbf{u}_z \frac{\mu_0}{4\pi} \int_{-l/2}^{l/2} \frac{[I_m]}{r} \, dz \tag{7.10}$$

Using Eq. 7.8 in Eq. 7.10 we have,

$$\mathbf{F} = \mathbf{u}_z F_z = \frac{\mu_0 I_{m0}}{4} \int_{-l/2}^{l/2} \frac{\exp\{j\omega(t - r/c)\}}{r} \tag{7.11}$$

If $r \gg 1$ and $\lambda \gg 1$, the phase difference of the contributions of the various current elements of length dz along the magnetic dipole can be neglected. Then,

$$F_z = \frac{\mu_0 I_{m0} l \exp\{j\omega(t - r/c)\}}{4\pi r} \tag{7.12}$$

Then the electric field E is obtained from

$$\mathbf{E} = \frac{1}{\mu_0} \nabla \times \mathbf{F} \tag{7.13}$$

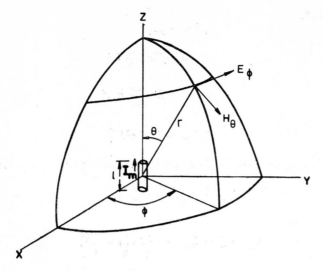

Fig. 7.2   Short magnetic dipole.

which gives

$$\mathbf{E} = \mathbf{u}_\phi E_\phi = \frac{[I_m]l \sin \theta}{4\pi}\left(\frac{j\omega}{cr} + \frac{1}{r^2}\right) \tag{7.14}$$

(converting F_z into spherical polar coordinates).

It can be noted that Eq. 7.14 is identical with the expression given in Eq. 3.24 for H_ϕ of a short electric dipole, with E replaced by H and I_m replaced by I.

For the radiation field only the $1/r$ term need be retained, and hence,

$$E_\phi = \frac{j[I_m]\omega l \sin \theta}{4\pi cr}$$

$$= \frac{j[I_m] \sin \theta}{2r}\frac{l}{\lambda} \tag{7.15}$$

Equation 7.15 is the far or radiation electric field of a short magnetic dipole of length l and carrying a fictitious magnetic current I_m. The far or radiation magnetic field H_θ is given by

$$H_\theta = \frac{E_\phi}{\eta_0} = \frac{j[I_m] \sin \theta}{240\pi r}\frac{l}{\lambda} \tag{7.16}$$

where $\eta_0 = (\mu_0/\epsilon_0)^{1/2}$.

Using Eq. 7.6 for the equivalent loop, the far-field components of the small electric current loop are given by

$$E_\phi = \frac{120\pi^2[I] \sin \theta}{r}\frac{A}{\lambda^2} \tag{7.17}$$

and

$$H_\theta = \frac{\pi[I] \sin \theta}{r}\frac{A}{\lambda^2} \tag{7.18}$$

Equations 7.17 and 7.18 are identical
with Eqs. 3.25 and 3.26 with I replaced by I_m and E replaced by H
and vice versa.

The far-field pattern of a small
loop is shown in Fig. 7.3.

7.3 Large Circular Loop Antenna with Uniform In-phase Current

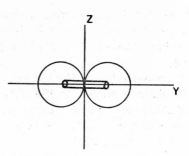

Fig. 7.3   Far-field pattern of small loop.

Consider a large circular loop of
radius a with uniform in-phase current as shown in Fig. 7.4, situated in
the X-Y plane. Although the condition of uniform in-phase current is
obtained easily on a small loop, for a large loop whose perimeter is $\geqslant \lambda/4$,
phase shifters must be introduced around the periphery in order to obtain
a uniform in-phase current on the loop.

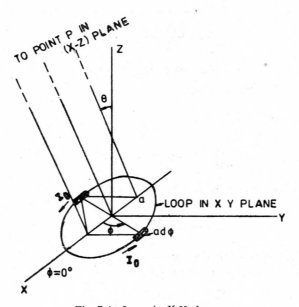

Fig. 7.4   Loop in X-Y plane.

Consider a pair of infinitesimally short diametrically opposite electric
dipoles of length $a\,d\phi$ as shown in Fig. 7.4. Since the current is along the
loop, it has a component only in the ϕ direction, and hence the magnetic
vector potential has only a ϕ component A_ϕ. The vector potential dA_ϕ at
a distant point P from the two diametrically opposed infinitesimally short
electric dipoles is

$$dA_\phi = \frac{\mu_0\, dM}{4\pi r} \tag{7.19}$$

where dM is the current moment due to this pair of electric dipoles carrying current I and of length $a\,d\phi$.

A cross-section of the loop in the X-Z plane is shown in Fig. 7.5. In this plane, $\phi = 0°$, and the ϕ component of the retarded current moment due to one dipole is

$$[I]a\,d\phi \cos \phi \qquad (7.20)$$

where $[I] = I_0 \exp \left\{ j\omega \left(t - \dfrac{r}{c} \right) \right\}$.

Due to the pair of diametrically opposed dipoles, the resultant moment dM is

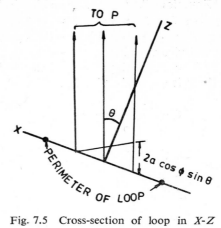

Fig. 7.5   Cross-section of loop in X-Z plane.

$$dM = 2j[I]a\,d\phi \cos \phi \sin \frac{\psi}{2} \qquad (7.21)$$

where $$\psi = 2\beta a \cos \phi \sin \theta$$

Hence,

$$dM = 2j[I]a \cos \phi[\sin (\beta a \cos \phi \sin \theta)]\,d\phi \qquad (7.22)$$

Therefore for the whole loop,

$$A_\phi = \int_{\phi=0}^{\pi} dA_\phi$$

$$= \frac{j\mu_0[I]a}{2\pi r} \int_0^\pi \sin (\beta a \cos \phi \sin \theta) \cos \phi\,d\phi \qquad (7.23)$$

by using Eq. 7.22 in Eq. 7.19.

Or

$$A_\phi = \frac{j\mu_0[I]a}{2r} J_1(\beta a \sin \theta) \qquad (7.24)$$

where J_1 is the Bessel function of the first order.

The far electric field has only a ϕ component given by

$$E_\phi = -j\omega A_\phi$$

$$= \frac{\mu_0\omega[I]a}{2r} J_1(\beta a \sin \theta)$$

$$= \frac{60\pi\beta a[I]}{r} J_1(\beta a \sin \theta) \qquad (7.25)$$

The magnetic field at a large distance has only a θ component given by

$$H_\theta = E_\phi/\eta_0$$

$$= \frac{\beta a[I]}{2r} J_1(\beta a \sin \theta) \qquad (7.26)$$

If $C_\lambda = \dfrac{2\pi a}{\lambda} = \beta a$, then E_ϕ or H_θ vary as $J_1(C_\lambda \sin \theta)$ and the far-field radiation pattern varies as $J_1(C_\lambda \sin \theta)$. Figure 7.6 shows $J_1(C_\lambda \sin \theta)$ as a

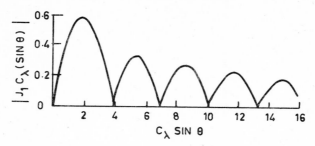

Fig. 7.6 $|J_1 C_\lambda(\sin \theta)|$ as a function of $C_\lambda \sin \theta$.

function of $C_\lambda \sin \theta$. This figure is useful in obtaining the radiation patterns of large loops. Figure 7.7 shows the radiation or far-field patterns of loops of different diameter.

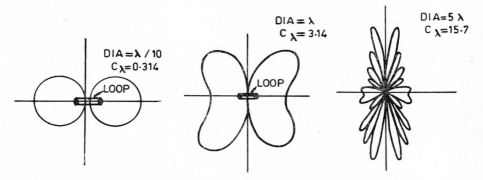

Fig. 7.7   Far field patterns of loops of different diameter.

As a special case of a small loop, $C_\lambda \sin \theta$ is very small and hence x is very small. Using the approximation

$$J_1(x) = x/2$$

for very small x, from Eq. 7.25

$$E_\phi = \frac{60\pi\beta a[I]}{2r} \beta a \sin \theta \qquad (7.27)$$

$$= \frac{120\pi^2[I] \sin \theta}{r} \frac{A}{\lambda^2} \qquad (7.28)$$

which is the same as Eq. 7.17.

7.4   Radiation Resistance of a Loop

The radiation resistance R_r of a loop is defined in the same manner as that of a linear antenna, i.e.,

$$R_r = 2W/I_0^2 \tag{7.29}$$

where W is the total power radiated and I_0 is the peak current in line on the loop. This radiation resistance appears at the input terminals of the loop, as shown in Fig. 7.8. The total radiated power W is given by

$$W \iint_S P_r \, da \tag{7.30}$$

where P_r = radial component of the Poynting vector and is given by

$$P_r = \tfrac{1}{2} |H|^2 Z_0 \tag{7.31}$$

where $|H|$ is the absolute value of the far magnetic field and equal to $|H_\theta|$ and Z_0 is the intrinsic impedance of free-space. S is a large sphere having the loop at its centre. Hence,

Fig. 7.8   Loop connected to a transmission line.

$$W = \iint_S P_r \, da = \iint_S \tfrac{1}{2} |H|^2 Z_0 \, da \tag{7.32}$$

$$= 15\pi(\beta a I_0)^2 \int_{\phi=0}^{2\pi} \int_{\theta=0}^{\pi} J_1^2(\beta a \sin \theta) \sin \theta \, d\theta \, d\phi \tag{7.33}$$

by using Eq. 7.26.
    Or

$$W = 30\pi^2(\beta a I_0)^2 \int_0^{\pi} J_1^2(\beta a \sin \theta) \sin \theta \, d\theta \tag{7.34}$$

For a small loop, Eq. 7.34 reduces to

$$W = \frac{15}{2} \pi^2(\beta a)^4 I_0^2 \int_0^{\pi} \sin^3 \theta \, d\theta$$

$$= 10\pi^2 \beta^4 a^4 I_0^2$$

$$= 10\beta^4 A^4 I_0^2 \tag{7.35}$$

(putting $A = \pi a^2$)
so that the radiation resistance R_r is given by

$$R_r = \frac{2W}{I_0^2} = \frac{20\beta^4 A^2 I_0^2}{I_0^2} = 10\beta^4 A^2 \tag{7.36}$$

or

$$R_r = 31,171(A/\lambda^2)^2 = 197 C_\lambda^4 \text{ ohms} \tag{7.37}$$

The value of R_r given by Eq. 7.37 is the radiation resistance of a small loop of area A. If the small loop consists of n number of turns, then

$$R_r = 31,171 \left(n \frac{A}{\lambda^2}\right)^2 \text{ Ohms} \tag{7.38}$$

For a larger loop of radius 'a', Eq. 7.34 has to be integrated. It is noted

that

$$\int_0^\pi J_1^2(x \sin \theta) \sin \theta \, d\theta = \frac{1}{x} \int_0^{2x} J_2(y) \, dy \qquad (7.39)$$

Using Eq. 7.39 in Eq. 7.34, we have

$$W = 30\pi^2 \beta a I_0^2 \int_0^{2\beta a} J_2(y) \, dy \qquad (7.40)$$

and hence

$$R_r = \frac{2W}{I_0^2} = 60\pi^2 C_\lambda \int_0^{2C_\lambda} J_2(y) \, dy \text{ Ohms} \qquad (7.41)$$

For large loops for which $C_\lambda \geqslant 5$, the following approximation can be used.

$$\int_0^{2C_\lambda} J_2(y) \, dy \simeq 1 \qquad (7.42)$$

so that

$$R_r = 60\pi^2 C_\lambda = 592 C_\lambda$$

$$= 3{,}720 \frac{a}{\lambda} \qquad (7.43)$$

For example, for a loop of perimeter 10, R_r is 5920 ohms.

For values of C_λ between 1/3 and 5, the integral in Eq. 7.34 can be evaluated using the transformation

$$\int_0^{2C_\lambda} J_2(y) \, dy = \int_0^{2C_\lambda} J_0(y) \, dy - 2J_1(2C_\lambda) \qquad (7.44)$$

Figure 7.9 shows the radiation resistance of single turn loops with uniform current as a function of C_λ.

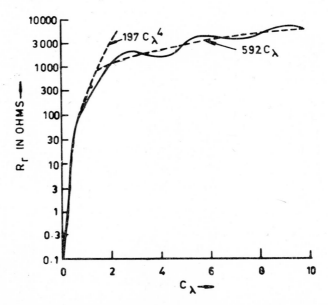

Fig. 7.9 Radiation resistance R_r of single-turn circular loop with uniform, in-phase current as a function of C_λ.

7.5   Directivity of Circular Loop Antennas with Uniform Current

The directivity D of an antenna is defined as the ratio of maximum radiation intensity to the average radiation intensity. The radiation intensity is power per unit angle. For the loop, the maximum radiation intensity is given by $r^2 P_r$ where P_r is given by Eq. 7.31 and the average radiation intensity is given by $W/4\pi$, where W is given by Eq. 7.33. Hence D is given by

$$D = \frac{2C_\lambda[J_1^2(C_\lambda \sin \theta)]_{max}}{\displaystyle\int_0^{2C_\lambda} J_2(y)\, dy} \tag{7.45}$$

For a small loop ($C_\lambda < 1/3$), D becomes

$$D = [\tfrac{3}{2} \sin^2 \theta]_{max} = \tfrac{3}{2} \tag{7.46}$$

This value of 3/2 is also the directivity of a short electric current element or dipole.

For a large loop, D becomes

$$D = 2C_\lambda J_1^2(C_\lambda \sin \theta) \tag{7.47}$$

For a loop with $C_\lambda \geqslant 1.84$, the maximum value of $J_1(C_\lambda \sin \theta)$ is 0.582. Hence for such a loop,

$$D = 0.68 C_\lambda \tag{7.48}$$

Figure 7.10 shows the directivity $\bar{D}$ as a function of C_λ.

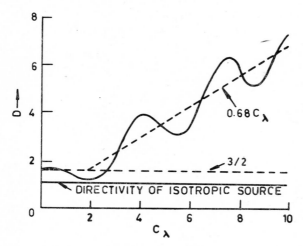

Fig. 7.10   Directivity of circular loop with uniform in-phase current as a function of C_λ.

7.6   Large Loop Antenna with Sinusoidal Current Distribution

If a circular conducting loop antenna is excited by a δ-function generator $V_0\delta(\phi)$ at $\phi = 0$ as shown in Fig. 7.11, and if the radius 'a' of the conduct-

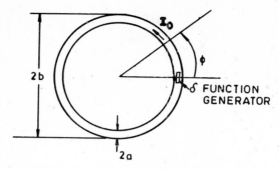

Fig. 7.11   Circular loop antenna excited by
δ-function generator.

ing wire is $\ll$ the radius b of the loop, it can be shown that the current distribution on the loop is sinusoidal and is of the form

$$I(\phi') = -\frac{jV_0}{\eta_0\pi}\left(\frac{1}{a_0} + 2\sum_1^\infty \frac{\cos{(n\phi')}}{a_n}\right) \tag{7.49}$$

where

$$\eta_0 = (\mu_0/\epsilon_0)^{1/2} \tag{7.50}$$

which is a Fourier series of an infinite number of terms. This can be proved by the integral equation method[1] similar to the method used for deriving the current distribution on a conducting cylindrical antenna as derived in Chap. 4. Alternatively the current $I(\phi')$ can be written in the form

$$I(\phi') = -\frac{jV_0}{\pi\eta_0}\sum_{-\infty}^\infty \frac{\exp{(jn\phi')}}{a_n} \tag{7.51}$$

The x and y components of the magnetic vector potential $\mathbf{A}$ at a distant point $P(r, \theta, \phi)$ (see Fig. 7.12) are given by

$$A_x = \frac{-b\mu_0}{4\pi}\frac{\exp{(-jk_0r)}}{r}\int_0^{2\pi} I(\phi')\exp{\{jk_0b\sin\theta\cos{(\phi-\phi')}\}}\sin\phi'\,d\phi' \tag{7.52}$$

and

$$A_y = \frac{b\mu_0}{4\pi}\frac{\exp{(-jk_0r)}}{r}\int_0^{2\pi} I(\phi')\exp{\{jk_0b\sin\theta\cos{(\phi-\phi')}\}}\cos\phi'\,d\phi' \tag{7.53}$$

The resulting integrals in Eqs. 7.52 and 7.53 for the nth term are

$$A_{xn} = \frac{jG}{2a_n}(K_1 - K_2) \tag{7.54}$$

and

$$A_{yn} = \frac{G}{2a_n}(K_1 + K_2) \tag{7.55}$$

---

[1]R. E. Collin and F. J. Zucker, *Antenna Theory*, Part I, Ch. 11.

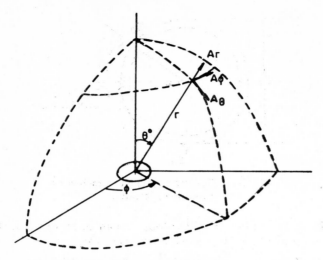

Fig. 7.12   Coordinate system for deriving radiation
field of large loop.

where

$$K_1 = \int_0^{2\pi} \exp\{j(n+1)\phi'\} \exp\{jk_0 b \sin\theta \cos(\phi - \phi')\}\, d\phi'$$

$$= 2\pi j^{n+1} \exp(j(n+1)\phi) J_{n+1}(k_0 b \sin\theta) \qquad (7.56)$$

$$K_2 = \int_0^{2\pi} \exp\{j(n-1)\phi'\} \exp(jk_0 b \sin\theta \cos(\phi - \phi'))\, d\phi'$$

$$= 2\pi j^{n-1} \exp(j(n-1)\phi) J_{n-1}(k_0 b \sin\theta) \qquad (7.57)$$

and

$$G = -\frac{jV_0}{4\pi^2\eta_0}\,\mu_0\,\frac{b\,\exp(-jk_0 r)}{r} \qquad (7.58)$$

The θ and ϕ components of A for the nth term are

$$A_{\theta n} = (A_{xn}\cos\phi + A_{yn}\sin\phi)\cos\theta \qquad (7.59)$$

and

$$A_{\phi n} = -A_{xn}\sin\phi + A_{yn}\cos\phi \qquad (7.60)$$

which give

$$A_{\theta n} = -\frac{2\pi G}{a_n}\exp\left(jn\left(\phi + \frac{\pi}{2}\right)\right)\frac{\eta J_n(k_0 b \sin\theta)}{k_0 b \sin\theta}\cos\theta \qquad (7.61)$$

and

$$A_{\phi n} = -\frac{j2\pi G}{a_n}\exp\left(jn\left(\phi + \frac{\pi}{2}\right)\right)J_n'(k_0 b \sin\theta) \qquad (7.62)$$

The resultant θ and ϕ components of $\mathbf{A}$ for all the terms in the series are

$$A_\theta = -2\pi G \sum_{-\infty}^{\infty}\frac{\exp(jn(\phi + \pi/2))}{a_n}\,\frac{n}{k_0 b}\,\frac{J_n(k_0 b \sin\theta)}{\sin\theta}\cos\theta \qquad (7.63)$$

and

$$A_\phi = -2\pi G \sum_{-\infty}^{\infty} \frac{\exp\{jn(\phi + \pi/2)\}}{a_n} J_n'(k_0 b \sin \theta) \tag{7.64}$$

The electric field $\mathbf{E}$ and the magnetic field $\mathbf{H}$ are then given by

$$\mathbf{E} = -j\omega(\mathbf{u}_\theta A_\theta + \mathbf{u}_\phi A_\phi) \tag{7.65}$$

and

$$H = -\frac{jk_0}{\mu_0}(\mathbf{u}_\phi A_\theta - \mathbf{u}_\theta A_\phi) \tag{7.66}$$

If $k_0 b \leqslant 1.0$, two terms in the series are adequate. If $k_0 b \leqslant 0.2$, then one term is enough.

For $n = 0$, that is, when the current is uniform,

$$A_{\theta_0} = 0 \tag{7.67}$$

and

$$A_{\phi_0} = \frac{j2\pi G}{a_0} J_1(k_0 b \sin \theta) \tag{7.68}$$

so that

$$E_{\theta_0} = H_{\phi_0} = 0 \tag{7.69}$$

$$E_{\phi_0} = -\frac{\omega}{k_0 \mu_0} H_{\theta_0} = \frac{2\pi \omega G}{a_0} J_1(k_0 b \sin \theta) \tag{7.70}$$

A similar result was proved in Section 7.3.

For $n = 1$,

$$A_{\theta_1} = \frac{4\pi G J_1(k_0 b \sin \theta)}{a_1 k_0 b \sin \theta} \cos \theta \sin \phi \tag{7.71}$$

and

$$A_\phi = \frac{4\pi G}{a_1} J_1'(k_0 b \sin \theta) \cos \phi \tag{7.72}$$

so that

$$E_{\theta_1} = \frac{-\omega}{k_0 \mu_0} H_{\phi_1} = \frac{-j4\pi \omega G}{a_1} \frac{J_1(k_0 b \sin \theta)}{k_0 b \sin \theta} \cos \theta \sin \phi \tag{7.73}$$

and

$$E_{\phi_1} = \frac{\omega}{k_0 \mu_0} H_{\theta_1} = \frac{-j4\pi \omega G}{a_1} J_1(k_0 b \sin \theta) \cos \phi \tag{7.74}$$

7.7   The Electrically Small Transmitting Loop

For a small loop for which $k_0 b < 1$,

$$I(\phi') \simeq I_0 + I_1(\phi') \tag{7.75}$$

For I_0 which is a uniform current, or what is called the 'zero' mode there are no accumulation of charges anywhere around the loop. The dipole mode is represented by $I_1(\phi')$, where $I_1(\phi') \alpha e^{j\phi'}$ the loop is oppositely charged at $\phi' = \pi/2$ and at $\phi' = -\pi/2$, and the current oscillates in synchronism on the two halves of the loop in a manner similar to two parallel dipoles that are driven in phase. The admittance of the loop is that of two circuits in parallel. Thus

$$Y = Y_0 + Y_1 \tag{7.76}$$

$$Z = \frac{Z_0 Z_1}{Z_0 + Z_1} \tag{7.77}$$

$$Y_0^{-1} = Z_0 = R_0 + j\omega L_0 = j\eta_0 \pi a_0 \tag{7.78}$$

and

$$Y_1^{-1} = Z_1 = R_1 - \frac{j}{\omega C_1} = \frac{j\eta_0 a_1 \pi}{2} \tag{7.79}$$

where it can be shown that[1]

$$R_0 \simeq \frac{\pi \eta_0}{6} k_0^4 b^4 \tag{7.80}$$

$$R_1 \simeq \frac{\pi \eta_0}{6} k_0^2 b^2 \tag{7.81}$$

$$L_0 \simeq \mu_0 b \left(\ln \frac{8b}{a} - 2 \right) \tag{7.82}$$

$$C_1 \simeq 2\epsilon_0 b \left(\ln \frac{8b}{a} - 2 \right)^{-1} \tag{7.83}$$

$R_1 > R_0$, R_1 is in series with C_1, and R_0 is in series with L_0. The dipole mode becomes significant when $k_0 b \geqslant 0.1$.

PROBLEMS

1. Calculate and plot the far-field pattern in a normal to the plane of a loop of diameter equal to λ and with uniform in-phase current distribution.

2. Calculate the total power radiated by the loop of Problem (1) and its radiation resistance and directivity.

3. If a small square loop is considered equivalent to 4 short dipoles, calculate the far-field pattern. Show that the pattern in the plane of the loop is a circle.

SUGGESTED READING

1. Collin, R. E. and Zucker, F. J., *Antenna Theory*, Parts 1 and 2, McGraw-Hill, New York, 1969.

2. Jasik, H., *Antenna Engineering Handbook*, McGraw-Hill, 1961.

3. Jordan, E. C. and Balmain, K. G., *Electromagnetic Waves and Radiating Systems*, Prentice-Hall of India, 1969.

4. Kraus, J. D., *Antennas*, McGraw-Hill, New York, 1950.

5. Schelkunoff, S. A. and Friis, H. T., *Antennas, Theory and Practice*, Chapman and Hall, London, 1952.

[1] R. E. Collin and F. J. Zucker, *Antenna Theory*, Part I, Chap. 11.

8

HELICAL ANTENNA

The helical antenna is the general form of antenna of which the linear and loop antennas are special cases. A helix of fixed diameter collapses to a loop as the spacing between turns approaches zero. On the other hand, a helix of fixed spacing between turns straightens out into a linear conductor as the diameter approaches zero.

An infinitely long helix is a transmission line which can support an infinite number of modes. Corresponding to these modes, a finite length helix can radiate in many modes. Two of these modes are important, namely (i) the axial mode of radiation, and (ii) the normal mode of radiation. In the axial mode of radiation, the field is maximum in the direction of the helix axis and is circularly polarized or nearly so. This is the most important mode of radiation, and occurs when the helix circumference is of the order of a wavelength. Further, it persists over a fairly wide frequency range. In the normal mode of radiation, the field is a maximum in a direction normal to the helix axis, and under particular conditions, the radiation field is circularly polarized. This normal mode occurs when the dimensions of the helix are small compared to a wavelength, and hence it has neither a wide-band nor a high efficiency.

8.1 Transmission Modes of Helices

The term 'transmission mode' is used to describe how an electromagnetic wave propagates along an infinitely long helix. In the lowest transmission mode, a helix has adjacent regions of positive and negative charge separated by many turns as shown in Fig. 8.1(a). This mode is called the T_0 mode, and it occurs when the length L of one turn of the helix is small as compared to the wavelength λ, i.e., $L \ll \lambda$ (see Fig. 8.2). This is the mode which occurs on low frequency inductances, and is also the important mode on the helix used in the travelling-wave tube. In the travelling-wave tube, this field interacts with the electron beam, because there is an appreciable axial distance between adjacent regions of positive and negative charge.

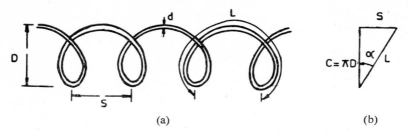

Fig. 8.1   Helix: (a) Dimension, (b) Relation between circumference, spacing, turn length, and pitch angle.

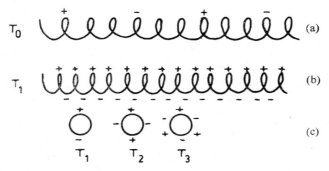

Fig. 8.2   Approximate instantaneous charge distributions on helices for different transmission modes: (a) T_0, (b) T_1, (c) T_1, T_2 and T_3.

When the helix circumference C_λ in free-space wavelengths is of the order of one wavelength, there is a first-order transmission mode called the T_1 mode on the helix. For small pitch angles, it has regions of adjacent positive and negative charge separated approximately by 1/2 a turn as shown in Fig. 8.1(b).

Higher order transmission modes are designated T_2, T_3, etc. and for small pitch angles, the approximate charge distribution around the helix for these modes are shown in Fig. 8.1(c).

8.2   Radiation Modes of Helices

For a short helix ($nL \ll \lambda$), the current may be assumed to be of uniform magnitude and in phase along the helix, and it may be assumed that the T_0 mode is propagating. Such a helix radiates with the maximum field in a direction normal to the helix axis as shown in Fig. 8.3(a). The space pattern is a figure of revolution, and this is called a 'normal radiation mode' and is designated as R_0 mode.

When C_λ is of the order of a wavelength and the T_1 transmission mode is propagating, then the helix radiates in the R_1 mode, with the maximum in the axial direction as shown in Fig. 8.3(b). A helix radiating in the

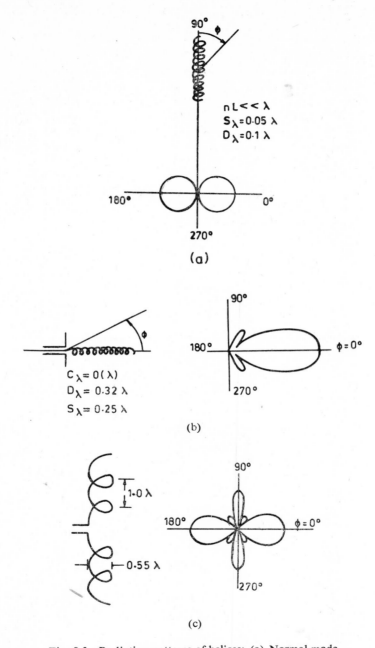

Fig. 8.3   Radiation patterns of helices: (a) Normal mode,
(b) Axial mode, (c) Multilobed pattern.

axial mode is usually called a helical beam antenna.

Figure 8.3(c) shows a multilobed pattern for a helix of larger dimensions.

8.3   Small Helix Radiating in the Normal Radiation Mode

If the dimensions of the helix are small ($nL \ll \lambda$), the maximum radiation is always in the normal direction. When the pitch angle is zero, the helix becomes a loop, and when the pitch angle is 90°, it straightens out into a linear antenna.

The helix may be considered as consisting of a number of small loops and short dipoles connected in series as shown in Fig. 8.4(a). The diameter D of the loops is the same as the helix diameter, and the length of the dipoles is the same as the spacing between turns. This assumption is quite good for a small helix. The current is assumed to be uniform in magnitude and in-phase over the entire length of the helix. For a small helix, the far-field is independent of the number of turns. Therefore, it is enough to derive the far-field pattern of a single loop and one short dipole connected in series as in Fig. 8.4(b).

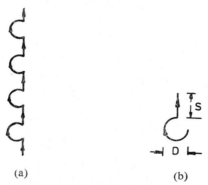

(a)   (b)

Fig. 8.4   Modified helix for normal mode
small helix,   (a) Several turns,
(b) Single turn.

The far-field of the small loop of Fig. 8.4(b) has an E_ϕ component given by

$$E_\phi = \frac{120\pi^2[I]\,\sin\,\theta}{r}\,\frac{A}{\lambda^2} \tag{8.1}$$

where the area of the loop is $A = \pi D^2/4$. The far-field of the short dipole of length S has an E_θ component given by

$$E_\theta = j\,\frac{60\pi[I]\,\sin\,\theta}{r}\,\frac{S}{\lambda} \tag{8.2}$$

It can be seen from Eqs. 8.1 and 8.2 that E_θ and E_ϕ are 90° out of phase. The ratio of the magnitudes of E_θ and E_ϕ is given by

$$AR = \frac{|E_\theta|}{|E_\phi|} = \frac{S\lambda}{2\pi A} = \frac{2S\lambda}{\pi^2 D^2} \tag{8.3}$$

where AR is the axial ratio of the polarization ellipse. The polarization ellipse is a vertical line when $E_\phi = 0$, which means vertical polarization, and the helix becomes a vertical dipole. When $E_\theta = 0$, it becomes a horizontal line indicating horizontal polarization, and the helix becomes a horizontal loop. When $|E_\theta| = |E_\phi|$, AR is unity and the polarization ellipse is a circle, indicating circular polarization. Therefore the radiation is circularly polarized when

$$\pi D = \sqrt{2S\lambda} \quad \text{or} \quad C_\lambda = \sqrt{2S_\lambda} \tag{8.4}$$

In general the radiation is elliptically polarized.

8.4 Measured Current Distribution on a Helix

When $C_\lambda < \frac{2}{3}$, the current distribution is nearly sinusoidal as shown in Fig. 8.5(a)[1]. If $C_\lambda \simeq 1$, the current distribution is shown in Fig. 8.5(b), and the helix radiates in the axial mode. In Fig. 8.5(a), the current distribution is caused by alternate reinforcement and cancellation of two oppositely directed travelling waves in the T_0 transmission mode of nearly equal amplitude. In Fig. 8.5b[1], we may assume two outgoing travelling-waves of different velocity. One of them (T_0 mode) is attenuated and the other (T_1 mode) is constant, and two smaller returning travelling-waves of different velocity, one (T_0 mode) attenuated and the other (T_1 mode) constant. In the central region of the helix only the relatively constant T_1 mode waves are of importance. If the helix is long, the T_0 waves can be neglected, and

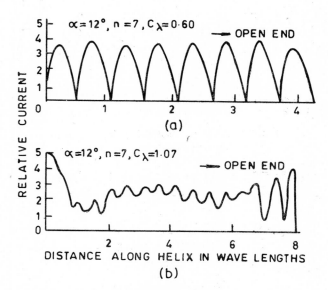

Fig. 8.5 Measured current distribution on helix,
(a) $C_\lambda < 0.6$, $C_\lambda = 1.07$.

[1]J.A. Marsh, 'Measured Current Distribution on Helical Antennas', *Proc. I.R.E.*, Vol. 39, pp. 668–675, June 1951.

the radiation pattern calculated on the basis of a single, outgoing T_1 wave of constant amplitude.

8.5   Terminal Impedance of Helix

When $C_\lambda < 2/3$, the measured terminal impedance of a helix is highly sensitive to frequency. When $3/4 < C_\lambda < 4/3$ and the helix is radiating in the axial mode, the terminal impedance is nearly constant as a function of frequency, provided that the pitch angle and number of turns are not too small. This makes the helix radiating in the axial mode a good broad-band antenna.

8.6   Radiation Pattern of Helix Radiating in the Axial Mode. The Phase Velocity of Wave Propagation on the Helix[1]

To derive the radiation pattern of a helix radiating in the axial mode, it is assumed that approximately a single travelling-wave of uniform amplitude is travelling along it. Using the principle of pattern multiplica-tion, the radiation pattern of such a helix is the product of the pattern of one turn and the pattern of an array of n isotropic point sources as shown in Fig. 8.6, where n is the number of turns. The spacing S between the sources is spacing between turns for a long helix, the array pattern is much sharper than the single-turn pattern and, therefore, determines the shape of the total far-field pattern fairly accurately.

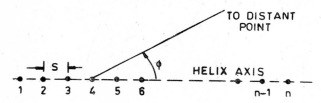

Fig. 8.6   Array of n isotropic point sources to represent the helix.

The array factor E for an array of n isotropic point sources is given by

$$E = \frac{\sin (n\psi/2)}{\sin (\psi/2)} \qquad (8.5)$$

where

$$\psi = S_r \cos \phi + \delta \qquad (8.6)$$

where

$$S_r = \frac{2\pi S}{\lambda} = 2\pi S_\lambda,$$

$$\delta = -\frac{2\pi L_\lambda}{p},$$

[1]J. D. Kraus, The Helical Antenna, *Proc. I.R.E.*, Vol. 37, pp. 263–272, March 1949.

$p = v/c$ = relative phase velocity of wave propagation along the helix,

v = phase velocity along the helix and

c = velocity of light in free space.

The radiation will be in the axial mode if fields from all the sources are in phase at a point on the helix axis ($\phi = 0°$), or ordinary endfire condition which requires that

$$\psi = -2\pi m \qquad (8.7)$$

where $m = 0, 1, 2, 3, \ldots$.

The minus sign in Eq. 8.7 is due to the fact that the phase of source 2 is retarded by $2\pi L_\lambda/p$ with respect to source 1. Similarly source 3 is retarded by the same amount with respect to source 2 etc. Putting $\phi = 0°$ in Eq. 8.6 and equating it to Eq. 8.7, we have

$$\frac{L_1}{p} = S_\lambda + m \qquad (8.8)$$

For $m = 1$, and $p = 1$, we have

$$L_\lambda - S_\lambda = 1 \text{ or } L - S = \lambda \qquad (8.9)$$

Equation 8.9 is an approximate relation between L and S for a helix radiating in the axial mode. However, $L^2 = \pi^2 D^2 + S^2$, and hence Eq. 8.9 can be rewritten as

$$D = \frac{\sqrt{2S_\lambda + 1}}{\pi}$$

or

$$C_\lambda = \sqrt{2S_\lambda + 1} \qquad (8.10)$$

The curve marked $C_\lambda = \sqrt{2S_\lambda + 1}$ in Fig. 8.7 represents Eq. 8.10 graphically. This curve defines the approximate upper limit of the axial or beam mode region. This describes the helix operating in the first order T_1 transmission mode. When $m = 2$, Eq. 8.8 becomes

$$L_\lambda = S_\lambda + 2$$

or

$$C_\lambda = 2\sqrt{S_\lambda + 1} \qquad (8.11)$$

which is also represented as a curve in Fig. 8.7, and corresponds to the T_2 transmission mode. Hence m corresponds to the order of the transmission mode on the helix radiating a maximum field in the axial direction. The important case is for $m = 1$.

When $m = 0$, it does not represent a realizable condition, unless $p > 1$, since when $m = 0$ and $p = 1$ in Eq. 8.8 we have $L = S$, when the helix becomes a straight line. However, the radiated field in the axial direction of a straight wire is zero. Therefore, there can be no axial field when $m = 0$. From Eq. 8.8,

$$p = L_\lambda/(S_\lambda + 1) \qquad (8.12)$$

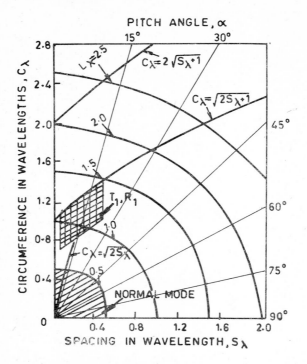

Fig. 8.7   Spacing-circumference chart of helix showing
regions for different modes of operation.

using the triangle of Fig. 8.1 and using Eq. 8.12

$$p = \frac{1}{\sin \alpha + (\cos \alpha / C_\lambda)} \tag{8.13}$$

Equations 8.12 and 8.13 give the relation between $p = v/c$ and the dimensions of the helix for in-phase fields in the axial direction. Figure 8.8 shows p as a function of C_λ for different values of α. The curves of p vs. C_λ show that in a helix radiating in the axial mode, p may be considerably less than unity, or that v is considerably less than c.

If Hansen and Woodyards' condition of increased directivity is used, then

$$\psi = -\left(2\pi m + \frac{\pi}{n}\right) \tag{8.14}$$

From Eqs. 8.14 and 8.6, we obtain,

$$p = \frac{L_\lambda}{S_\lambda + m + (1/2n)} \tag{8.15}$$

When $m = 1$,

$$p = \frac{L_\lambda}{S_\lambda + (2n + 1)/2n} \tag{8.16}$$

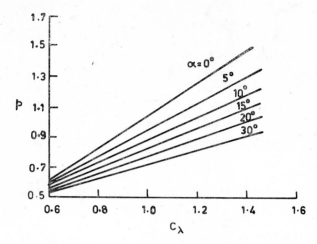

Fig. 8.8 p as a function of C_λ for different values of α.

For large values of n, Eq. 8.16 reduces to Eq. 8.12. Equation 8.16 can also be written in the form

$$p = \frac{1}{\sin \alpha + [(2n+1)/2n][\cos \alpha / C_\lambda]}$$ (8.17)

Using the value of p as given by Eq. 8.17, the array factor pattern calculated by using Eq. 8.5 is in agreement with the measured pattern. The value of p given by Eq. 8.17 is also in agreement with the measured value. Therefore, it can be concluded that the increased directivity condition is approximated as a natural condition on a helix radiating in the axial mode.

8.7   Axial Ratio and Conditions for Circular Polarization for Helix Radiating in the Axial Mode

Consider the helix shown in Fig. 8.9. The axial ratio of the radiated field

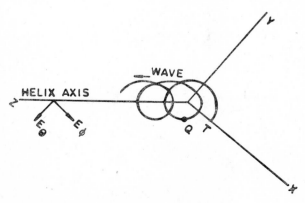

Fig. 8.9   Helix.

in the direction of the axis of the helix is to be determined, and the conditions for circular polarization will be examined.

The electric field components E_θ and E_ϕ at a distant point on the axis will be determined. It is assumed that a single uniform travelling wave is supported on the helix, and let p be the relative phase velocity. Let D be the diameter of the helix and S be the spacing between turns. If the helix is unrolled in the X-Z plane, Fig. 8.10 is obtained, and Fig. 8.11 shows the

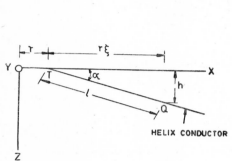

Fig. 8.10   Unrolled helix.

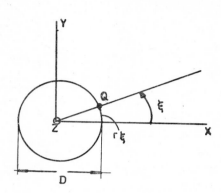

Fig. 8.11   Helix as viewed from the direction of the positive z-axis.

helix as viewed from a point on the Z-axis. The angle ξ is measured from the X-Z plane. The coordinates of a point Q on the helix can be specified as r, ξ, z, and its distance from the terminal point T is l as measured along the helix. From Figs. 8.10 and 8.11, we have

$$h = l \sin \alpha, \; z_p - h = z_p - l \sin \alpha$$

$$\alpha = \tan^{-1} \frac{5}{\pi D} = \cos^{-1} (r\xi/l) \tag{8.18}$$

$$r\xi = l \cos \alpha$$

where z_p is the z coordinate of the distant point on the z-axis.

The component E_ϕ of the electric field at point P for a helix of n turns is

$$E_\phi = E_0 \int_0^{2\pi n} \sin \xi \exp [j\omega\{t - (z_p/c) + (l \sin \alpha/c) - (l/pc)\}] \, d\xi \tag{8.19}$$

where E_0 is a constant which depends on the magnitude of the current on the helix. In Eq. 8.19

$$\frac{l \sin \alpha}{c} - \frac{l}{pc} = \frac{r\xi}{c} \left(\tan \alpha - \frac{1}{p \cos \alpha} \right) = \frac{r\xi q}{c} \tag{8.20}$$

where

$$q = \tan \alpha - \frac{1}{p \cos \alpha} \tag{8.21}$$

Equation 8.19 reduces to

$$E_\phi = E_0 \exp \{j(\omega t - \beta z_p)\} \int_0^{2\pi n} \sin \xi \exp (jk\xi) \, d\xi \qquad (8.22)$$

where

$$\beta = \frac{\omega}{c} = \frac{2\pi}{\lambda}$$

and

$$k = \beta rq = L_\lambda \left(\sin \alpha - \frac{1}{p} \right) \qquad (8.23)$$

Equation 8.22 can be integrated to give

$$E_\phi = \frac{E_0 \exp \{j(\omega t - \beta z_p)\}}{(k^2 - 1)} (\exp (j2\pi nk) - 1) \qquad (8.24)$$

Similarly the expression for the θ component E_θ of the electric field is

$$E_\theta = E_0 \int_0^{2\pi n} \cos \xi \exp \left\{ j\omega \left(t - \frac{zp}{c} + \frac{l \sin \alpha}{c} - \frac{1}{pc} \right) \right\} d\xi \qquad (8.25)$$

which reduces to

$$E_\theta = \frac{jE_{1k}}{(k^2 - 1)} (\exp (j2\pi nk) - 1) \qquad (8.26)$$

The field is circularly polarized at a point on the z-axis if

$$\frac{E_\phi}{E_\theta} = \pm j \qquad (8.27)$$

or

$$\frac{E_\phi}{E_\theta} = \frac{1}{jk} = -\frac{j}{k} \qquad (8.28)$$

or

$$k = \pm 1 \qquad (8.29)$$

The axial ratio AR is given by

$$AR = \frac{|E_\phi|}{|E_\theta|} = \left| \frac{1}{jk} \right| = \frac{1}{k} \qquad (8.30)$$

Using Eq. 8.23 in Eq. 8.30, we obtain

$$AR = \frac{1}{|L_\lambda (\sin \alpha - 1/p)|} \qquad (8.31)$$

when $k = -1$ for circular polarization Eq. 8.30 gives

$$L_\lambda \left(\sin \alpha - \frac{1}{p} \right) = -1 \qquad (8.32)$$

from which p is obtained as

$$p = \frac{L_\lambda}{S_\lambda + 1} \qquad (8.33)$$

which is the ordinary end-fire condition (Eq. 8.12).

Using Eq. 8.17 which is the condition for increased directivity, we can see that

$$AR = \frac{2n + 1}{2n} \qquad (8.34)$$

If n is a large, AR approaches unity and the polarization is circular.

Thus, it has been shown that both the end-fire condition and the increased directivity condition make the field on the axis of the helix almost circularly polarized.

This circularly polarized field of the helix is of great advantage in using the helical antenna in receiving signals of arbitrary polarization.

8.8   Feed Arrangements for Helical Antennas

Figure 8.12 shows different methods of feeding axial mode helical antennas, while Fig. 8.13 shows tapered helical antennas which are more

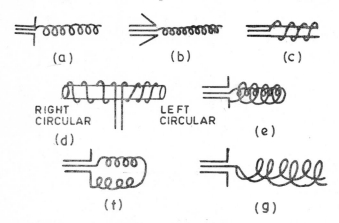

Fig. 8.12   Axial mode helices showing various constructional and feed arrangements.

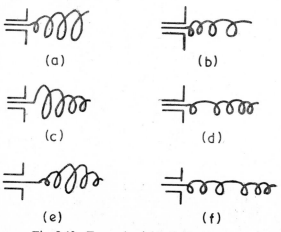

Fig. 8.13   Tapered axial helical antennas.

broad-band than untapered helices. Figure 8.14 shows slotted metal cylinder and slotted metal conical structures supporting helical antennas.

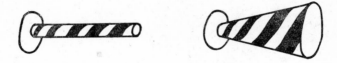

Fig. 8.14 Slotted helical antennas, (a) On cylinder,
(b) On cone.

Figure 8.15 shows arrangements for producing linear polarization with axial mode helical antennas.

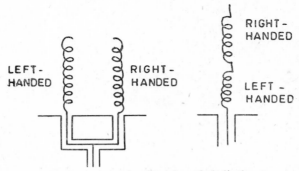

Fig. 8.15 Arrangements of axial mode helical antennas
for producing linear polarization.

PROBLEMS

1. A helical beam antenna has $\alpha = 14°$, $n = 10$, $D = 20$ cm.
   (a) What is the value of p at 500 Mc for (i) in-phase fields, (ii) increased directivity?
   (b) Calculate and plot the field pattern, assuming that each turn is an isotropic source for $p = 1$, 0.8 and 0.5.
   (c) If each loop has a cosine field pattern, calculate and plot the field pattern for $p = 1$, 0.8.

2. A helix has 8 turns. The diameter is 2.5 cm. and the turn spacing is 15 cm. Assume increased directivity and calculate the phase velocity. Calculate the field pattern at 600 Mc and plot the pattern.

SUGGESTED READING

1. Collin, R. E. and Zucker, F. J., Antenna Theory, Parts 1 and 2, McGraw-Hill, New York, 1969.
2. Jasik, H., Antenna Engineering Handbook, McGraw-Hill, 1961.
3. Jordan, E. C. and Balmain, K. G., Electromagnetic Waves and Radiating Systems, Prentice-Hall of India, 1969.
4. Kraus, J. D., Antennas, McGraw-Hill, New York, 1950.
5. Schelkunoff, S. A. and Friis, H. T., Antennas, Theory and Practice, Chapman and Hall, London, 1952.

9

SLOT AND MICROSTRIP
ANTENNAS

9.1 Slot Antennas

At higher frequencies like microwave frequencies, slot-antennas, cut in a metal surface such as the skin of an aircraft or a moving vehicle or the wall of a metal waveguide, are found to be very convenient radiators. The dimensions of such slots are very much smaller than the dimensions of the metal surface in which they are cut, and are usually less than a wavelength. The slot may be excited by means of an energized cavity placed behind it, or through a waveguide, or by a transmission line connected across it. The electromagnetic field distribution in the slot may be obtained by using Babinets Principle as applied by Booker[1] to slot antennas and to complementary wire antennas.

9.2 Babinets' Principle

In optics, Babinets' Principle states that the sum of the fields beyond any two complementary absorbing screens add to produce the field that would exist there without any screen. If S_1 is a thin absorbing screen pierced with apertures of any shape or size, then the screen S_2 obtained by interchanging the region of absorbing screen space and aperture space is called the complementary screen. If U_1 is the ratio at every point of the field on the right of screen S_1 to the field without the screen, and if U_2 is the similar ratio for screen S_2, then Babinets' Principle in optics states that

$$U_1 + U_2 = 1$$

In electromagnetics, thin perfectly absorbing screens are not available even approximately. But an extension of the principle has been established

---

[1] H. G. Booker, 'Slot Aerials and Their Relation to Complementary Wire Aerials (Babinets' Principle),' *J.I.E.E.*, vol. III A, pp. 620–626 (1946).

by Booker to conducting screens and polarized fields. This generalized Babinets' Principle can be enunciated as follows.

Let a source s_1 to the left of an infinite screen S_1 produce a field on the right of S_1, and let U_1 be the ratio of this field to the field that would exist there in the absence of the screen. Similarly, let U_2 be the ratio of the field to the right of the complementary screen S_2 produced by a source s_2 which is conjugate to s_1 to the left of screen S_2, to the field that would exist there in the absence of the screen. Then

$$U_1 + U_2 = 1 \tag{9.1}$$

A source s_2 which is conjugate to source s_1 is defined as one for which the distribution of electric and magnetic source currents and charges is replaced by the corresponding distribution of magnetic and electric currents and charges. This replacement has the effect of interchanging $\mathbf{E}$ and $\mathbf{H}$ for the incident fields. This is illustrated in Fig. 9.1. If $U_1 = E_1/E^i$, then $U_2 = H_2/H^i$.

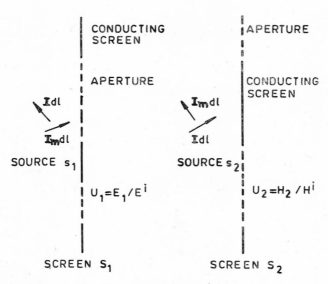

Fig. 9.1   Illustrating Babinets' Principle $U_1 + U_2 = 1$.

The basis of Babinets' principle is the duality property of the electromagnetic field. Maxwell's equations in a source-free region are

$$\nabla \times \mathbf{E} = -j\omega\mu\mathbf{H} \tag{9.2}$$

and

$$\nabla \times \mathbf{H} = j\omega\epsilon\mathbf{H} \tag{9.3}$$

Equations 9.2 and 9.3 exhibit a high degree of symmetry in the expressions for $\mathbf{E}$ and $\mathbf{H}$. If a field $(\mathbf{E}_1, \mathbf{H}_1)$ satisfies these equations, then a second field (E_2, H_2) obtained from it by the transformations

$$E_2 = \pm(\mu/\epsilon)^{1/2}H_1 \tag{9.4}$$

and

$$H_2 = \pm(\epsilon/\mu)^{1/2}E_1 \tag{9.5}$$

is also a solution of Eqs. 9.2 and 9.3. This is a statement of the duality property of the electromagnetic field. The transformations given in Eqs. 9.4 and 9.5 also affect the boundary conditions to be satisfied by the new field. Clearly, since the roles of E and H are interchanged, all electric walls $E_t = 0$ have to be interchanged with magnetic walls $H_t = 0$ and vice versa. The new boundary conditions are still the continuity of the tangential components, and this changes the relative amplitudes of the fields on the two sides of the discontinuity surface.

9.3 Slot and Dipole Antennas as Dual Problems

A slot in a conducting plane sheet and the complementary flat dipole antenna is shown in Fig. 9.2. Considering these two antennas are boundary-value problems, it is necessary to find the appropriate solutions to Maxwell's equations or the wave equations to satisfy the proper boundary conditions. Either one of the wave equations

$$\nabla^2 E = \mu\epsilon E \tag{9.6}$$

or

$$\nabla^2 H = \mu\epsilon H \tag{9.7}$$

have to be solved to satisfy the proper boundary conditions for both the problems. For the dipole, Eq. 9.7 may be solved to satisfy the following boundary conditions:

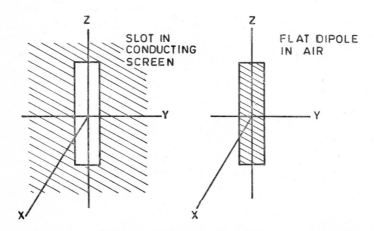

Fig. 9.2   Slot antenna in a conducting screen and the complementary flat dipole.

(i) The tangential components of the magnetic field, namely, H_y and H_z are zero in the y-z plane and outside the perimeter of the dipole.

(ii) The normal component H_x of the magnetic field is zero in the y-z plane and within the perimeter of the dipole.

For the slot, Eq. 9.6 may be solved to satisfy the following boundary conditions:

(i) The tangential components E_y and E_z of the electric field are zero in the y-z plane outside the perimeter of the slot.

(ii) The normal component E_x of the electric field is zero in the y-z plane within the perimeter of the slot.

It can be seen that mathematically the two problems are identical except that E and H are interchanged. Hence the two problems are dual problems. Therefore, except for a constant, the solution for E for the slot is the same as the solution for H for the dipole. Hence, we can formulate

$$\mathbf{E}_s = k_1 \mathbf{H}_d \tag{9.8}$$

and

$$\mathbf{H}_s = k_2 \mathbf{E}_d \tag{9.9}$$

where $\mathbf{E}_s$ and $\mathbf{H}_s$ are the electric and magnetic fields of the slot; and $\mathbf{E}_d$ and $\mathbf{H}_d$ are the electric and magnetic fields of the dipole. From Eqs. 9.8 and 9.9, it can be seen that the electric field distribution of the slot is the same as the magnetic field distribution of the dipole and vice-versa. Also the impedance of the dipole is proportional to the admittance of the slot and vice-versa.

To find the relationships between the impedance and admittance properties of the two antennas, consider Fig. 9.3, in which a slot and a flat dipole with small finite gaps are shown. The impedance of the dipole is the ratio of the voltage across the gap to the current through the generator into one arm of the dipole, while the admittance of the slot is the ratio of the current into one edge to the voltage across the gap.

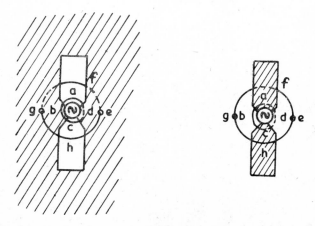

Fig. 9.3 Slot and flat dipole with small gaps.

For the dipole,

$$V = -\int_{abc} \mathbf{E}_d \cdot d\mathbf{s} \tag{9.10}$$

and

$$I = \oint_{efghe} \mathbf{H}_d \cdot d\mathbf{s} = 2 \int_{efg} \mathbf{H}_d \cdot d\mathbf{s} \tag{9.11}$$

and hence the impedance Z_d of the dipole is

$$Z_d = -\frac{\int_{abc} \mathbf{E}_d \cdot d\mathbf{s}}{2 \int_{efg} \mathbf{H}_d \cdot d\mathbf{s}} \tag{9.12}$$

For the slot,

$$I = \oint_{abcda} \mathbf{H}_s \cdot d\mathbf{s} = 2 \int_{abc} \mathbf{H}_s \cdot d\mathbf{s} \tag{9.13}$$

and

$$V = \int_{efg} \mathbf{E}_s \cdot d\mathbf{s} \tag{9.14}$$

and hence the admittance Y_s of the slot is

$$Y_s = \frac{2 \int_{abc} \mathbf{H}_s \cdot d\mathbf{s}}{\int_{efg} \mathbf{E}_s \cdot d\mathbf{s}} \tag{9.15}$$

Using Eqs. 9.8 and 9.9 in Eqs. 9.12 and 9.15, it can be seen that

$$\frac{1}{Z_s} = Y_s = -4 \frac{k_2}{k_1} Z_d \tag{9.16}$$

or

$$Z_s Z_d = -\frac{k_1}{4k_2} \tag{9.17}$$

To find out the ratio k_1/k_2, the distant fields of the slot and the dipole are considered. At corresponding distant points,

$$\mathbf{E}_s = \eta_0 \mathbf{H}_s \tag{9.18}$$

and

$$\mathbf{E}_d = \eta_0 \mathbf{H}_d \tag{9.19}$$

where $\eta_0 = (\mu_0/\epsilon_0)^{1/2}$ = intrinsic impedance of free-space.

Using Eqs. 9.18 and 9.19 with Eqs. 9.8 and 9.9, we obtain

$$-\frac{k_1}{k_2} = \eta_0^2 = 377^2 \tag{9.20}$$

and hence

$$Z_s Z_d = \eta_0^2/4 = 377^2/4 \tag{9.21}$$

For the theoretical half-wave dipole, $Z_d \simeq (73 + j43)$ ohms, and therefore for the theoretical half-wave slot,

$$Z_s \simeq \frac{377^2}{4(73 + j43)} \simeq 418 \ \underline{/-30.5°} \ \text{ohms} \qquad (9.22)$$

For a resonant-length dipole, the input impedance is a resistance. It depends on the thickness of the dipole, and is usually of the order of 65 ohms for a practical half-wave dipole. Hence the impedance of a practical resonant-length slot is of the order of 550 ohms. For a folded half-wave dipole as shown in Fig. 9.4, the input impedance is approximately four times that of a half-wave dipole. This is because the two half-wave dipoles of the folded dipole are in parallel and, therefore, the total effective current is $2I$. Hence the power radiated is four times the power radiated by a simple half-wave dipole. However, the current that is delivered by the generator is only I. The input resistance, therefore, is four times that of the simple dipole. Thus, the input impedance of a folded half-wave slot as shown in Fig. 9.4 is approximately $550/4 = 138$ ohms.

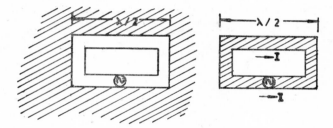

Fig. 9.4   Folded slot and folded dipole.

9.4   Radiation from Rectangular Aperture in Infinite Conducting Plane

Figure 9.5 represents a rectangular aperture of dimensions $2a$ by $2b$ in

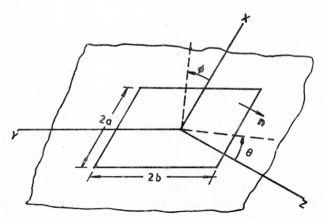

Fig. 9.5   Rectangular aperture in infinite conducting plane.

an infinite conducting plane coinciding with the x-y plane. Let $\mathbf{E}_0$ and $\mathbf{H}_0$ be the tangential electric and magnetic fields in the aperture. Since the tangential component of the electric field is zero while the tangential component of the magnetic field may exist on the infinite conducting plane, it is convenient to determine the radiated field in terms of the tangential electric field E_0 in the aperture surface.

In the region $z > 0$ which is source-free, the electric field $\mathbf{E}$ is a solution of the equation

$$(\nabla^2 + k_0^2)\mathbf{E}(x, y, z) = 0 \qquad (9.23)$$

and is also solenoidal so that

$$\nabla \cdot \mathbf{E} = 0 \qquad (9.24)$$

taking the Fourier transform of E with respect to x and y, we have

$$\mathbf{E}(k_x, k_y, z) = \int_{-\infty}^{\infty} \int_{-\infty}^{\infty} \mathbf{E}(x, y, z) \exp{(jk_x x + jk_y y)} \, dx \, dy \qquad (9.25)$$

where $k_0^2 = k_x^2 + k_y^2 + k_z^2, k_x, k_y, k_z$ are the wave numbers in the x, y and z directions and k_0 the intrinsic wave number $2\pi/\lambda_0$. Taking the Fourier transform of Eq. 9.23, we have

$$\left[\frac{\partial^2}{\partial z^2} + k_0^2 - (k_x^2 + k_y^2)\right] \mathbf{E}(k_x, k_y, z) = 0 \qquad (9.26)$$

Equation 9.26 may be solved to obtain

$$\mathbf{E}(k_x, k_y, z) = \mathbf{f}(k_x, k_y) \exp{(-jk_z z)} \qquad (9.27)$$

where $\mathbf{f}$ is a vector function of k_x and k_y. When k_x is real, the positive real root is chosen, and when k_x is imaginary, the root with a negative imaginary part is chosen when k_z is imaginary so as to satisfy the radiation condition. Then the electric field in the region $z > 0$ can be put in the form

$$E(x, y, z) = \frac{1}{4\pi^2} \int_{-\infty}^{\infty} \int_{-\infty}^{\infty} \mathbf{f}(k_x, k_y) \exp{(-j\mathbf{k} \cdot \mathbf{r})} \, dk_x \, dk_y \qquad (9.28)$$

where

$$\mathbf{k} = k_x \mathbf{u}_x + k_y \mathbf{u}_y + k_z \mathbf{u}_z \qquad (9.29)$$

and

$$|\mathbf{k}| = k_0 \qquad (9.30)$$

According to Eq. 9.28 the field in the region $z > 0$ can be regarded as made up of a superposition of plane waves of the form $\mathbf{f} \exp{(-j\mathbf{k} \cdot \mathbf{r})}$. If k_z is real the wave is propagating and will contribute to the energy flow, but if k_z is purely imaginary which happens when $(k_x^2 + k_y^2) > k_0^2$, the wave is evanescent. Therefore, only if $(k_x^2 + k_y^2) < k_0^2$ and hence k_z is real, the plane waves contribute to the far-zone radiation field. The components k_x and k_y of $\mathbf{k}$ are called spatial frequencies.

Since $\nabla \cdot \mathbf{E}(x, y, z) = 0$, taking the divergence of Eq. 9.28, we get

$$\mathbf{k} \cdot \mathbf{f} = 0 \qquad (9.31)$$

because $\nabla \cdot \mathbf{f} \exp(-j\mathbf{k}\cdot\mathbf{r}) = \mathbf{f}\cdot\nabla \exp(-j\mathbf{k}\cdot\mathbf{r}) = -j\mathbf{f}\cdot\mathbf{k} \exp(-j\mathbf{k}\cdot\mathbf{r})$. From Eq. 9.31, we can see that only two components of $\mathbf{f}$ are independent. Hence, we can infer that $\mathbf{E}(x, y, z)$ can be uniquely determined in terms of the tangential electric field on the aperture plane.

In the $z = 0$ plane

$$\mathbf{E}_t(x, y, 0) = \mathbf{E}_0(x, y)$$

$$= \frac{1}{4\pi^2} \int_{-\infty}^{\infty} \int_{-\infty}^{\infty} \mathbf{f}_t(k_x, k_y) \exp -j(k_x x + k_y y)\, dk_x\, dk_y$$

(9.32)

where t stands for tangential and denotes the x and y components on the aperture plane. Taking the inverse Fourier transform of Eq. 9.32, we get

$$\mathbf{f}_t = \iint_{S_0} \mathbf{E}_0(x, y) \exp[j(k_x x + k_y y)]\, dx\, dy \qquad (9.33)$$

where the integration is over the aperture S_0 only, since E_0 is zero on the conducting screen. From Eq. 9.31, the z component of f is given by

$$f_z = -\frac{\mathbf{k}_t \cdot \mathbf{f}}{k_z} \qquad (9.34)$$

The magnetic field $\mathbf{H}$ may be found from the electric field by using the Maxwell's equation $\nabla \times \mathbf{E} = -j\omega\mu_0\mathbf{H}$. Using Eq. 9.28 for $\mathbf{E}$, we obtain,

$$\mathbf{H}(x, y, z) = \frac{1}{4\pi^2 k_0 \eta_0} \int_{-\infty}^{\infty} \int_{-\infty}^{\infty} \mathbf{k} \times \mathbf{f} \exp(-j\mathbf{k}\cdot\mathbf{r})\, dk_x\, dk_y \quad (9.35)$$

(where $\eta_0 = (\mu_0/\epsilon_0)^{1/2}$), since

$$\nabla \times \mathbf{f} \exp(-j\mathbf{k}\cdot\mathbf{r}) = -\mathbf{f} \times \nabla \exp(-j\mathbf{k}\cdot\mathbf{r}) = j\mathbf{f} \times \mathbf{k} \exp(-j\mathbf{k}\cdot\mathbf{r})$$

The complex radiated power is obtained by integrating the complex Poynting vector over the aperture plane S_0. Thus,

$$P + 2j\omega(W_m - W_e) = \text{complex radiated power}$$

$$= \tfrac{1}{2} \iint_{S_0} \mathbf{E} \times \mathbf{H}^* \cdot \mathbf{u}_z\, dx\, dy$$

$$= \frac{1}{8\pi^2 \eta_0 k_0} \int_{-\infty}^{\infty} \int_{-\infty}^{\infty} (k_0^2 - k_t^2)\mathbf{f}_t \cdot \mathbf{f}_t^* + |\mathbf{k}_t \cdot \mathbf{f}_t|^2 \frac{dk_x\, dk_y}{k_z^*}$$

(9.36)

For $k_t^2 \leqslant k_0^2$, k_z is real and hence, the integrand is real. Thus, the real radiated power comes only from the portion of the plane wave spectrum for which k_x and k_y are located in the region $k_t^2 \leqslant k_0^2$ which is a circle. This region of k_x and k_y is called the visible region. The evanescent waves contribute to the average reactive energy $(W_m - W_e)$ corresponding to the invisible portion $k_t > k_0$ of the k_x, k_y space.

Booker and Clemmow[1] were the first to represent any arbitrary electromagnetic field as a superposition of propagating and evanescent plane waves and its application to the problem of radiation from an aperture in an infinite conducting plane.

The Fourier transform (Eq. 9.28) can be evaluated to obtain the far-field asymptotically by the method of stationary phase as discussed by Jeffreys and Jeffreys[2]. The basis of this method is that when r is very large, a function like $\exp(-j\mathbf{k}\cdot\mathbf{r}) = \cos(\mathbf{k}\cdot\mathbf{r}) - j\sin(\mathbf{k}\cdot\mathbf{r})$ oscillates very rapidly between equal positive and negative values except for certain values of k_x and k_y for which $\mathbf{k}\cdot\mathbf{r}$ remains stationary. The word 'stationary' means that $\mathbf{k}\cdot\mathbf{r}$ does not change for first-order changes in k_x and k_y. Therefore, when a slowly varying function of k_x and k_y is multiplied by $\exp(-j\mathbf{k}\cdot\mathbf{r})$ and integrated over k_x and k_y, only those values of k_x and k_y for which the phase remains stationary contribute to the integral. This is because for those values of k_x and k_y that make the phase $\mathbf{k}\cdot\mathbf{r}$ vary rapidly, the integrand oscillates rapidly between positive and negative values and the contribution to the value of the integral is small. As r tends to infinity, the leading term in the asymptotic expansion of the integral is given exactly by the contributions arising from the stationary phase points only.

The integral in Eq. 9.28 is evaluated as follows:

$$\mathbf{E}(r) = \frac{1}{4\pi^2} \int_{-\infty}^{\infty} \int_{-\infty}^{\infty} \mathbf{f}(k_x, k_y) \exp(-j\mathbf{k}\cdot\mathbf{r}) dk_x\, dk_y \tag{9.28}$$

$$\mathbf{k}\cdot\mathbf{r} = k_x x + k_y y + k_z z = r[k_x \sin\theta \cos\phi$$
$$+ k_y \sin\theta \sin\phi + (k_0^2 - k_x^2 - k_y^2)^{1/2} \cos\theta] \tag{9.37}$$

The stationary phase points are given by

$$\frac{\partial(\mathbf{k}\cdot\mathbf{r})}{\partial k_x} = 0 \quad \text{and} \quad \frac{\partial(\mathbf{k}\cdot\mathbf{r})}{\partial k_y} = 0 \tag{9.38}$$

which give

$$k_x = k_z \frac{\sin\theta \cos\phi}{\cos\theta}$$

and

$$k_y = k_z \frac{\sin\theta \sin\phi}{\cos\theta} \tag{9.39}$$

Since $k_z = k_0 \cos\theta$, the stationary values of k_x and k_y are

$$k_x = k_1 = k_0 \cos\phi \sin\theta = k_0 u$$

and

$$k_y = k_2 = k_0 \sin\phi \sin\theta = k_0 v \tag{9.40}$$

[1]Booker, H. G. and P. C. Clemmow, 'The Concept of an Angular Spectrum of Plane Waves and Its Relation to that of Polar Diagram and Aperture Distribution', *Proc. IEE*, vol. 97, Part III, pp. 11–17, January 1950.
[2]Jeffreys H. and B. S. Jeffreys, 'Methods of Mathematical Physics', 2nd Ed., Chap. 17, Cambridge University Press, London, 1950.

$\mathbf{k} \cdot \mathbf{r}$ can be approximated by the first few terms in a Taylor series expansion, and will be given by

$$\mathbf{k} \cdot \mathbf{r} = \mathbf{k} \cdot \mathbf{r}\Big|_{k_1, k_2} + \frac{1}{2} \frac{\partial^2 (\mathbf{k} \cdot \mathbf{r})}{\partial k_x^2}\Big|_{k_1, k_2} (k_x - k_1)^2 + \frac{1}{2} \frac{\partial^2 (\mathbf{k} \cdot \mathbf{r})}{d k_y^2}\Big|_{k_1, k_2} (k_y - k_2)^2$$

$$+ \frac{\partial^2 (\mathbf{k} \cdot \mathbf{r})}{\partial k \partial k_{yx}}\Big|_{k_1, k_2} (k_x - k_1)(k_y - k_2)$$

$$= k_0 r - [A(k_x - k_1)^2 + B(k_y - k_2)^2 + C(k_x - k_1)(k_y - k_2)] \quad (9.41)$$

where A, B and C are constants. In the vicinity of the stationary phase point, $\mathbf{f}(k_x, k_y)$ may be replaced by $\mathbf{f}(k_1, k_2)$. Then the asymptotic value of $\mathbf{E}(\mathbf{r})$ is given by

$$\mathbf{E}(\mathbf{r}) \approx f(k_0 \cos \phi \sin \theta, k_0 \sin \phi \cos \theta)$$

$$\cdot \exp(-jk_0 r) \frac{1}{4\pi^2} \iint\limits_{S_0'} \exp[j\{A(k_x - k_1)^2 + B(k_y - k_2)^2$$

$$+ C(k_x - k_1)(k_y - k_2)\}] d(k_x - k_1) d(k_y - k_2) \quad (9.42)$$

where S_0' is a small circle centred on the stationary phase point which contributes the major portion to the integral. If $u' = (k_x - k_1)$ and $v' = (k_y - k_2)$ are the new variables, since the phase factor becomes large for large r when u' and v' differ slightly from zero, the integration may be extended over the whole $u'v'$ plane with little error, because the integrand oscillates rapidly for non-zero values of u' and v'. Then $\mathbf{E}(\mathbf{r})$ is given by

$$\mathbf{E}(\mathbf{r}) \approx -j \frac{\exp(-jk_0 r)}{2\pi r} k_0 \cos \theta \, \mathbf{f}(k_0 \cos \phi \sin \theta, k_0 \sin \phi \cos \theta) \quad (9.43)$$

If Eq. 9.43 is expressed in terms of $\mathbf{f}_t$, we have

$$\mathbf{E}(\mathbf{r}) \approx j \frac{\exp(-jk_0 r)}{2\pi r} k_0 \cos \theta \Big[\mathbf{f}_t(k_0 \cos \phi \sin \theta, k_0 \sin \phi \cos \theta)$$

$$- \frac{\sin \theta}{\cos \theta} (f_x \cos \phi + f_y \sin \phi) \mathbf{u}_z \Big]$$

$$= jk_0 \frac{\exp(-jk_0 r)}{2\pi r} [\mathbf{u}_\phi (f_y \cos \phi - f_x \sin \phi) \cos \theta$$

$$+ \mathbf{u}_\theta (f_x \cos \phi + f_y \sin \phi)] \quad (9.44)$$

9.5 Half-Wavelength Slot in Infinite Ground Plane

Consider a half-wavelength slot cut in a perfectly conducting plane as shown in Fig. 9.6. Such a slot may be fed either by a coaxial line connected to an r-f source as shown in Fig. 9.7(a), or in the side wall or across the end of a waveguide as shown in Fig. 9.7(b), or by a cavity as shown in Fig. 9.7(c). For any of these feeding arrangements, the tangential

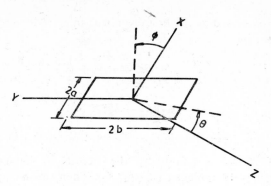

Fig. 9.6 Slot in ground conducting plane.

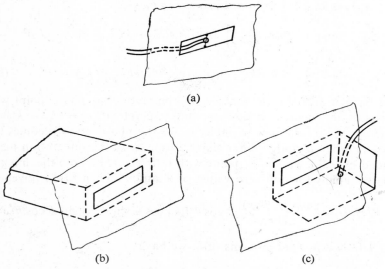

(a)

(b)   (c)

Fig. 9.7   Methods of feeding slots in ground conducting plane: (a) Coaxial
folding, (b) Fed by end of waveguide, (c) Cavity-fed slot.

electric field in the aperture may be assumed to be given by

$$\mathbf{E}_0 = \mathbf{u}_x E_{00} \cos \frac{\pi y}{2a} \quad \text{for } |x| < a, \ |y| \leqslant b \qquad (9.45)$$

and the tangential electric field in the xy plane outside the slot may be
assumed to be zero. This assumption is quite valid, because this is the
distribution of the TE_{10} mode in the waveguide for the slot fed by the end
of a waveguide. It is also quite reasonable for the cavity-fed slot which is
excited in its fundamental mode as well as for the coaxial-fed slot.

Using the results of Section 9.4, the field in the region $z > 0$ may now
be derived. The transform $f_t(k_x, k_y)$ will be given by

$$\mathbf{f}_t(k_x, k_y) = \int_{x=-a}^{a} \int_{y=-b}^{b} \mathbf{E}_0(x, y) \exp \left[j(k_x x + k_y y) \right] dx \, dy \qquad (9.46)$$

so that the field at any point in the half-space $z > 0$ is given by

$$E(x, y, z) = \frac{1}{4\pi^2} \int_{k_x=-\infty}^{\infty} \int_{k_y=-\infty}^{\infty} \mathbf{f}(k_x, k_y) \exp(-j\mathbf{k}\cdot\mathbf{r})\, dk_x dk_y \quad (9.47)$$

which can be evaluated to be equal to

$$E(x, y, z) = jk_0 \frac{\exp(-jk_0 r)}{2\pi r} [\mathbf{u}_\phi(f_y \cos\phi - f_x \sin\phi)\cos\theta$$
$$+ \mathbf{u}_\theta(f_x \cos\phi + f_y \sin\phi)] \quad (9.48)$$

where f_x and f_y are evaluated at $k_x = k_0 \cos\phi \sin\theta$ and $k_y = k_0 \sin\phi \sin\theta$.
Using Eq. 9.45 in Eqs. 9.46 to 9.48, we obtain

$$\mathbf{f}_t(k_x, k_y) = \mathbf{u}_x \left\{ -\frac{2E_{00}\pi}{b} \frac{\sin(k_x a)}{k_x} \frac{\cos(k_y b)}{k_y^2 - (\pi/2b)^2} \right\} \quad (9.49)$$

so that

$$f_x(k_0 \cos\phi \sin\theta, k_0 \sin\phi \sin\theta)$$
$$= -\frac{2E_{00}\pi}{b} \frac{\sin(k_0 a \cos\phi \sin\theta)}{k_0 \cos\phi \sin\theta} \frac{\cos(k_0 b \sin\phi \sin\theta)}{[(k_0 \sin\phi \sin\theta)^2 - (\pi/2b)^2]} \quad (9.50)$$

Therefore, the far-field components of the electric field are

$$E_\phi = jk_0 E_0 ab \frac{\exp(-jk_0 r)}{r} \sin\phi \cos\theta \frac{\sin(k_0 a \cos\phi \sin\theta)}{k_0 a \cos\phi \sin\theta}$$
$$\times \frac{\cos(k_0 b \sin\phi \sin\theta)}{[(k_0 b \sin\phi \sin\theta)^2 - (\pi/2)^2]} \quad (9.51)$$

$$E_\theta = -jk_0 E_{00} ab \frac{\exp(-jk_0 r)}{r} \cos\phi \frac{\sin(k_0 a \cos\phi \sin\theta)}{k_0 a \cos\phi \sin\theta}$$
$$\times \frac{\cos(k_0 b \sin\phi \sin\theta)}{[(k_0 b \sin\phi \sin\theta)^2 - (\pi/2)^2]} \quad (9.52)$$

If the width $2a$ is small so that $k_0 a \ll 1$, and if the length $2b$ of the slot is a half-wavelength so that $k_0 b = \pi/2$, then the field components E_ϕ and E_θ become

$$E_\phi = -j \frac{V_0}{\pi} \frac{\exp(-jk_0 r)}{r} \sin\phi \cos\theta \frac{\cos(\frac{1}{2}\pi \sin\phi \sin\theta)}{(1 - \sin^2\phi \sin^2\theta)} \quad (9.53)$$

and

$$E_\theta = j \frac{V_0}{\pi} \frac{\exp(-jk_0 r)}{r} \cos\phi \frac{\cos(\frac{1}{2}\pi \sin\phi \sin\theta)}{(1 - \sin^2\phi \sin^2\theta)} \quad (9.54)$$

where $V_0 = 2aE_{00}$ may be taken as the voltage across the slot. In the $\phi = 0°$ plane, E_ϕ is zero and E_θ is constant, so that the pattern in this plane is a circle. In the $\phi = \pi/2$ ($x = 0$) plane, E_θ is zero and

$$E_\phi = -j \frac{V_0}{\pi} \frac{\exp(-jk_0 r)}{r} \frac{\cos(\frac{1}{2}\pi \sin\theta)}{\cos\theta} \quad (9.55)$$

The pattern for E_ϕ in this plane $\phi = \pi/2$ is the same as that of a half-wavelength electric dipole oriented along the y-axis. The aperture of the narrow half-wavelength slot produces exactly the same fields as a magnetic dipole with magnetic current given by

$$I_m = -4aE_{00} \cos(\pi y/2b)\mathbf{u}_y \quad \text{for } |y| \leqslant b \quad (9.56)$$

radiating in free space (with no ground plane). Therefore the slot is just the dual of an electric dipole as we tried to show in a simple manner in Section 9.3.

The total power radiated by the slot is obtained by integrating $\frac{1}{2}$ Re $(\mathbf{E} \times \mathbf{H}^*)$ over a hemisphere of a very large radius centred on the aperture, and is given by

$$
\begin{aligned}
P &= \int_{\theta=0}^{\pi/2} \int_{\phi=0}^{2\pi} \frac{\eta_0^{-1}}{2} \frac{V_0^2}{\pi^2 r^2} \left(\sin^2 \phi \cos^2 \theta + \cos^2 \phi\right) \\
&\quad \times \frac{\cos^2 \left[(\pi/2) \sin \phi \sin \theta\right]}{(1 - \sin^2 \phi \sin^2 \theta)^2} r^2 \sin \theta \, d\theta \, d\phi \\
&= \frac{\eta_0^{-1}}{2} \frac{V_0^2}{\pi^2} \int_{\theta=0}^{\pi/2} \int_{\phi=0}^{2\pi} \frac{\cos^2 \left[(\pi/2) \sin \phi \sin \theta\right]}{(1 - \sin^2 \phi \sin^2 \theta)} \sin \theta \, d\theta \, d\phi \\
&= \frac{\eta_0^{-1}}{2} \frac{V_0^2}{\pi^2} \int_{\psi=0}^{\pi} \int_{\xi=-\pi/2}^{\pi/2} \frac{\cos^2 \left[(\pi/2) \cos \psi\right]}{\sin^2 \psi} \sin \psi \, d\psi \, d\xi
\end{aligned}
\tag{9.57}
$$

where $\sin \theta \sin \phi = \cos \psi$ and ξ is the angle made by the projection of r in the xz plane with the z axis.

For a half-wavelength thin slot, $k_0 a \ll 1$ and $k_0 b = \pi/2$, so that

$$
\rho = \frac{\eta_0^{-1} V_0^2}{4\pi} \left[\ln (2\gamma\pi) - Ci (2\pi)\right] = 2.436 \frac{\eta_0^{-1} V_0^2}{4\pi}
\tag{9.58}
$$

As V_0 is the voltage across the slot, a radiation conductance G_r may be defined as the conductance which, if placed across the voltage V_0, would dissipate the same power as that radiated by the slot, that is

$$
\tfrac{1}{2} V_0^2 G_r = P
$$

which gives

$$
G_r = 2.436 \frac{\eta_0^{-1}}{2\pi} = 1.03 \times 10^{-3} \text{ mhos}
\tag{9.59}
$$

9.6 Slots on Conducting Cylinders

A slot may be formed by a waveguide opening through the side of a rocket body or aircraft wing. An approximate model of this slot antenna is a slot on a conducting cylinder. The slot may be a circumferential slot or an axial slot as shown in Figs. 9.8(a) and 9.8(b).

If the electric field in the aperture of the slot is known, the field in the region outside the cylinder is found in the following manner. The general expressions involving cylindrical mode functions for an arbitrary field in a cylindrical coordinate system

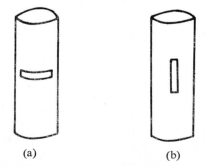

(a)   (b)

Fig. 9.8   Slots on a cylinder: (a) Circumferential slot, (b) Axial slot.

are written, and the unknown coefficients are evaluated by setting the general expressions equal to the assumed fields on the surface of the cylinder.
Silver and Saunders[1,2] have used this method for an aperture of arbitrary
shape, and have used their results to study a slot of rectangular shape on
a conducting cylinder. They have applied their general results to the
special case of a narrow axial half-wavelength slot having a cosinusoidal
distribution for the ϕ component of the electric field on the aperture, and
a tabular data has been given to calculate the radiation patterns. The
radiation patterns are not exactly omnidirectional in the plane normal to
the slot, and are not quite symmetrical in the plane of the slot as shown
in Fig. 9.9.

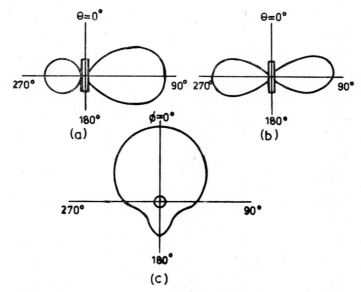

Fig. 9.9   Radiation patterns of narrow axial $\lambda/2$ slot on a conducting
cylinder: (a) $|E_\phi|$ vs. θ for $\phi=0°$, (b) $|E_\phi|$ vs. θ for $\phi=\pm\pi/2$,
(c) $|E_\phi|$ vs. ϕ for $\theta = \pi/2$.

9.7   Slots on Conducting Spheres

Consider a slot on a perfectly conducting sphere as shown in Fig. 9.10.
The theoretical problem is studied in a manner similar to the slot on a conducting cylinder. The resulting fields outside the sphere are written as a
sum of spherical mode functions chosen to satisfy the radiation condition at infinity. Figure 9.11 shows some radiation patterns of half-wave
slots on a conducting sphere for different values of k_0a, where 'a' is the
radius of the sphere.

[1]Silver, S. and W. K. Saunders, The External Field Produced by a Slot in an Infinite
Circular Cylinder, *J. Appl. Phys.*, vol. 21, pp. 153–158, February 1950.
[2]Silver, S. and W. K. Saunders, The Radiation from a Transverse Rectangular Slot in a
Circular Cylinder, *J. Appl. Phys.*, vol. 21, pp. 745–749, August 1950.

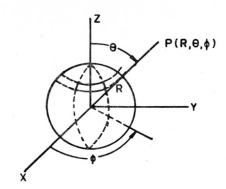

Fig. 9.10   A slot on a conducting sphere.

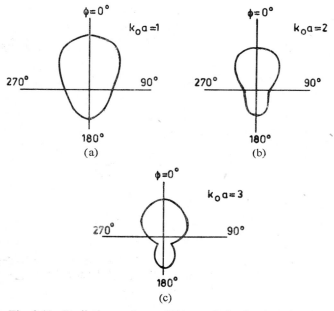

Fig. 9.11   Radiation patterns ($|E_\theta|$ vs. ϕ) in the $\theta = \pi/2$ plane for a half-wave slot on a conducting sphere.

9.8   Slot Arrays on Waveguides

Arrays of slots are used to obtain higher directivity and lower side lobes. In order to optimize the array it is necessary to control the excitation level of each slot in the array. This can be achieved by cutting the slots in a waveguide wall with the array axis extending along the length of the waveguide. A usual method of doing this is by using an array of slots in the broadwall of a waveguide.

Figure 9.12 shows a longitudinal slot in the broadwall of a rectangular waveguide, excited in the dominant TE_{10} mode. The current flowing in the

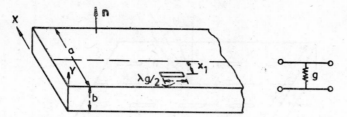

Fig. 9.12   Longitudinal $\lambda_0/2$ slot in the broadwall of a rectangular waveguide and its equivalent circuit.

upper broadwall of the waveguide is

$$\mathbf{J}_s \propto \mathbf{u}_x \cos\frac{\pi x}{a} - j\frac{\beta a}{\pi}\mathbf{u}_z \sin\frac{\pi x}{a} \qquad (9.60)$$

At the centre where $x = a/2$, the current is entirely in the z-direction. Hence, a narrow slot at the centre does not cut any current flow lines and is not excited. However, if the slot is placed away from the centre, say, at $x = x_1$, current flow lines are intercepted and the slot is excited. The excitation increases with increasing x_1. The normalized conductance of the slot is given by

$$g = 2.09\,\frac{\lambda_g}{\lambda_0}\frac{a}{b}\cos^2\frac{\pi\lambda_0}{2\lambda_g}\sin^2\frac{\pi x_1}{a} \qquad (9.61)$$

where λ_g is the guide wavelength and λ_0 is the free-space wavelength, and this acts as a shunt conductance across the waveguide.

There are two types of arrays of slots on a waveguide, namely, the resonant array and the non-resonant array. The resonant array is designed on the broadside wall and uses a slot spacing of $\lambda_g/2$, where λ_g is the guide wavelength. The *non-resonant* array can be designed to have the main lobe at any angle with respect to the normal to the broadwall of the guide but in the plane containing the array axis and the normal $\mathbf{n}$.

In the *resonant array*, for radiation in the broadside direction, the slots must all be excited in phase, which can be done by making the spacing between consecutive slots equal to $\lambda_g/2$. In order to avoid more than one main lobe in visible space, the spacing must be less than λ_0 for a broadside array. However, in a waveguide, $\lambda_g > \lambda_0$, and hence the spacing cannot be $\lambda_g/2$. This difficulty is overcome in the following manner. It is noted that the surface current in the x direction is of opposite sign on adjacent sides of the centre-line of the broadwall of the rectangular waveguide. The slot excitation is reversed in sign by off-setting it on the adjacent side of the centre-line. Hence a slot array as shown in Fig. 9.13(a), where every other slot is on the opposite side of the centre-line, and with a slot spacing of $\lambda_g/2$, is used. Figure 9.13(b) represents the equivalent circuit of such an array, and $g_e = \sum\limits_{n=1}^{N} g_n$ is the equivalent shunt conductance.

If a_n is the desired excitation coefficient for the nth-element of the array,

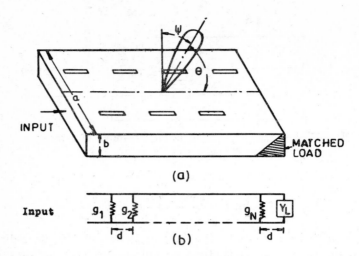

(a)

(b)

Fig. 9.13   Resonant array of slots on waveguide and its equivalent circuits.

for example, the coefficient to give a Chebyshev or some other form of array, then the conductance g_n is given by

$$g_n = Ka_n^2 \tag{9.62}$$

where K is chosen so that $g_e = 1$, i.e.,

$$K \sum_{n=1}^{N} a_n^2 = 1 \tag{9.63}$$

in order that the array is matched to the input waveguide for maximum radiated power. Once g_n is specified, the slot off-set x_1 is found for the nth slot by using Eq. 9.61.

The resonant broadside array on the broadwall of a waveguide is a narrow-band antenna, because the short-circuit spacing $\lambda_g/4$ from the last slot changes with the frequency and as frequency changes it does not simulate an open circuit any more and reflects a wave with appreciable amplitude. A *non-resonant array* is usually designed for operations other than broadside. The array consists of resonant longitudinal broadwall slots, and the waveguide is terminated in a normalized load Y_L. The array and its equivalent circuit are shown in Fig. 9.14.

Let d be the slot spacing and x_n the off-set measured from the centre-line $x = a/2$. Then the location of the n-th slot is given by

$$\mathbf{r}_n = nd\mathbf{u}_z + x_n\mathbf{u}_x \tag{9.64}$$

Let the excitation of the n-th slot be a_n (not including the possible 180° phase change due to off-set on the opposite side of the centre). Then the array factor is given by

$$f = \sum_{n=1}^{N} a_n \exp\left(jk_0\mathbf{r}\cdot\mathbf{r}_n - j\beta nd\right) \tag{9.65}$$

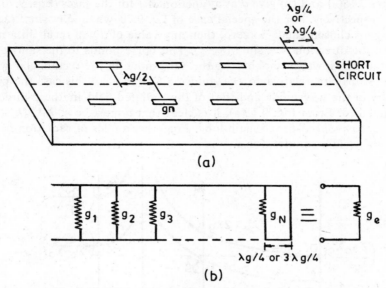

Fig. 9.14   Non-resonant array and its equivalent circuit.

for the case of the slots all off-set on one side of the centre-line, and

$$f = \sum_{n=1}^{N} a_n \exp (jk_0 \mathbf{r} \cdot \mathbf{r}_n + jn\pi - j\beta nd) \qquad (9.66)$$

for the case where every other slot is displaced on adjacent sides of the centre-line. In Eqs. 9.65 and 9.66, $\beta = 2\pi/\lambda_g$ and r is given by

$$\mathbf{r} = \mathbf{u}_x \sin \theta \cos \phi + \mathbf{u}_y \sin \theta \sin \phi + \mathbf{u}_z \cos \theta \qquad (9.67)$$

where θ is the polar angle and ϕ is the azimuthal angle in spherical polar coordinates.

To have maximum radiation in the direction $\phi = \pi/2$, $\theta = \pi/2 - \psi$, the condition

$$k_0 d \sin \psi - \beta d = 2m\pi \qquad (9.68)$$

has to be satisfied for the case (Eq. 9.65), and

$$k_0 d \sin \psi - \beta d + \pi = 2m\pi \qquad (9.69)$$

for the case (Eq. 9.66), where m is an integer.

From Eqs. 9.68 and 9.69, we obtain

$$d = \frac{m\lambda_0\lambda_g}{\lambda_g \sin \psi - \lambda_0} \qquad (9.70)$$

and

$$d = \frac{(2m - 1)\lambda_0\lambda_g}{2(\lambda_g \sin \psi - \lambda_0)} \qquad (9.71)$$

Equations 9.70 and 9.71 should be satisfied such that only one main lobe occurs in visible space, i.e., second order beams should not occur.

Figures 9.15(a) and (b) give d as a function of ψ for the cases (Eqs. 9.70 and 9.71) respectively. For the special case of Eq. 9.70 when $\psi = \sin^{-1}(\lambda_0/\lambda_g)$, when m is chosen equal to zero, then any value of d will result in a main lobe in the direction $\psi = \sin^{-1}(\lambda_0/\lambda_g)$. This corresponds to the same angle that the TEM waves making up the TE_{10} mode in the rectangular waveguide propagate relative to the guide axis. Hence, in this case the phase velocity of the waveguide and that of the radiated field in the z-direction are matched. From Fig. 9.15(a), it is clear that d must be $< \lambda_0\lambda_g/(\lambda_0 + \lambda_g)$ in order to avoid a second main lobe. For other angles of radiation $m = 0$ does not yield a solution for d.

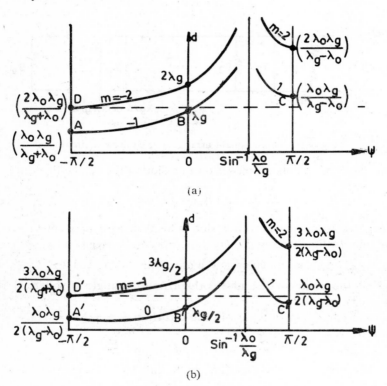

(a)

(b)

Fig. 9.15 d vs. ψ, (a) For Eq. 9.70,   (b) For Eq. 9.71.

Many solutions for d are possible depending on the relative value of λ_g/λ_0, but $\lambda_g/2\lambda_0$ should be kept so that the guide will not be too dispersive, because high dispersion will make the radiation very frequency sensitive. For $\lambda_g < 2\lambda_0$ the value of d must be so chosen that

$$\frac{\lambda_0\lambda_g}{(\lambda_0 + \lambda_g)} < d < \frac{2\lambda_0\lambda_g}{(\lambda_0 + \lambda_g)} \tag{9.72}$$

and

$$\sin \psi = \frac{\lambda_0}{\lambda_g} - \frac{\lambda_0}{d} \tag{9.73}$$

which will give a single main lobe in any direction between $-\pi/2$ and $\sin^{-1}[(2\lambda_0 - \lambda_g)/\lambda_g]$ with all of the slots off-set on one side of the centre-line.

Curves of Fig. 9.15(b) apply in the case of alternate slots being off-set on opposite sides of the centre-line. For this case if λ_g is chosen $< 2\lambda_0$, then

$$\frac{\lambda_0\lambda_g}{(\lambda_0 + \lambda_g)} < 2d < \frac{3\lambda_0\lambda_g}{(\lambda_0 + \lambda_g)} \qquad (9.74)$$

and

$$\sin\psi = \frac{\lambda_0}{\lambda_g} - \frac{\lambda_0}{2d} \qquad (9.75)$$

and the angle of radiation is then limited to the range between $-\pi/2$ and $\sin^{-1}[(2\lambda_0 - \lambda_g)/3\lambda_g]$.

For angles greater than the angle ψ given by Eq. 9.75, it is not possible to get a single lobe in the forward direction without having another main lobe appearing at some other angle, since points C, C' can never be lower than points A, A' because $\lambda_g - \lambda_0 < (\lambda_g + \lambda_0)$.

The fact that the main lobe depends on the relative value of λ_g/λ_0 for a given value of d can be made use of in scanning the lobe by changing the frequency. A highly dispersive waveguide is useful for frequency scanning, and this can be done by the use of waveguides bent into a sinusoidal shape so that the electrical length between the slots is increased.

9.9 Microstrip Antennas

A microstrip is a type of open waveguiding structure which is a state-of-art transmission line that can be used as a circuit component like a filter, coupler, resonator, etc. The microstrip antenna is a much later development and its assembly is physically very simple and flat. This has made it very popular. Figure 9.16 shows a simple microstrip patch antenna. The upper surface of the dielectric substrate supports the printed conducting

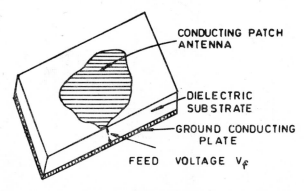

CONDUCTING PATCH ANTENNA

DIELECTRIC SUBSTRATE

GROUND CONDUCTING PLATE

FEED VOLTAGE V_f

Fig. 9.16 Microstrip patch antenna driven against ground plane.

strip which is suitably contoured while the entire lower surface of the substrate is backed by the conducting ground plate. Such an antenna is sometimes called a printed antenna because of the manufacturing process. The microstrip antenna is similar to a slot antenna not only because it is a flat antenna, but also because its radiation characteristics can be derived by considering it to be a slot antenna with magnetic walls.

Many types of microstrip antennas have been evolved which are variations of the basic sandwich structure. The common property of all these antennas is that they are electrically driven with respect to the ground plane. A wide range of dielectric substrate thicknesses and permittivities are used including air dielectric. Microstrip antennas can be designed as very thin planar printed antennas and they are very useful elements of different types of arrays, especially conformal arrays which can be designed on a surface of any type and shape.

9.10 The Microstrip as a Transmission Line

Figure 9.17 shows a microstrip line. Usually the substrate is a dielectric of relative permittivity ϵ_r and relative permeability $\mu_r = 1$. A simple approximate analysis assumes that the line supports a TEM wave and the characteristic line impedance is

$$Z_d = \frac{1}{v_d C_d} \text{ ohms} \tag{9.76}$$

where v_d is the velocity of propagation along the line and C_d is the line capacity per unit length. If Z_0, v_0 and C_0 are the corresponding parameters for an air-spaced line, then,

$$v_d = \frac{v_0}{\sqrt{\epsilon_r}}; \; C_d = C_0 \epsilon_r; \; v_0 = \frac{1}{\sqrt{\mu_0 \epsilon_0}}; \; Z_d = \frac{Z_0}{\sqrt{\epsilon_r}} \tag{9.77}$$

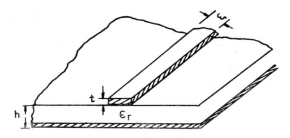

Fig. 9.17 Microstrip line.

Since the dielectric material only partially fills the space above the ground plane in the microstrip, it is reasonable to expect the corresponding values of velocity, capacity and impedance denoted by v_m, C_m and Z_m respectively, to lie somewhere between the values corresponding to the two extreme cases, namely, the completely dielectric-filled case and the case of empty space. Thus

$$v_d < v_m < v_0$$

$$C_0 < C_m < C_d \tag{9.78}$$

and

$$Z_d < Z_m < Z_0$$

Hence, it is convenient to introduce the concept of a filling factor q $(0 \leqslant q \leqslant 1)$ and effective relative permittivity ϵ_e $(1 \leqslant \epsilon_e \leqslant \epsilon_r)$, where

$$\epsilon_e = 1 + q(\epsilon_r - 1), \quad v_m = \frac{v_0}{\sqrt{\epsilon_e}}$$

$$C_m = C_0 \epsilon_e, \qquad Z_m = \frac{Z_0}{\sqrt{\epsilon_e}} \tag{9.79}$$

$$\lambda_m = \frac{\lambda_0}{\sqrt{\epsilon_e}}, \qquad \beta = \frac{2\pi}{\lambda_m}$$

where λ_0 is the wavelength in the air-filled line and is identical to the wavelength in free space. C_0 and Z_0 can be calculated by a conformal transformation method as shown by Wheeler[1-3].

From Eq. 9.79 the microstrip properties can be obtained approximately. The structure can support not only the TEM mode but also a set of discrete hybrid modes having non-zero E_z and H_z components, where z is the direction of propagation. However, the lowest-order hybrid mode strongly resembles a TEM mode and it is called a quasi-TEM mode or wave.

Wheeler's basic expression for Z_m is given by

$$Z_m = \frac{377}{\pi \sqrt{2} \sqrt{\epsilon_{r+1}}} \left\{ \ln\left(\frac{8h}{w}\right) + \frac{1}{32}\left(\frac{w}{h}\right)^2 \right.$$

$$\left. - \frac{1}{2}\left(\frac{\epsilon_r - 1}{\epsilon_{r+1}}\right)\left(\ln\frac{\pi}{2} + \frac{1}{\epsilon_r}\ln\frac{4}{\pi}\right)\right\} \text{ ohms} \quad \text{for } w/h < 1 \tag{9.80}$$

and

$$Z_m = \frac{377}{2\sqrt{\epsilon_r}}\left[\frac{w}{2h} + 0.441 + 0.082\left(\frac{\epsilon_r - 1}{\epsilon_r^2}\right)\right.$$

$$\left. + \left(\frac{\epsilon_{r+1}}{2\pi\epsilon_r}\right)\left\{1.451 + \ln\left(\frac{w}{2h} + 0.94\right)\right\}\right]^{-1} \text{ ohms} \quad \text{for } w/h > 1 \tag{9.81}$$

[1] H. A. Wheeler, 'Transmission Properties of Parallel Wide Strips by a Conformal Mapping Approximation', *IEEE Trans. MTT*-12, pp. 280–289, 1964.

[2] H. A. Wheeler, 'Transmission-line Properties of a Strip on a Dielectric Sheet', *IEEE Trans. MTT*-13, pp. 172–185, 1965.

[3] H. A. Wheeler, 'Transmission-line Properties of a Strip on a Dielectric Sheet on a Plane', *IEEE Trans. MTT*-25, pp. 631–647, 1977.

Schneider[1] has given an expression for ϵ_e as

$$\epsilon_e = \frac{1}{2}\left\{\epsilon_r + 1 + (\epsilon_r - 1)\left(1 + \frac{10h}{w}\right)^{-1/2}\right\} \qquad (9.82)$$

which may be used with Eqs. 9.80, 9.81 and 9.79 by putting $\epsilon_r = 1$ in Eqs. 9.80 and 9.81. The strip thickness t can be corrected for by substituting w' for w where

$$w' = w + \frac{t}{\pi}\left\{1 + \ln\left(\frac{2x}{t}\right)\right\} \qquad (9.83)$$

with $x = h$ for $w > (h/2\pi) > 2t$ and $x = 2\pi w$ for $(h/2\pi) > w > 2t$.

One of the important limitations of microstrip antennas are dissipative losses. These losses occur in the dielectric substrate material and in the conducting sheets. If the substrate material has a complex relative permittivity $\epsilon_r - j\epsilon_r$, then the attenuation constant α_{du} due to this is

$$\alpha_{du} = \frac{\omega}{2}(\mu_0\epsilon_0\epsilon_r)^{1/2}\tan\delta \qquad (9.84)$$

where $\tan\delta = \epsilon_r'/\epsilon_r$, assuming TEM wave propagation. It is reasonable to use the effective relative permittivity ϵ_e instead of ϵ_r to give better accuracy. Schneider[2] has given the expression

$$\alpha_d = 27.3\,\frac{\epsilon_r\,(\epsilon_e - 1)}{\epsilon_e\,(\epsilon_r - 1)}\,\frac{\tan\delta}{\lambda_m}\ \text{dB/cm} \qquad (9.85)$$

The conductor loss gives rise to an attenuation constant α_c given by

$$\alpha_c \simeq \frac{8.68R_s}{wZ_m}\ \text{dB/cm} \qquad (9.86)$$

where $R_s = \left(\dfrac{\omega\mu_0}{2\sigma_c}\right)^{1/2}$, R_s is the surface resistivity in (ohms per square) of both sides of the strip, and σ_c is the conductivity of the strip.

9.11   Radiation from Microstrip Antennas

Any discontinuity in a microstrip transmission line radiates, and it is an undesirable effect in a transmission line. However finite length microstrips have very desirable characteristics when used as antennas. Even the open-circuit microstrip termination can be used as a radiating element.

We shall consider, in particular, the radiation characteristics of a microstrip patch antenna. The patch can be of any shape. Figure 9.18 represents some types of feed arrangements for such antennas.

The electromagnetic field distribution of a patch radiator depends on the different modes it can support. If the spacing between the patch and the ground plane is $<\lambda_0/10$ which is true for typical microstrip substrates,

[1]M. V. Schneider, 'Microstrip Lines for Microwave Integrated Circuits', *Bell Syst. Tech. J.*, 48, pp. 1421–1444, 1969.
[2]M. V. Schneider, 'Dielectric Loss in Integrated Microwave Circuits', *Bell Syst. Tech. J.*, 48, pp. 2325–2332, 1969.

the field is usually concentrated underneath the patch. The fields leak out into the air through the substrate surrounding the patch, and this is a complex boundary-value problem. This problem can be dealt with

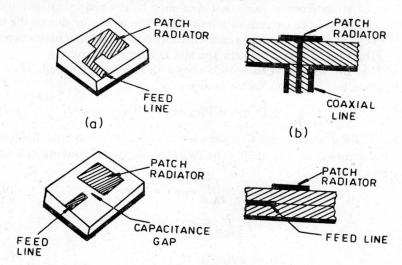

Fig. 9.18   Feed arrangements for microstrip patch antennas: (a) Direct connection to surface microstrip line, (b) Coaxial-line feed, (c) Connection to surface microstrip line with matched capacitance coupling, (d) Double layer microstrip transmission-line feed.

approximately by applying an open-circuit boundary condition at the edges of the microstrip patch, which is justified by assuming a TEM wave. The open-circuit boundary condition means a high impedance at the patch periphery, and hence, the electric field parallel to the edge has maximum value and the magnetic field parallel to the edge is minimum. This can be modelled by placing a magnetic wall at the effective edge position as shown in Fig. 9.19. The boundary conditions for the patch are expressed as

$$\mathbf{H} \cdot \mathbf{u}_z = 0, \quad \mathbf{E} \times \mathbf{u}_z = 0 \text{ at } z = 0 \text{ and } h \tag{9.87}$$

and

$$\mathbf{E} \cdot \mathbf{u}_n = 0, \quad \mathbf{H} \times \mathbf{u}_n = 0 \text{ on the magnetic wall} \tag{9.88}$$

Therefore, a good approximation of the microstrip patch antenna is that of an enclosed cavity which can support an infinite set of resonant modes of different frequencies. Since the height of the cavity is very small as compared to a wavelength, we may assume that $\partial/\partial z(\mathbf{E}, \mathbf{H}) = 0$. Then Eqs. 9.87 and 9.88 make

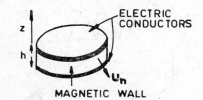

Fig. 9.19   Magnetic wall model of patch antenna.

$$E_x = E_y = H_z = 0 \tag{9.89}$$

Under the conditions given in Eq. 9.89, the field distribution in the cavity is the dual of the field structure of a TE mode in a rectangular metal waveguide at cut-off.

The radiation from a patch antenna may be derived by assuming that either it is due to the current distribution on the patch, or due to the field on the edge slot which may be represented by equivalent magnetic currents. The method based on the field in the slot is easier to handle.

Let V be the voltage at the patch edge with respect to ground. Then the far-field electric radiation vector is given by

$$\mathbf{L} = -2 \int_C \mathbf{u}_z \times \mathbf{u}_n V(c) \exp\left(jk_0 r \cos \psi\right) dc \qquad (9.90)$$

where the integration range C is the closed loop around the patch effective edge as depicted in Fig. 9.20. The radiation field components E_θ and E_ϕ are given by

$$E_\theta = -j \frac{L_\phi}{2\lambda_0 R} \exp\left(-jk_0 R\right) \qquad (9.91)$$

$$E_\phi = j \frac{L_\theta}{2\lambda_0 R} \exp\left(-jk_0 R\right) \qquad (9.92)$$

and

$$H_\theta = -E_\phi/Z_0; \quad H_\phi = E_\theta/Z_0 \qquad (9.93)$$

Applying this to a rectangular patch as shown in Fig. 9.21, the field components are assumed to be the duals of the well-known rectangular waveguide TE modes and given by

$$E_z = \frac{V_0}{h} \cos \frac{m\pi x}{a} \cos \frac{n\pi y}{b}$$

$$H_x = -\frac{j\omega\epsilon_0}{k_0^2} \frac{n\pi}{b} \frac{V_0}{h} \cos \frac{m\pi x}{a} \sin \frac{n\pi y}{b} \qquad (9.94)$$

$$H_y = \frac{j\omega\epsilon_0}{k_0^2} \frac{m\pi}{a} \frac{V_0}{h} \sin \frac{m\pi x}{a} \cos \frac{n\pi y}{b}$$

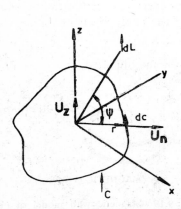

Fig. 9.20 Coordinate system for calculating the radiation from a patch.

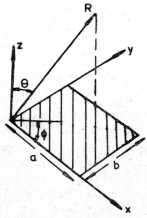

Fig. 9.21 Rectangular patch antenna.

where V_0 may be considered as the peak voltage at the corner of the patch. The resonant frequency of the patch is the same as the cut-off frequency of the rectangular waveguide and is given by

$$f_r = \frac{1}{2\sqrt{\mu_0 \epsilon_0 \epsilon_r}} \left\{ \left(\frac{m}{a}\right)^2 + \left(\frac{n}{b}\right)^2 \right\}^{1/2} \tag{9.95}$$

The voltages at the edges of the patch are given by

$$V(y) = V_0 \cos (n\pi y/b) \quad \text{at } x = 0; \ 0 \leqslant y \leqslant b$$

$$V(y) = V_0 \cos (n\pi y/b) \cos (m\pi) \quad \text{at } x = a; \ 0 \leqslant y \leqslant b$$

$$V(x) = V_0 \cos (m\pi x/a) \quad \text{at } y = 0; \ 0 \leqslant x \leqslant a \tag{9.96}$$

$$V(x) = V_0 \cos (m\pi x/a) \cos (n\pi) \quad \text{at } y = b; \ 0 \leqslant x \leqslant a$$

Using Eq. 9.96 in Eq. 9.90, we have

$$L = 2jV_0\{1 - \cos (m\pi) \exp (jk_0 a \cos \phi \sin \theta)\}$$

$$\times \{1 - \cos (n\pi) \exp (jk_0 b \sin \phi \sin \theta)\}$$

$$\times \left[\mathbf{u}_x \frac{k_0 \cos \phi \sin \theta}{k_0^2 \cos^2 \phi \sin^2 \theta - \dfrac{m^2\pi^2}{a^2}} - \mathbf{u}_y \frac{k_0 \sin \phi \sin \theta}{k_0^2 \sin^2 \phi \sin^2 \theta - \dfrac{n^2\pi^2}{b^2}} \right]$$
$$\tag{9.97}$$

Converting Eq. 9.97 to spherical polar coordinates, and substituting in Eqs. 9.91 and 9.92, we obtain

$$E_\theta = -\frac{\exp (-jk_0 R)}{2\lambda_0 R} 2V_0\{1 - \cos (m\pi) \exp (jk_0 a \cos \phi \sin \theta)\}$$

$$\times \{1 - \cos (n\pi) \exp (jk_0 b \sin \phi \sin \theta)\} k_0 \sin \theta \cos \phi \sin \phi$$

$$\times \left\{ \frac{1}{k_0^2 \cos^2 \phi \sin^2 \theta - \dfrac{m^2\pi^2}{a^2}} + \frac{1}{k_0^2 \sin^2 \phi \sin^2 \theta - \dfrac{n^2\pi^2}{b^2}} \right\} \tag{9.98}$$

and

$$E_\phi = \frac{\exp (-jk_0 R)}{2\lambda_0 R} 2V_0\{1 - \cos (m\pi) \exp (jk_0 a \cos \phi \sin \theta)\}$$

$$\times \{1 - \cos (n\pi) \exp (jk_0 b \sin \phi \sin \theta)\} k_0 \sin \theta \cos \theta$$

$$\times \left\{ \frac{\sin^2 \phi}{k_0^2 \sin^2 \phi \sin^2 \theta - \dfrac{n^2\pi^2}{b^2}} - \frac{\cos^2 \phi}{k_0^2 \cos^2 \phi \sin^2 \theta - \dfrac{m^2\pi^2}{a^2}} \right\} \tag{9.99}$$

The lowest-order mode for which $m = 0$ and $n = 1$ is the most important case. For this mode the radiation fields in the $\phi = 0°$ and $\phi = 90°$ planes are given by

$$E_{\theta}(\phi = 0°) = 0$$

$$E_{\phi}(\phi = 0°) = -j \frac{\exp{(-jk_0R)}}{2\lambda_0 R} 4V_0 a \exp\left(j \frac{k_0 a}{2} \sin\theta\right) \cos\theta$$

$$\times \frac{\sin\{(k_0a/2)\sin\theta\}}{(k_0a/2)\sin\theta}$$

$$(9.100)$$

$$E_{\theta}(\phi = 90°) = -j \frac{\exp{(-jk_0R)}}{2\lambda_0 R} 4V_0 a \exp\left(j \frac{k_0 b}{2} \sin\theta\right)$$

$$\times \cos\left(\frac{k_0 b}{2}\sin\theta\right)$$

$$E_{\phi}(\phi = 90°) = 0$$

Figure 9.22 shows the theoretical radiation patterns of a square patch antenna in $\phi = 45°/225°$ plane. It can be seen that the radiation patterns are not very directional. However, by using arrays of patch antennas, highly directional patterns can be obtained.

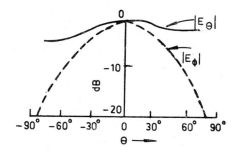

Fig. 9.22   Theoretical radiation patterns of square patch
antenna in 45°/225° plane $\epsilon_r = 2.0$.

9.12 Limitations of Microstrip Antennas

The bandwith of a square or a circular patch antenna for a VSWR S is given by

$$\text{Bandwidth} = \frac{100(S-1)}{\sqrt{S}} \frac{8}{\epsilon_r} \frac{h}{\lambda_0} \%$$

$$(9.101)$$

which shows that the bandwidth decreases with decreasing h. Thus, thinner antennas have lesser bandwith.

The feed structure of these antennas is often printed on the substrate surface together with the radiating elements. The feeder lines introduce additional loss and hence the efficiency is reduced.

The mechanical tolerances of thin microstrip antennas usually place a limit on the precision with which the aperture phase and amplitude distribution can be controlled in manufacture.

PROBLEMS

1. What are the dimensions of a slot antenna whose terminal impedance is $(70 + j0)$ ohms? The slot is open on both sides.

2. Calculate and draw the far-field pattern in the $\phi = 0°$ and $\phi = 90°$ planes of a rectangular slot of dimensions $(3\lambda/2) \times 0.3\lambda$ in a conducting plane at a wavelength of 3 cms.

3. Calculate and draw the radiation patterns of a patch antenna of dimensions $2\lambda \times 3\lambda$ at a wavelength of 5 cms in the $\phi = 0°$ and $\phi = 90°$ planes.

4. Calculate the characteristic impedance of a microstrip line of width $w = \lambda/2$, height $h = 0.1\lambda$, and ϵ_r of the substrate $= 2.5$.

SUGGESTED READING

1. Collin, R. E. and Zucker, F. J., *Antenna Theory*, Parts 1 and 2, McGraw-Hill, New York, 1969.

2. Jasik, H., *Antenna Engineering Handbook*, McGraw-Hill, 1961.

3. Jordan, E. C. and Balmain, K. G., *Electromagnetic Waves and Radiating Systems*, Prentice-Hall of India, 1969.

4. Kraus, J. D., *Antennas*, McGraw-Hill, New York, 1950.

5. Schelkunoff, S. A. and Friis, H. T., *Antennas, Theory and Practice*, Chapman and Hall, London, 1952.

6. James, J. R., Hall, P. S. and Wood, C., *Microstrip Antenna*, Peter Peregrinus Ltd. (on behalf of Institution of Electrical Engineers, U.K.), 1981.

10

HORN ANTENNAS

Electromagnetic horn antennas have the desirable property of effecting a smooth transition from a medium like a waveguide which supports a finite number of propagating modes to a medium like free space which can support a large or infinite number of modes. They provide a fairly ideal wavefront for good directivity, and also have the advantage of being able to accommodate a very broad band of frequencies. Horns of different shapes are used to control properties like gain, radiation pattern and impedance.

10.1 Different Types of Horns

Some of the more frequently used types of horns are pyramidal, sectoral H-plane, sectoral E-plane, conical and biconical as shown in Fig. 10.1. The pyramidal horn of Fig. 10.1(a) is usually used as a standard primary gain antenna, because its gain can be calculated to within a tenth of a decibel if it is accurately constructed. This antenna is also very useful in obtaining specified beamwidths independently in the two principal planes. The H-plane and E-plane sectoral horns of Figs. 10.1(b) and (c) are useful in obtaining fan-shaped beams in the plane containing the flare, while the pattern is very broad in the other plane. Because of its axial symmetry, the conical horn of Fig. 10.1(d) can handle any polarization of the dominant TE_{11} mode. The biconical horn of Fig. 10.1(e) gives an omnidirectional radiation pattern in the plane normal to the axis, and is useful in the VHF-UHF band for broadcasting purposes. Other types of horns are shown in Fig. 10.2. The broadband horn of Fig. 10.2(a) may be used to give a good broadband impedance match to free space by properly designing the E-plane flare length and angle. The box horn in Fig. 10.2(b) shows a narrower pattern in the H-plane than a sectoral horn of the same aperture. The other horns shown in Figs. 10.2(c) to (h) have H-plane patterns either broader or narrower than the open-ended waveguide. The hog horn of Fig. 10.2(d) is used to illuminate 'cheese' or 'pill-box' antennas. It combines a 90°-direction change, aperture enlargement in the H-plane, and phase correction in the H-plane. The asymmetrical horn of Fig. 10.2(c) is used as an off-set feed for parabolic cylindrical antennas.

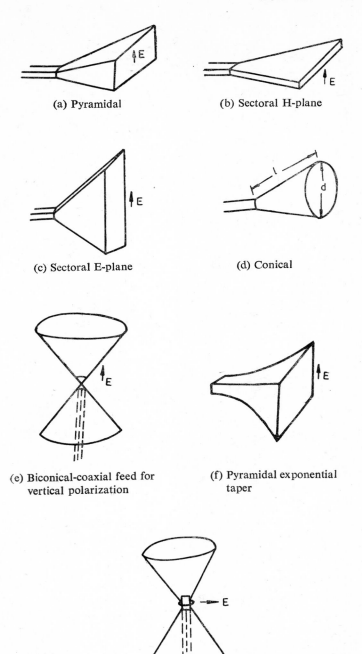

(a) Pyramidal

(b) Sectoral H-plane

(c) Sectoral E-plane

(d) Conical

(e) Biconical-coaxial feed for
vertical polarization

(f) Pyramidal exponential
taper

(g) Biconical-loop feed for
horizontal polarization

Fig. 10.1   Frequently used types of horns.

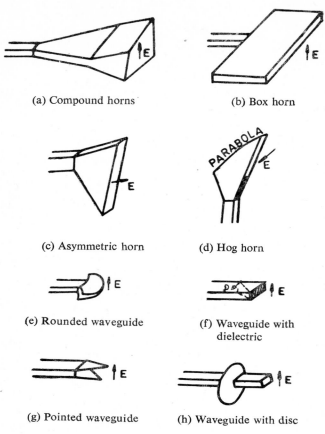

(a) Compound horns   (b) Box horn

(c) Asymmetric horn   (d) Hog horn

(e) Rounded waveguide   (f) Waveguide with
                                             dielectric

(g) Pointed waveguide   (h) Waveguide with disc

Fig. 10.2   Other types of horns.

10.2   Rectangular Horn

If a rectangular horn fed by a rectangular waveguide carrying the dominant TE_{01} mode is flared out in the plane of the magnetic field lines, it is
called an H-plane sectoral horn as shown in Fig. 10.1(b), and if it is flared
out in the plane of the electric field, it is called an E-plane sectoral horn
as shown in Fig. 10.1(c). If it is flared out in both planes, it is called a
pyramidal horn as shown in Fig. 10.1(a). To minimize reflections of the
guided wave, the transition region of the horn between the waveguide at
the throat and the free-space at the aperture can be given a gradual exponential taper as shown in Fig. 10.1(f).

Neglecting edge effects, the radiation pattern of a horn antenna can be
determined by the application of the Equivalence Principle as enunciated
in Section 2.21, if the aperture distribution is known.

The electromagnetic field in a semi-infinitely long rectangular horn is
determined by solving Maxwell's equations to satisfy the proper boundary

conditions at the conducting walls of the horn.[1,2]

Figure 10.3 represents the fields inside the finite length H-plane sectoral

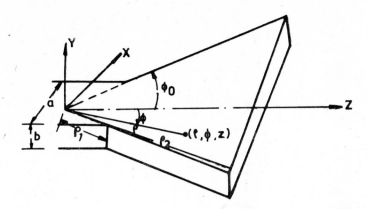

Fig. 10.3 H-plane sectoral horn.

horn for the dominant TE_{01} mode. These are

$$E_y = A \cos (p\phi)[H_p^{(2)}(k_0\rho) + \alpha H_p^{(1)}(k_0\rho)] \exp (j\omega t) \qquad (10.1)$$

$$H_\rho = \frac{pA}{j\omega\mu} \frac{\sin (p\phi)}{\rho} [H_p^{(2)}(k_0\rho) + \alpha H_p^{(1)}(k_0\rho)] \exp (j\omega t) \qquad (10.2)$$

$$H_\phi = \frac{kA}{j\omega\mu} \cos (p\phi)[H_p^{(2)}(k_0\rho) + \alpha H_p^{(1)}(k_0\rho)] \exp (j\omega t) \qquad (10.3)$$

using cylindrical coordinates ρ, ϕ, z. In Eqs. 10.1 to 10.3, for E_y, H_ρ and H_ϕ, A is an amplitude constant, $k_0 = 2\pi/\lambda_0 = \omega\sqrt{\mu_0\epsilon_0}$, $p = \pi/2\phi_0$, where ϕ_0 is the half-flare angle in the H-plane. $H_p^{(2)}(k_0\rho)$ represents a forward going wave and $\alpha H_p^{(1)}(k_0\rho)$ represents the reflected backward wave. H_ρ being proportional to $(1/\rho)[H_p^{(2)}(k\rho) + \alpha H_p^{(1)}(k\rho)]$ becomes very small for large ρ compared to E_y and H_ϕ. Hence the TE_{01} mode becomes very nearly a TEM mode near the mouth of the horn. Also $E_y/H_\phi \to \eta_0$ at the mouth of the horn, where η_0 is the intrinsic impedance $(\mu_0/\epsilon_0)^{1/2}$ of free-space. Near the throat of the horn $E_y/H_\phi = Z =$ characteristic impedance of the waveguide, if $\lambda > \lambda_c$, where λ_c is the cut-off wavelength of the waveguide. Hence the impedance match of the H-plane sectoral horn is good both at the throat and the mouth and, therefore, effects a smooth transition from the rectangular waveguide to the free-space.

The fields inside an E-plane sectoral horn as shown in Fig. 10.4 are given by

$$E_\phi = A \cos (\pi x/a)[H_1^{(2)}(\beta\rho) + \alpha H_1^{(1)}(\beta\rho)] \exp (j\omega t) \qquad (10.4)$$

[1]Barrow, W. L. and L. J. Chu, 'Theory of the Electromagnetic Horn', *Proc. I. R. E.*, vol. 27, pp. 51–64, Jan. 1939.

[2]Silver, S., *Microwave Antenna Theory and Design*, McGraw-Hill Book Co., Inc., New York, 1949, Chap. 10.

$$H_\rho = \frac{j\pi A}{\omega\mu a} \sin(\pi x/a)[H_1^{(2)}(\beta\rho) + \alpha H_1^{(1)}(\beta\rho)] \exp(j\omega t) \qquad (10.5)$$

$$H_x = \frac{j\beta A}{\omega\mu} \cos(\pi x/a)[H_0^{(2)}(\beta\rho) + \alpha H_0^{(1)}(\beta\rho)] \exp(j\omega t) \qquad (10.6)$$

using cylindrical coordinates ρ, ϕ, x, and where

$$\beta^2 = k_0^2 - \frac{\pi^2}{a^2} = \left[\frac{2\pi}{\lambda_0}\left\{1 - \left(\frac{\lambda}{2a}\right)^2\right\}\frac{1}{2}\right]^2 \text{ and } k_0^2 = \omega^2\mu_0\epsilon_0$$

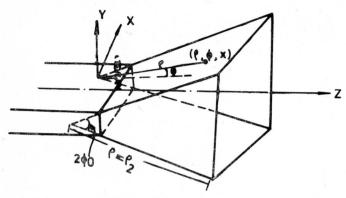

Fig. 10.4   E-plane sectoral horn.

In this horn, the guide-wavelength in the horn is the same as the guide-wavelength λ_g in the waveguide. Further, E_ϕ/H_x in the horn is also the characteristic impedance of the waveguide. Hence, the impedance match at the throat of the horn is very good, though it may not be so good at the mouth.

10.3   Radiation Pattern of Rectangular Horn

Using the expressions for the field components at the mouth of the horn obtained by substituting $\rho = \rho_2$ in Eqs. 10.1 to 10.6, for the H-plane and E-plane horns, the radiation patterns of these horns can be obtained by the application of the Equivalence Principle. The surface electric and magnetic current densities at the mouth of the horn are evaluated as

$$\mathbf{J} = \mathbf{n} \times \mathbf{H}^0 \qquad (10.7)$$

$$\mathbf{M} = -\mathbf{n} \times \mathbf{E}^0 \qquad (10.8)$$

where $\mathbf{E}^0$ and $\mathbf{H}^0$ are the electric and magnetic fields at the mouth of the horn and $\mathbf{n}$ is the unit normal. In terms of $\mathbf{J}$ and $\mathbf{M}$, the vector magnetic and electric potentials $\mathbf{A}$ and $\mathbf{F}$ at a distant point P are given by

$$\mathbf{A} = \iint_S \frac{\mathbf{J} \exp(-j\beta r)}{4\pi r} \, da \qquad (10.9)$$

and

$$\mathbf{F} = \iint\limits_{S} \frac{\mathbf{M} \exp{(-j\beta r)}}{4\pi r}\, da \qquad (10.10)$$

where S is a closed surface consisting of the mouth S_1 of the horn and a large surface S_2 enclosing the distant point as shown in Fig. 10.5. Since the surface currents on S_2 are negligible, only the surface currents on S_1 need to be considered in the integration of Eqs. 10.9 and 10.10. The radiation patterns thus obtained for the E-plane, H-plane and pyramidal horns are highly directional as shown in Fig. 10.6. The integration is usually done numerically or on the computer. To obtain an expression for the absolute gain or directivity of a horn, the field intensity at a distant point on the axis of the horn is obtained.

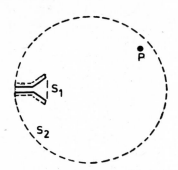

Fig. 10.5 Surface of integration for the determination of the radiation from a horn.

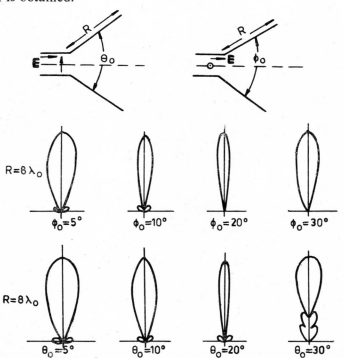

Fig. 10.6 Radiation patterns of sectoral horns: (a) E-plane sectoral horn, (b) H-plane sectoral horn, (c) Radiation patterns of E-plane sectoral horn, (d) Radiation patterns of H-plane sectoral horn.

10.4 Approximate Expression for Radiation Pattern of Rectangular Horn, using Huyghens' Principle

The principle proposed by Christian Huyghens (1629–1695), now called Huyghens' Principle[1], is very useful in determining the far-field of an antenna. According to this principle, each point of a wavefront can be considered as the source of a secondary spherical wave. These secondary spherical waves combine to form a new wavefront which is the envelope of the secondary wavelets. Thus, a spherical wave from a single point source propagates as a spherical wave, while an infinite plane wave has a planar wavefront. The wave at a field point is obtained by the super-position of these elementary secondary wavelets, with due regard to their phase differences when they reach the point in question.

When applied to electromagnetic waves, the point source is replaced by what is called a Huyghens' source which consists of a combination of an electric current element and a magnetic current element which are related to the tangential magnetic field intensity and tangential electric field intensity respectively on the wavefront by the equivalence theorem and given by Eqs. 10.7 and 10.8. The field at a distant point is evaluated as the summation of the electromagnetic fields due to all the secondary Huyghens' sources on the wavefront. Hence, Huyghens' Principle is essentially the same as the Equivalence Principle.

For a rectangular horn as shown in Fig. 10.7 (sectoral or pyramidal), the wavefront is cylindrical, but may be considered to be approximately plane at the mouth of the horn. The electric field intensity may be assumed to be approximately E_{0y} in the y-direction and the magnetic field to be approximately H_{0x} in the x-direction. The moment of a typical electric current element over the area $dx\,dy$ in the wavefront is $(\mathbf{n} \times H_{0x}\mathbf{u}_x)\,dx\,dy = \mathbf{u}_y H_{0x}\,dx\,dy$. Similarly, the moment of a typical magnetic current element

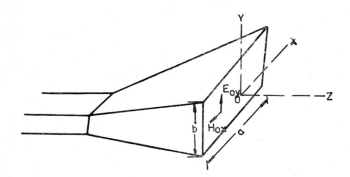

Fig. 10.7   Rectangular horn.

---

[1]Baker, B. B. and E. T. Copson, *The Mathematical Theory of Huyghens' Principle*, Oxford Unviersity Press, New York, 1939.

over the area $dx\,dy$ in the wavefront is $-(\mathbf{n}\times E_{0y}\mathbf{u}_y)\,dy\,dx = -E_{0y}\mathbf{u}_x\,dx\,dy$. This pair of electric and magnetic current elements form the Huyghens' source.

Since the electric current element is in the y-direction, it has a y-component A_y of the magnetic vector potential at a distant point $P(r, \theta, \phi)$, given by

$$A_y = \frac{H_{0x}\,dx\,dy}{4\pi r}\exp\,(-j\beta r) \tag{10.11}$$

so that

$$E_y = -j\omega\mu_0 A_y = \frac{-j60\pi H_{0x}\,dx\,dy}{\lambda r}\exp\,(-j\beta r) \tag{10.12}$$

because

$$\frac{\omega\mu_0}{4\pi} = \frac{2\pi f}{4\pi}\sqrt{\frac{\mu_0}{\epsilon_0}}\sqrt{\mu_0\epsilon_0} = \frac{f\cdot 120\pi}{2}\cdot\frac{1}{c} = \frac{60\pi}{\lambda}$$

Hence

$$\begin{aligned} E_{\theta_1} &= E_y \cos\theta \sin\theta \\ &= \frac{j60\pi H_{0x}\,dx\,dy}{\lambda r}\cos\theta \sin\phi \exp\,(-j\beta r) \end{aligned} \tag{10.13}$$

and

$$E_{\phi_1} = E_y \cos\phi = -\frac{j60\pi H_{0x}\,dx\,dy}{\lambda r}\cos\phi \exp\,(-j\beta r) \tag{10.14}$$

The electric vector potential at the distant point due to the magnetic current element $E_{0y}\,dx\,dy$ has an x component given by

$$F_x = \frac{-E_{0y}\,dx\,dy}{4\pi r}\exp\,(-j\beta r) \tag{10.15}$$

so that

$$H_x = j\omega\epsilon F_x = \frac{-jE_{0y}\,dx\,dy}{240\pi\lambda r}\exp\,(-j\beta r) \tag{10.16}$$

Hence

$$H_{\theta_2} = H_x \cos\theta \cos\phi = \frac{-jE_{0y}\,dx\,dy}{240\pi\lambda r}\cos\theta \cos\phi \exp\,(-j\beta r) \tag{10.17}$$

and

$$H_{\phi_2} = -H_x \sin\phi = \frac{jE_{0y}\,dx\,dy}{240\pi\lambda r}\sin\phi \exp\,(-j\beta r) \tag{10.18}$$

and the corresponding components of the electric field at the distant point are

$$E_{\theta_2} = 120\pi H_{\phi_2} = \frac{jE_{0y}\,dx\,dy}{2\lambda r}\sin\phi \exp\,(-j\beta r) \tag{10.19}$$

and

$$E_{\phi_2} = -120\pi H_{\theta_2} = \frac{jE_{0y}\,dx\,dy}{2\lambda r}\cos\theta \cos\phi \exp\,(-j\beta r) \tag{10.20}$$

Therefore, the total far-field of Huyghens' source is given by $E_\theta = E_{\theta_1} + E_{\theta_2}$ and $E_\phi = E_{\phi_1} + E_{\phi_2}$, which are the sums of the θ and ϕ components of the electric and magnetic current elements.

For a uniform plane wave in free-space,

$$E_{0x} = 120\pi H_{0y} \tag{10.21}$$

and

$$E_{0y} = -120\pi H_{0x} \tag{10.22}$$

By examining Eqs. 10.1 to 10.6, we can see that approximately the fields represented by Eqs. 10.21 and 10.22 can represent the fields at the mouth of the rectangular horn which can be considered as a wavefront. Therefore, we can assume that the components of the electromagnetic field at a distant point due to a Huyghens' source at the mouth of the rectangular horn are given by

$$E_\theta = j\frac{E_{0y}\,dx\,dy}{2\lambda r}(1 + \cos\theta)\sin\phi\,\exp\,(-j\beta r) \tag{10.23}$$

and

$$E_\phi = \frac{jE_{0y}\,dx\,dy}{2\lambda r}(1 + \cos\theta)\cos\phi\,\exp\,(-j\beta r) \tag{10.24}$$

The electric field on the mouth of the horn for the dominant mode is given by

$$E_{0y} = E_0\cos(\pi x/a) \tag{10.25}$$

where E_0 is an amplitude constant, assuming that the wavefront is plane. Equation 10.25 is more accurate if a lens is placed in the mouth of the horn which transforms the cylindrical wavefront to a plane wavefront. Here x is measured from the centre of the aperture and a and b are the dimensions of the aperture. The field at any distant point is then given by the product of the field due to a Huyghens' source and the array factor of a continuous rectangular array of Huyghens' sources covering the whole aperture. Therefore,

$$\begin{aligned}
E_\theta = {} & j\frac{(1 + \cos\theta)\sin\phi}{2\lambda r}\exp\,(-j\beta r)\int_{x=-a/2}^{+a/2}\int_{y=-b/2}^{b/2}E_0\{\cos(\pi x/a) \\
& \times \exp\,[j\beta(x\sin\theta\cos\phi + y\sin\theta\sin\phi)]\}\,dx\,dy \\
= {} & \frac{j(1 + \cos\theta)\sin\phi}{2\lambda r}\left[\frac{\exp\,(-j\beta r)E_0 a\sin\{(\pi b/\lambda)\sin\theta\sin\phi\}}{\sin\theta\sin\phi\,(\pi^2 - \beta^2 a^2\sin^2\theta\cos^2\phi)}\right] \\
& \times\left(\cos\frac{\pi a}{\lambda}\sin\theta\cos\phi\right)
\end{aligned} \tag{10.26}$$

and

$$\begin{aligned}
E_\phi = {} & \frac{j(1 + \cos\theta)\cos\phi}{2\lambda r}\frac{\exp\,(-j\beta r)E_0 a\sin\{(\pi b/\lambda)\sin\theta\sin\phi\}}{(\pi^2 - \beta^2 a^2\sin^2\theta\cos^2\phi)} \\
& \times\cos\frac{\pi a}{\lambda}(\sin\theta\cos\phi)
\end{aligned} \tag{10.27}$$

Equations 10.26 and 10.27 also give the approximate components of the far electric field of the open end of a rectangular waveguide carrying the dominant TE_{10} mode, assuming that there is no impedance mismatch at the mouth. However, in practice, for the open end of a waveguide there is mismatch so that E_y is relatively large and H_x is relatively small. Hence, the magnetic currents predominate over the aperture and the radiation pattern is, therefore, more omnidirectional. But in the case of a horn with a large aperture, the impedance match at the mouth is very good and thus, a highly directional radiation pattern results.

10.5   Gain of Rectangular Horn

To obtain an expression for the gain of a rectangular horn, it is necessary to find the maximum radiation intensity which is in the direction of the axis of the horn. Consider the sectoral E-plane horn shown in Fig. 10.8.

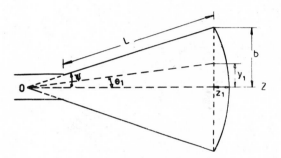

Fig. 10.8   Sectoral E-plane horn.

Let a and b be the dimensions of the aperture, x and y the rectangular coordinates on the aperture and z-axis the axis of the horn. The strengths of the Huyghens' sources on the aperture are given by

$$E^0 = E_y^0 \cos (\pi x/a) \exp (j\omega t) \tag{10.28}$$

for the TE_{10} mode. The total field at a distant point on the z-axis is

$$E = \frac{E_y^0}{\lambda r} \int_{y=-b/2}^{b/2} \int_{x=-a/2}^{a/2} \cos (\pi x/a) \exp (j\beta z_1) \, dx \, dy \tag{10.29}$$

The reference phase is taken as that of a source in the plane of the aperture, and the total field at the distant point is taken as the summation of the contributions from all the Huyghens' sources distributed over the cylindrical wavefront. From the geometry of Fig. 10.8, we have

$$z_1 = L \cos \theta_1 - L \cos \psi$$

$$\simeq \frac{b^2}{8L} - \frac{y_1^2}{2L} \tag{10.30}$$

because $\theta_1 \simeq \sin \theta_1 = y_1/L$, and $\psi \simeq \sin \psi = b/2L$. Therefore,

$$L \cos \theta_1 - L \cos \psi \simeq L \left\{ 1 - \frac{\theta_1^2}{2!} - 1 + \frac{\psi^2}{2!} \right\}$$

$$= L \left\{ \frac{\psi^2}{2!} - \frac{\theta_1^2}{2!} \right\} = \frac{b^2}{8L} - \frac{y_1^2}{2L} \qquad (10.31)$$

Therefore,

$$|E| = 2E_y^0 \left| \int_{-a/2}^{a/2} \int_{-b/2}^{b/2} \cos (\pi x/a) \exp (-j\beta\epsilon) \, dx \, dy \right|$$

$$= 2E_y^0 \left[\frac{a}{\pi} \sin \frac{\pi x}{a} \right]_{-a/2}^{a/2} \left| \int_0^{b/2} \exp (-j\beta\epsilon) \, dy \right|$$

$$= \frac{4aE_y}{\pi \lambda r} \left| \int_0^{b/2} \exp (-j\beta\epsilon) \, dy \right| \qquad (10.32)$$

where $\epsilon = y^2/2L$.

Putting $\beta y^2/2L = \pi/2$, $v^2 = \beta\epsilon$, we have

$$|E| = \frac{4a}{\pi} \frac{E_y^0}{\lambda r} \left\{ \frac{\lambda L}{2} \right\}^{1/2} \left| \int_0^{b/(2\lambda L)^{1/2}} \exp \left(-j \frac{\pi}{2} v^2 \right) dv \right|$$

$$= (2L/\lambda)^{1/2} (2aE_y^0/\pi r)[C\{b/(2\lambda L)^{1/2}\} - jS\{b/(2\lambda L)^{1/2}\}] \qquad (10.33)$$

where

$$C(v) = \int_0^v \cos (\pi v^2/2) \, dv \qquad (10.34)$$

and

$$S(v) = \int_0^v \sin (\pi v^2/2) \, dv \qquad (10.35)$$

are the Fresnel integrals. Therefore,

$$|E|^2 = (2L/\lambda)(2aE_y^0/\pi r)[C^2\{b/(2\lambda L)^{1/2}\} + S^2\{b/(2\lambda L)^{1/2}\}] \qquad (10.36)$$

Figure 10.9 shows the Cornu's spiral which is obtained by plotting S against C. From Fig. 10.9 and Eq. 10.36, it can be seen that if b is increased for a given L, the forward signal will first increase to maximum and then decrease, increasing again to secondary maxima, etc. A similar variation results if b and L are increased keeping the horn angle ψ constant.

The total radiated power from the horn can be taken approximately to be

$$W = \frac{E_y^{02}}{\eta} \int_{y=-b/2}^{b/2} \int_{x=-a/2}^{a/2} \cos^2 (\pi x/a) \, dx \, dy$$

$$= \frac{ab}{2} \frac{E_y^{02}}{\eta} \qquad (10.37)$$

Therefore, the absolute gain D or directivity of the horn is

$$D = \frac{\text{maximum radiation intensity}}{\text{average radiation intensity}}$$

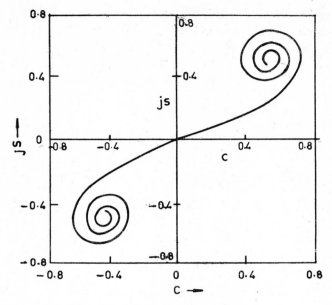

Fig. 10.9   Cornu's spiral.

$$= \frac{\dfrac{2L\eta}{2}\dfrac{2aE_y^{0^2}}{\pi}\;[C^2\{b/(2\lambda L)^{1/2}\} + S^2\{b/(2\lambda L)^{1/2}\}]}{\dfrac{abE_y^{0^2}}{8\pi\,\eta}}$$

[Using Eq. 10.36]

$$= \frac{64a}{b}\frac{L}{\lambda}\,[C^2\{b/(2\lambda L)^{1/2}\} + S^2\{b/(2\lambda L)^{1/2}\}] \qquad (10.38)$$

Figure 10.10 shows some typical gain versus b curves for an E-plane sectoral horn. These curves are helpful in designing a horn for optimum

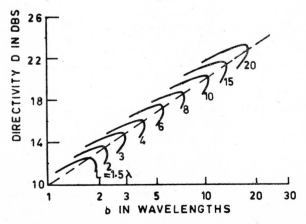

Fig. 10.10   Gain vs. b of E-plane sectoral horn.

gain. Similar curves can be obtained for H-plane and pyramidal horns.[1,2]

An approximate expression for the directivity of a rectangular horn is given by

$$D \simeq 7.5 A_E \lambda A_H \lambda \qquad (10.39)$$

where $A_E \lambda$ and $A_H \lambda$ are the aperture lengths in wavelengths in the E and H planes respectively.

10.6   Conical Horns

The properties of a conical horn excited in the dominant TE_{11} mode are similar to those of the pyramidal horn. The electromagnetic field of the conical horn has been derived rigorously by Schorr and Beck[3]. They have also calculated the radiation patterns for moderate values of aperture phase deviation. The gain of a conical antenna is optimum for a given slant length of flare l when the diameter d of the aperture is related to l by

$$d = [3l\lambda]^{1/2} \qquad (10.40)$$

This corresponds to a maximum phase deviation of $3/8\lambda$ in the aperture wavefront and is the same criterion as for the H-plane flare in a sectoral or pyramidal horn. The effective area of an optimum conical horn is approximately 52 per cent of its actual area, 2 per cent more than that of the optimum pyramidal horn. The gain of an optimum conical horn is

$$\text{Gain (opt. in dB)} = 20 \log \frac{C}{\lambda} - 2.82 \qquad (10.41)$$

10.7   Biconical Horn

The biconical horn may be excited in either vertical or horizontal polarization as shown in Figs. 10.1(e) and (g). The radiation pattern of the biconical horn is uniform (omnidirectional) in the horizontal plane for either vertical or horizontal polarization. In the vertical plane it is similar to that of the pyramidal or conical horn.

PROBLEMS

1. Calculate and plot the radiation patterns in the $\phi = 0°$ and $\phi = 90°$ planes of a pyramidal horn of dimensions $a = 12$ cms., $b = 9$ cms., $l = 12$ cms. at frequencies of 9000 MHz and 12000 MHz. Calculate the gain of the horn at these frequencies.

[1]Southworth, G. C., *Principles and Applications of Waveguide Transmission*, D. Van Nostrand, Princeton, 1950.

[2]Schelkunoff, S. A. and H. T. Fries, *Antennas, Theory and Practice*, John Wiley & Sons, Inc., New York, Chap. 16, 1952.

[3]Schorr, M. G. and F. J. Beck, 'Electromagnetic Field of the Conical Horn', *J. Appl. Phys.*, vol. 21, pp. 795–801, August 1950.

2. Calculate the gain of the horn antenna of length $l = 12$ cms., $a = 12$ cms. and b varying from 5 cms. to 10 cms. in steps of 1 cm. Plot gain versus b.

SUGGESTED READING

1. Collin, R. E. and Zucker, F. J., *Antenna Theory*, Parts 1 and 2, McGraw-Hill, New York, 1969.

2. Jasik, H., *Antenna Engineering Handbook*, McGraw-Hill, 1961.

3. Jordan, E. C. and Balmain, K. G., *Electromagnetic Waves and Radiating Systems*, Prentice-Hall of India, 1969.

4. Kraus, J. D., *Antennas*, McGraw-Hill, New York, 1950.

5. Schelkunoff, S. A. and Friis, H. T., *Antennas, Theory and Practice*, Chapman and Hall, London, 1952.

6. Fradin, A. Z., *Microwave Antennas*, Pergamon Press, Oxford, 1961.

7. Schelkunoff, S. A., *Electromagnetic Waves*, D. Van Nostrand and Co., Inc., Princeton, N.J., 1943.

8. Silver, S., *Microwave Antenna Theory and Design*, McGraw-Hill Book Co., New York, 1949.

9. Fry, D. W. and Goward, F. K., *Aerials for Centimeter Wavelengths*, Cambridge University, 1950.

10. Love, A. W., *Electromagnetic Horn Antennas*, IEEE Press, The Institution of Electrical and Electronics Engineers, Inc., New York, 1976.

11

REFLECTOR ANTENNAS

Reflector antennas are used to modify the radiation pattern of a radiating element. Since another antenna is required to excite a reflector antenna, it is usually called a secondary antenna. Similarly a lens antenna is also called a secondary antenna.

11.1 Different Types of Reflectors

Figure 11.1 shows a number of different types of reflector antennas together with the primary feed antennas. The flat reflector shown in Fig. 11.1(a) and the reflecting element shown in Fig. 11.1(b) reduces the backward radiation and increases the gain in the forward direction. The large flat sheet is less frequency sensitive than the thin element. The corner reflector shown in Fig. 11.1(c) gives a sharper radiation pattern and higher directivity. The passive corner reflector shown in Fig. 11.1(d) is a good radar

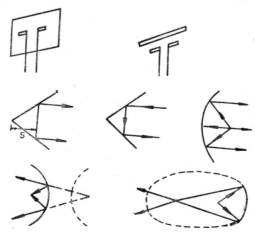

Fig. 11.1 Reflector-antennas: (a) large flat sheet reflector for dipole, (b) reflector element for dipole, (c) corner reflector, (d) passive corner reflector, (e) parabolic reflector, (f) elliptic reflector, (g) hyperbolic reflector.

target. The parabolic reflector shown in Fig. 11.1(e) gives a highly directional radiation pattern and is an important reflector antenna used in radar and communication systems. The elliptic reflector shown in Fig. 11.1(f) and the hyperbolic reflector in Fig. 11.1(g) produce beams from one focus if the source is at the other focus.

The gains of flat reflectors and corner reflector are calculated using the method of images[1]. Figure 11.2 shows the gain of a corner reflector antenna over a λ/2 dipole in free-space with the same power input as a function of the antenna to corner spacing and of the corner angle.

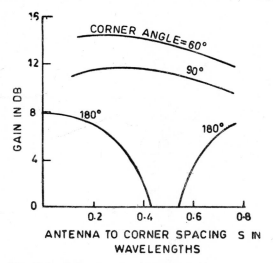

Fig. 11.2 Gain of corner antenna versus antenna to corner spacing S.

A grid of parallel wires or conductors is sometimes used instead of a flat reflector to reduce the wind resistance as shown in Fig. 11.3. The spacing between the wires of the grid should be < 0.1λ.

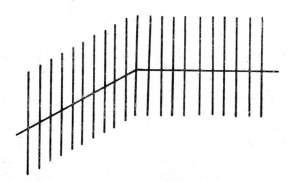

Fig. 11.3 Corner reflector consisting of grid of parallel wires.

[1]Brown, G. H., 'Directional Antennas', *Proc. I.R.E.*, vol. 25, p. 122, January 1937.

11.2   Corner Reflector Antenna

The corner reflector antenna, with a dipole or horn as feed, has been used extensively as a moderate-gain reflector antenna, because of the simplicity of its construction. The mathematical analysis of a corner reflector antenna is usually done by assuming perfectly conducting intersecting planes. A simpler analysis is by assuming a current element feed on the bisector plane of the corner angle π/N, where N is an integer. This analysis has been found to be very useful (Fig. 11.4(a)). The application of the image

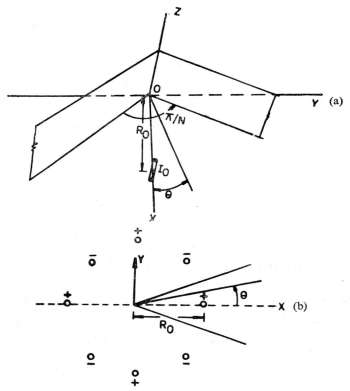

Fig. 11.4   Corner reflector: (a) Coordinate system, (b) Equivalent circular array produced by multiple images.

principle shows that the field between the planes is the same as would result from an array of $2N$ current elements equispaced on a circle of radius R_0 with adjacent elements having opposite polarities for $-\pi/2N \leqslant \theta \leqslant \pi/2N$. Figure 11.4(b) shows a cross-section in the $\phi = 0°$ plane of the array for $N = 4$, in which case the corner angle is 45°. Let the corner be illuminated by a current element I_0 at $R = R_0$ on the bisector line. Then the field E parallel to the z-axis is of the form

$$E = NI_0 \int_{m=0}^{\infty} j^{(2m+1)N} J_{(2m+1)N}(k_0 R_0 \cos \phi) \cdot \cos (2m + 1)N\theta \quad (11.1)$$

which becomes zero for $\theta = \pm\pi/N$, i.e., at the reflector planes. The pattern given by Eq. 11.1 must be multiplied by the element pattern of the feed which may be a dipole or a horn or any other antenna.

11.3 Parabolic Reflector

It is to be examined how a plane wave front over a large aperture can be produced by a reflector and a point source. In Fig. 11.5(a), the equality

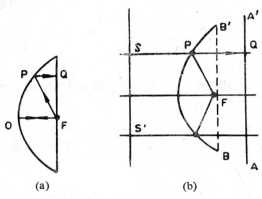

(a)   (b)

Fig. 11.5 Parabolic reflector.

of the two paths FOF and FPQ from the point source at P to the plane front QF gives

$$2L = R(1 + \cos \theta) \qquad (11.2)$$

so that

$$R = \frac{2L}{(1 + \cos \theta)} \qquad (11.3)$$

which is the equation of the required surface. Equation 11.3 is the equation of a parabola with the focus at F and focal length L. The parabola is also defined as a curve on which the distance from any point P (see Fig. 11.5(b)) to a fixed point F which is the focus, is equal to the perpendicular distance of P to a fixed line SS' called the directrix. Therefore,

$$PF = PS \qquad (11.4)$$

Therefore if AA' is a line normal to the axis at an arbitrary distance QS from the directrix, then

$$FP + PQ = PS + PQ = SQ \qquad (11.5)$$

Therefore, a property of a parabola is that all rays from an isotropic source at the focus that are reflected from the parabola arrive at a line AA' with equal phase. The image of the focus is the directrix and the reflected field along the line AA' appears as though it originated at the directrix as a plane wave. The aperture of the parabola is BB'.

An in-phase line source radiates a cylindrical wave and hence a parabolic cylinder is necessary to convert such a wave into a plane wave at the aperture as shown in Fig. 11.6(a). A spherical wave from an isotropic source is converted into a plane wave by a paraboloid of revolution as shown in Fig. 11.6(b).

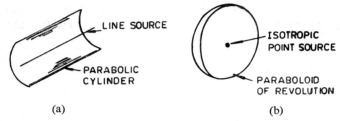

(a)   (b)

Fig. 11.6   Parabolic antennas: (a) Parabolic cylinder, (b) Paraboloid of revolution.

If an isotropic source is placed at the focus of a paraboloid of revolution as shown in Fig. 11.7, and if the distance L between the source and the

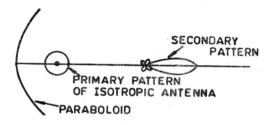

Fig. 11.7   Paraboloid excited by isotropic source.

paraboloid, i.e., the focal length is an even multiple of $\lambda/4$, the direct radiation in the axial direction from the source will be in opposite phase to the reflected radiation and hence will cancel it. If it is an odd multiple of $\lambda/4$, then the direct radiation will be in phase with the reflected radiation and will enhance it. Therefore, it is desirable to eliminate the direct radiation from the source by using a directional source like a horn as shown in Fig. 11.8. The presence of the primary antenna in the path of the

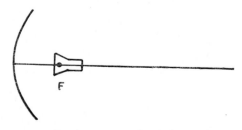

Fig. 11.8   Paraboloid excited by horn.

reflected wave has two disadvantages, namely, (i) the waves reflected
from the parabola back to the primary antenna produce interaction
and mismatching, and (ii) the primary antenna acts as an obstruction,
blocking out the central portion of the
aperture. To avoid both these effects,
only a portion of the paraboloid can
be used and the primary pattern displaced as shown in Fig. 11.9

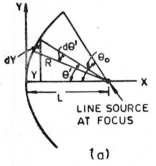

Fig. 11.9   Small portion of paraboloid with
horn as feed to avoid inter-
ference.

To obtain an expression for the
field distribution across the aperture
of a parabolic cylinder and a paraboloid of revolution, consider
Figs. 11.10(a) and (b).

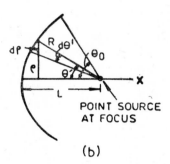

Fig. 11.10   Parabolic antennas: (a) Parabolic cylinder, (b) Paraboloid of
revolution.

For the parabolic cylinder, the line source is isotropic in a plane perpendicular to its axis. For a unit distance in the z-direction perpendicular
to the x-y plane, the power W in a strip of width dy is

$$W = dy\, P_y \tag{11.6}$$

where P_y = power density at y. W is also given by

$$W = d\theta'\, U \tag{11.7}$$

where U is the power per unit angle per unit length in the z-direction.
Therefore,

$$dy\, P_y = d\theta'\, U \tag{11.8}$$

so that

$$P_y/U = d\theta'/dy$$

$$= \frac{1}{\dfrac{d}{d\theta}(R \sin \theta')} \tag{11.9}$$

where

$$R = \frac{2L}{(1 + \cos \theta')} \tag{11.10}$$

Therefore,

$$P_y = \frac{(1 + \cos \theta')}{2L} U \tag{11.11}$$

Therefore, if $P_{\theta'}$ is the power density at θ' and P_0 is the power density at $\theta' = 0$, then

$$\frac{P_{\theta'}}{P_0} + \frac{(1 + \cos \theta')}{2} \tag{11.12}$$

Therefore,

$$\frac{E_{\theta'}}{E_0} = (P_{\theta'}/P_0)^{1/2} = \left\{ \frac{1 + \cos \theta'}{2} \right\}^{1/2} \tag{11.13}$$

where $E_{\theta'}$ and E_0 are the field densities at $\theta' = \theta'$ and $\theta = 0°$ respectively.

For the paraboloid of revolution shown in Fig. 11.10(b), the source is an isotropic point source, and the total power W through the annular section of radius ρ and width $d\rho$ is

$$W = 2\pi\rho \, d\rho P_\rho \tag{11.14}$$

where P_ρ is the power density at a distance ρ from the axis. This power is also equal to the power radiated by the isotropic source over the solid angle $2\pi \sin \theta' \, d\theta'$. Therefore,

$$W = 2\pi \sin \theta' \, d\theta' \, U \tag{11.15}$$

where U is the radiation intensity or power per unit solid angle. Then,

$$\rho \, d\rho \, P_\rho = \sin \theta' \, d\theta' \, U \tag{11.16}$$

or

$$\frac{P_\rho}{U} = \frac{\sin \theta'}{\rho \, d\rho/d\theta'} \tag{11.17}$$

and

$$\rho = R \sin \theta' = \frac{2L \sin \theta'}{(1 + \cos \theta')} \tag{11.18}$$

so that

$$P_\rho = \frac{(1 + \cos \theta')^2}{4L^2} U \tag{11.19}$$

Therefore, the ratio of the power density $P_{\theta'}$ at $\theta' = \theta'$ to the power density P_0 at $\theta' = 0°$ is

$$\frac{P_{\theta'}}{P_0} = \frac{(1 + \cos \theta')^2}{4} \tag{11.20}$$

Hence

$$\frac{E_{\theta'}}{E_0} = \frac{1 + \cos \theta'}{2} \tag{11.21}$$

From Eqs. 11.13 and 11.21 we can see that for small θ_0, $E_{\theta'}/E_0$ at any point on a parabolic cylinder or on a paraboloid of revolution is nearly equal to 1, because $\cos \theta' \simeq 1$. In such a case the illumination may be assumed to be nearly uniform over the aperture.

11.4   Radiation Pattern and Gain of Paraboloid of Revolution

A paraboloid with uniform illuminated aperture may be considered equivalent to a circular aperture of the same diameter D in an infinitely extending absorbing screen with a uniform plane wave incident on it as shown in Fig. 11.11. The radiation pattern of such a circular aperture may be derived by applying Huyghens' Principle.

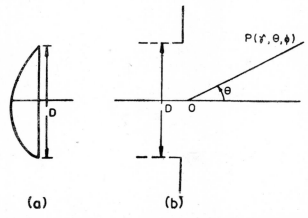

(a)   **(b)**

Fig. 11.11   (a) Large paraboloid with uniform illuminated aperture, (b) Equivalent uniformly illuminated circular aperture in infinitely extending absorbing screen.

The array factor is given by

$$S(\theta, \phi) = \int_{\phi'=0}^{2\pi} \int_{\rho=0}^{D/2} F(\rho, \phi') \exp \left[jk\rho \sin \theta \cos (\phi - \phi') \right] \rho \, d\rho \, d\phi'$$

$$(11.22)$$

where $F(\rho, \phi')$ is the illumination at point (ρ, ϕ') on the aperture. For uniform illumination, $F(\rho, \phi') = 1$. Therefore,

$$S(\theta, \phi) = \int_{\phi'=0}^{2\pi} \int_{\rho=0}^{D/2} \exp \{ jk\rho [\sin \theta \cos (\phi - \phi')] \} \rho \, d\rho \, d\phi'$$

$$= \frac{\pi D^2}{2} \int_0^1 x J_0(ux) \, dx \qquad (11.23)$$

where $x = 2\rho/D$, $u = (\pi D/\lambda) \sin \theta$. Therefore,

$$S(\theta, \phi) = \frac{\pi D^2}{2} \frac{J_1\left(\dfrac{\pi D}{\lambda} \sin \theta\right)}{\dfrac{\pi D}{\lambda} \sin \theta} \qquad (11.24)$$

Equation 11.24 has to be multiplied by the expressions for the field components of a Huyghens' source as given by Eqs. 10.23 and 10.24 to obtain the expressions for the field components of the circular aperture or equivalent paraboloid with uniform illumination. Since the array factor given by Eq. 11.24 is highly directional, it mainly determines the total radiation pattern.

The angle θ_0 of the first two nulls of the pattern is the first zero of $J_1((\pi D/\lambda) \sin \theta)$ or $(\pi D/\lambda) \sin \theta_0 = 3.83$, so that

$$\theta_0 = \sin^{-1} (3.83\lambda/\pi D) = \sin^{-1} (1.22\lambda/D) \qquad (11.25)$$

If D/λ is very large, θ_0 is quite small, and

$$\theta_0 \simeq \frac{1.22}{D_\lambda} = \frac{70}{D_\lambda} \text{ degrees} \qquad (11.26)$$

where $D_\lambda = D/\lambda$. The beamwidth between these two nulls is $2\theta_0 = 140/D_\lambda$ degrees. The beamwidth between half-power points is $58/D_\lambda$.

The directivity of the large aperture or large paraboloid with uniform distribution can be shown to be given by

$$D = \frac{4\pi(\text{area})}{\lambda^2} = \frac{4\pi \cdot \pi D^2}{4\lambda^2} = 9.87 D_\lambda^2 \qquad (11.27)$$

The gain or power gain over a $\lambda/2$ dipole is then given by

$$G = 6 D_\lambda^2 \qquad (11.28)$$

because the directivity of a $\lambda/2$ dipole is 3/2. For example, a circular aperture or a paraboloid of diameter 10λ has a gain of 600 or nearly 28 dB with respect to a $\lambda/2$ dipole.

For a paraboloid with tapered illumination, the directivity and gain will be slightly less. For such a tapered illumination the expression for the radiation pattern will be more involved and the integration can be done by using numerical methods.

11.5   Radiation Pattern of Cylindrical Parabolic Reflector

The cylindrical parabolic reflector is usually fed by a line-source, which may consist of a line array of in-phase $\lambda/2$ antennas as shown in Fig. 11.12(a) or a driven stub antenna with a reflector element as shown in Fig. 11.12(b). The stub antenna may be of length less than $\lambda/2$. The source also may be an open-ended waveguide or small horn.

Neglecting edge effects, the radiation pattern of such an antenna is the same as the pattern of a rectangular aperture of dimension L by H.

A long linear array of discrete closely spaced source may be approximated by a continuous array with the same amplitude and phase distribution. A uniform distribution yields the maximum directivity.[1] Tapering

---

[1]Taylor, T. T., "A Discussion of the Maximum Directivity of an Antenna", *Proc. I.R.E.*, vol. 36, p. 1135, Sept. 1948.

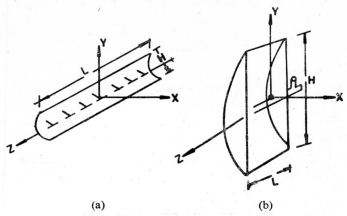

Fig. 11.12   Cylindrical parabolic reflector: (a) Fed by linear array of in-phase λ/2 antennas, (b) Fed by stub with reflector element ('Pillbox' or 'Cheese' antenna).

of the amplitude from a maximum at the centre to a smaller value at the edges reduces the side-lobe level but gives less directivity. A distribution with an inverse taper results in a sharper main lobe but gives increased side-lobe level.

The Fourier transform method is convenient for finding the radiation patterns of certain aperture distributions. Consider a continuous linear in-phase source of length L on a rectangular aperture of height L as shown in Fig. 11.13. Let $L \gg \lambda$, and let the amplitude distribution be known and

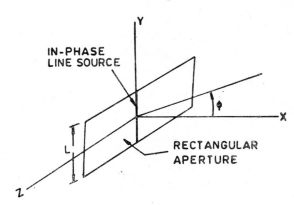

Fig. 11.13   Line source on rectangular aperture.

let the phase distribution be uniform. If $F(y)$ is the amplitude distribution, then the radiation field pattern as a function of ϕ in the x-y plane is given by

$$E(\phi) = \int_{-L/2}^{L/2} F(y) \exp (j\beta y \sin \phi) \, dy \qquad (11.29)$$

Thus $E(\phi)$ is the Fourier transform of $F(y)$. If the amplitude distribution is uniform, $F(y) = 1$, and the radiation pattern is

$$E(\phi) = \frac{2 \sin [(L_r \sin \phi)/2]}{\beta \sin \phi} \qquad (11.30)$$

where $L_r = 2\pi L/\lambda$.

The field patterns for other types of amplitude distributions are given by Ramsay.[1] These are given as follows:

For $F(y) = 1$, $(1 - 2y)/L$, $\cos (\pi y/L)$, and $\cos^2 (\pi y/L)$, the normalized field patterns are

$$\frac{\sin [(L_r \sin \phi)/2]}{[(L_r \sin \phi)/2]} \, , \quad \left\{ \frac{\sin (L_r \sin \phi)/4}{L_r \sin \phi} \right\}^2 , \quad \left\{ \frac{(\pi/2)^2 \cos [(L_r \sin \phi)/2]}{(\pi/2)^2 - [(L_r \sin \phi)/2]^2} \right\}$$

and

$$\left\{ \frac{\sin [(L_r \sin \phi)/2]}{L_r \sin \phi} \frac{2}{1 - [(L_r \sin \phi)^2/4\pi^2]} \right\}$$

respectively.

11.6   Cassegrain Antenna

In a radar antenna, sometimes it is necessary to locate the feed system and the associated waveguide circuits behind the antenna, because these items are quite sophisticated and big in size. To achieve this, the Cassegrain system is employed. Figure 11.14 shows a Cassegrain antenna, where the

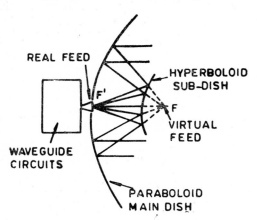

Fig. 11.14   Cassegrain antenna system.

main dish antenna is a paraboloid, while the auxiliary reflector or 'subdish' is a hyperboloid. If the feed is located at the near focus of the hyperboloid, and the focus of the paraboloid coincides with the far focus of the hyperboloid, the antenna radiates a collimated beam. The virtual feed is smaller

[1]Ramsay, J. F., 'Fourier Transforms in Aerial Theory', *Marconi Rev.*, vol. 9, p. 139, October-December 1946.

than the real feed, and this property is used to advantage in the design of certain types of antennas like those involving a monopulse system.

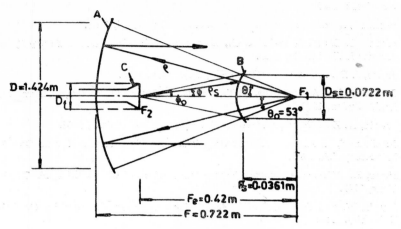

Fig. 11.15   Cassegrain antenna geometry.

A small subdish reduces aperture blocking, but it has to be supported at a greater distance from the main reflector-dish. Thus the choice of sub-reflector dish must be a compromise. The magnification of the hyperbolic dish $= (e + 1)/(e - 1)$, where e is the eccentricity and is given by

$$e = \frac{\text{distance between the two conjugate foci}}{\text{constant difference between the two focal radii}}$$

The focal radii of a point on the locus of the hyperbola are the straight lines which join the point to the foci. The difference between focal radii is a constant, irrespective of the point chosen. The effective focal length of the Cassegrain antenna is the distance between F and F' multiplied by $(e + 1)/(e - 1)$. The magnifying property of the Cassegrain antenna permits the use of a parabola of conventional f/D ratio to obtain the same effect as a parabola with a larger f/D.

PROBLEMS

1. Calculate and plot the radiation patterns of a paraboloid reflector with uniformly illuminated aperture when the diameter is (i) 4λ and (ii) 12λ. Also calculate the directivities.

2. Calculate and plot the radiation pattern of a cylindrical parabolic reflector of aperture $16\lambda \times 12\lambda$ when the illumination is (i) uniform and (ii) follows a cosine variation with maximum intensity at the centre and zero at the edges.

3. A square corner reflector has a driven $\lambda/2$ dipole antenna spaced $\lambda/2$ from the corner. Calculate and plot the radiation pattern in a plane at right angles to the driven element, assuming perfectly conducting sheets which are infinitely extending.

SUGGESTED READING

1. Collin, R. E. and Zucker, F. J., *Antenna Theory*, Parts 1 and 2, McGraw-Hill, New York, 1969.

2. Jasik, H., *Antenna Engineering Handbook*, McGraw-Hill, 1961.

3. Jordan, E. C. and Balmain, K. G., *Electromagnetic Waves and Radiating Systems*, Prentice-Hall of India, 1969.

4. Kraus, J. D., *Antennas*, McGraw-Hill, New York, 1950.

5. Schelkunoff, S. A. and Friis, H. T., *Antennas, Theory and Practice*, Chapman and Hall, London, 1952.

6. Fradin, A. Z., *Microwave Antennas*, Pergamon Press, Oxford, 1961.

7. Schelkunoff, S. A., *Electromagnetic Waves*, D. Van Nostrand and Co., Inc., Princeton, N.J., 1943.

8. Silver, S., *Microwave Antenna Theory and Design*, McGraw-Hill Book Co., New York, 1949.

9. Fry, D. W. and Goward, F. K., *Aerials for Centimeter Wavelengths*, Cambridge University, 1950.

10. Love, A. W., *Electromagnetic Horn Antennas*, IEEE Press, The Institution of Electrical and Electronics Engineers, Inc., New York, 1976.

11. Love, A. H., *Reflector Antennas*, IEEE Press, New York, 1978.

12. Rusdh, W. V. T. and Potter, P. D., *Analysis of Reflector Antennas*, Academic Press, New York, 1970.

12

LENS ANTENNAS

Curved reflector antennas and lens antennas are usually used as collimating elements in high-gain narrow beam microwave antennas. The lens antenna is also a secondary antenna like the reflector antenna and needs a primary antenna to excite it.

There are two main types of lens antennas, namely, (i) those in which the electrical path length is increased by the lens medium, and (ii) those in which the electrical path length is decreased by the medium. The first type is also called a delay lens because the wave is retarded in the lens medium. Examples of delay lenses are dielectric lenses and H-plane metal plate lenses. E-plane metal plate lenses belong to the second type.

12.1 Real Dielectric and Artificial Dielectric Lenses

Dielectric lenses can be constructed of real dielectrics like lucite or polystyrene, or of artificial dielectrics which are space arrays of discrete metal bodies of different shapes like discs, spheres, strips, etc. Table 12.1 gives the effective relative permittivity, effective relative permeability and refractive index of some important types of artificial dielectrics. Here $N=$ number of spheres or discs per cubic metre, $a=$ radius of sphere or disc in metres, $N'=$ number of strips per square metre, $w=$ width of strips in metres. The size of the metal bodies should be small, compared to the wavelength to avoid resonance effects, usually $< \lambda/4$, and the spacing between the

Table 12.1   Properties of Artificial Dielectrics

Type of metal body	Relative permittivity ϵ_r	Relative permeability μ_r	Refractive index, n
Sphere	$(1 + 4\pi Na^3)$	$(1 - 2\pi Na^2)$	$[(1 + 4\pi Na^3)(1 - 2\pi Ns^2)]^{1/2}$
Disc	$(1 + 5.3Na^3)$	$\simeq 1$	$(1 + 5.3Na^3)^{1/2}$
Strip	$(1 + 7.8N'w^2)$	$\simeq 1$	$(1 + 7.8N'w^2)^{1/2}$

bodies must be $< \lambda$ to avoid diffraction effects. The effective ϵ_r and the effective μ_r can be either measured or calculated approximately[1]. For example, the calculation of ϵ_r for an artificial dielectric consisting of a space array of metal spheres as shown in Fig. 12.1(a), will now be discussed.

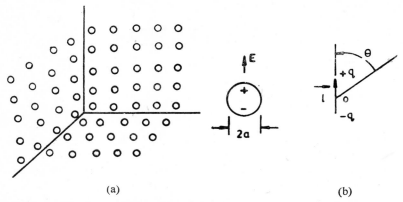

(a)   (b)

Fig. 12.1   (a) Artificial dielectric of space array of metal spheres, (b) single metal sphere in electric field.

Let an uncharged conducting sphere of radius a be placed in an electric field E as shown in Fig. 12.1(b). The field induces positive and negative charges as shown in the figure so that the sphere is equivalent to an electric dipole of moment ql. At a distant point (r, θ), the potential due to the dipole is

$$V = \frac{ql \cos \theta}{4\pi\epsilon_0 r^2} \qquad (12.1)$$

The polarization P of the artificial dielectric is given by

$$\mathbf{P} = N q \mathbf{l} \qquad (12.2)$$

where $N =$ number of spheres per cubic meter, and $\mathbf{l} =$ vector joining the two charges $+q$ and $-q$ of the dipole. Then the displacement density $\mathbf{D}$ is given by

$$\mathbf{D} = \epsilon \mathbf{E} = \epsilon_0 \mathbf{E} + \mathbf{P} \qquad (12.3)$$

Therefore, the effective permittivity of the artificial dielectric is

$$\epsilon = \epsilon_0 + \frac{\mathbf{P}}{\mathbf{E}} = \epsilon_0 + \frac{N q \mathbf{l}}{\mathbf{E}} \qquad (12.4)$$

In a uniform field $\mathbf{E}$, the potential V is

$$V = -\int_0^r E \cos \theta \, dr = -Er \cos \theta \qquad (12.5)$$

The potential V_0 outside the sphere placed in an originally uniform field

[1]Kock, W. E., 'Metallic Delay Lens' ,B.S.T.J., vol. 27, pp. 58–82, January 1948.

is then

$$V_0 = -Er \cos \theta + \frac{ql \cos \theta}{4\pi\epsilon_0 r^2} \qquad (12.6)$$

At the surface of the sphere $r = a$, $V_0 = 0$, and hence

$$0 = -Ea \cos \theta + \frac{ql \cos \theta}{4\pi\epsilon_0 a^2} \qquad (12.7)$$

so that $ql/E = 4\pi\epsilon_0 a^3$.

Therefore,

$$\epsilon = \epsilon_0 + 4\pi N\epsilon_0 a^3 = \epsilon_0(1 + 4\pi Na^3) \qquad (12.8)$$

so that

$$\epsilon_r = (1 + 4\pi Na^3) \qquad (12.9)$$

12.2   Delay Lens

A delay lens may be made of a real dielectric or an artificial dielectric. It is very similar to an optical lens and can be designed by ray analysis methods of geometrical optics. Figure 12.2 shows a plano-convex delay

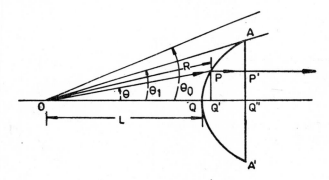

Fig. 12.2   Plano-convex dielectric lens.

lens whose shape is to be determined. The spherical wavefront from an isotropic point source at O is to be transformed into a plane wavefront at AA'. This requires that all paths from the source O to the plane AA' are of equal electrical length. Therefore, the electrical length of path OPP' should be equal to the electrical length of path $OQQ'Q''$. Therefore,

$$\frac{R}{\lambda_0} = \frac{L}{\lambda_0} + \frac{R \cos \theta - L}{\lambda_d} \qquad (12.10)$$

where λ_0 and λ_d are the wavelengths in free space and in the dielectric respectively. Equation 12.10 can be rewritten as

$$R = L + n(R \cos \theta - L) \qquad (12.11)$$

where $n = \lambda_0/\lambda_g$ = refractive index of the dielectric medium.

Solving for R from Eq. 12.11, we have

$$R = \frac{(n-1)L}{(n \cos \theta - 1)} \tag{12.12}$$

which is the equation of a hyperbola. Hence, the shape of the dielectric lens for an isotropic point source should be a hyperboloid of revolution, and for a linear in-phase source it should be a cylindrical hyperbola.

To obtain the distribution of illumination on the aperture of the lens, consider Fig. 12.3. The total power W through the annular section of

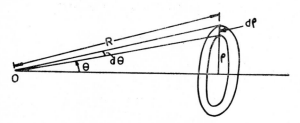

Fig. 12.3   Annular zone of delay lens.

radius ρ and width $d\rho$ is given by

$$W = 2\pi\rho \, d\rho P_\rho \tag{12.13}$$

where P_ρ is the power per unit area at radius ρ. This power is also equal to the power radiated by the isotropic source over the solid angle $2\pi \sin \theta \, d\theta$. Therefore,

$$W = 2\pi \sin \theta \, d\theta \, U \tag{12.14}$$

where U is the radiation intensity of the isotropic source. Equating Eqs. 12.13 and 12.14, we have

$$\rho \, d\rho P_\rho = \sin \theta \, d\theta \, U \tag{12.15}$$

so that

$$\frac{P_\rho}{U} = \frac{\sin \theta}{\rho \, d\rho/d\theta} \tag{12.16}$$

However $\rho = R \sin \theta$, and using Eq. 12.11, we obtain,

$$P_\rho = \frac{(n \cos \theta - 1)^2}{(n-1)^2(n - \cos \theta)L^2} U \tag{12.17}$$

Therefore,

$$\frac{P_\theta}{P_0} = \frac{\text{Power density at } \theta = \theta}{\text{Power density at } \theta = 0} = \frac{(n \cos \theta - 1)^2}{(n-1)^2(n - \cos \theta)} \tag{12.18}$$

Hence,

$$\frac{E_\theta}{E_0} = \left[\frac{P_\theta}{P_0}\right]^{1/2} = \frac{1}{(n-1)} \frac{(n \cos \theta - 1)}{(n - \cos \theta)^{1/2}} \tag{12.19}$$

For example, for $n = 1.5$,

$$\frac{E_\theta}{E_0} = 0.7 \text{ at } \theta = 20° \text{ and } = 0.14 \text{ at } \theta = 40°$$

Therefore, to obtain a nearly uniform illumination over the aperture, the angle θ_1 to the edge of the lens should be $\leqslant 20°$. However, in practice, the source or primary antenna is not isotropic and may have a directional pattern and the illumination becomes even more nonuniform, unless the pattern of the primary antenna has less intensity in the axial direction than in the direction off the axis. A tapered illumination is preferable to a uniform illumination to suppress minor lobes. Therefore, a lens with a larger θ, or shorter focal length is preferable as shown in Fig. 12.4(a). But such a lens becomes very bulky. This can be avoided by having a zonal lens as shown in Fig. 12.4(b), whose weight is reduced by removing sections or zones, in a manner such that the thickness z of a zone step is such that the electrical length of z in the dielectric is an integral number of wavelengths longer (usually 1) than the electrical length of z in air. For one wavelength difference,

$$\frac{Z}{\lambda_d} - \frac{Z}{\lambda_0} = 1$$

or

$$Z = \frac{\lambda_0}{(n-1)} \tag{12.20}$$

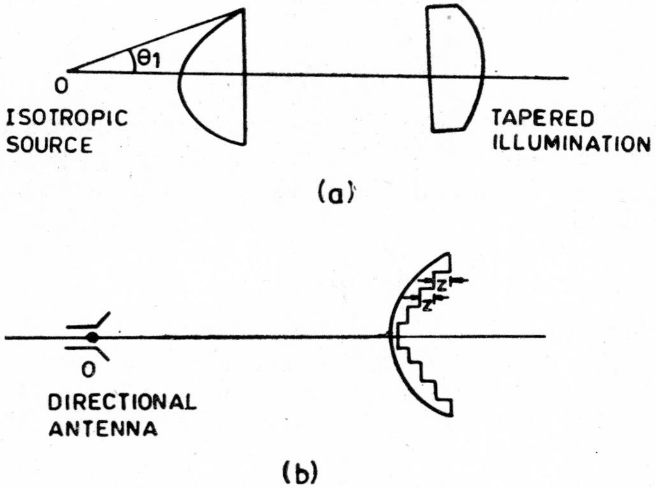

Fig. 12.4   (a) Short focal lens, (b) Zoned lens.

However, the zoned lens is more frequency sensitive than the unzoned lens.

The use of an artificial dielectric lens reduces the weight considerably. Figure 12.5 shows some such lenses. The shapes of these lenses are also hyperbolical just like that of the real dielectric lenses.

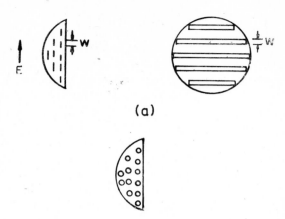

(a)

Fig. 12.5   Artificial dielectric lenses made of (a) metal
strips, (b) metal spheres.

12.3   E-Plane Metal Plate Lens

In the E-plane metal plate lens, the wave is accelerated by the lens medium. The metal plates are parallel to the E-plane. The velocity v of propagation of a TE_{10} wave in a parallel plane waveguide consisting of two parallel conducting planes of large extent is

$$v = \frac{v_0}{\{1 - (\lambda/2b)^2\}^{1/2}} \tag{12.21}$$

where v_0 and λ are the velocity and wavelength in free space, and b is the spacing between the plates, $\lambda = 2b$ is the cut-off wavelength so that waves of wavelengths $< 2b$ are transmitted. The equivalent refractive index of such a medium is

$$n = \frac{v_0}{v} = \left\{1 - \left(\frac{\lambda}{2b}\right)^2\right\}^{1/2} \tag{12.22}$$

which is always < 1.

A metal plate lens can be made from such a medium as shown in Fig. 12.6. The shape is plano-convex and the width t of the plates which is the thickness of the lens is such that a spherical wave from an isotropic source is transformed to a plane wave on the plane side of the lens. The electric field is parallel to the plates. The electrical path length OPP'

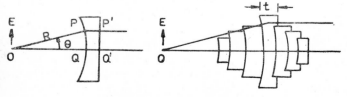

Fig. 12.6   E-plane metal plate lens.

should be equal to the electrical path length OQQ', so that

$$\frac{L}{\lambda} = \frac{R}{\lambda} + \frac{L - R \cos \theta}{\lambda_g} \qquad (12.23)$$

where λ and λ_g are the free-space wavelength and guide wavelength respectively. Equation 12.23 can be rewritten as

$$R = \frac{(1 - n)L}{(1 - n \cos \theta)} \qquad (12.24)$$

where $n = \dfrac{v_0}{v} = \dfrac{\lambda}{\lambda_g} < 1$. Since $n < 1$, Eq. 12.24 represents an ellipse.

Hence, the three-dimensional concave surface of the lens is generated by rotating the ellipse given by Eq. 10.24.

The E-plane metal plate lens is more frequency sensitive than the dielectric delay lens, but zoning the E-plane metal plate lens decreases the frequency sensitivity. A zoned E-plane metal plate lens used with a horn, as shown in Fig. 12.7, permits a much shorter length of the horn for the same size aperture.

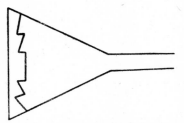

Fig. 12.7   Zoned E-plane metal plate lens with rectangular metal horn.

12.4   H-Plane Metal Plate Lens

If a wave enters a stack of metal plates oriented parallel to the H-plane as in Fig. 12.8(a), its velocity is not affected. However, if it is a stack of

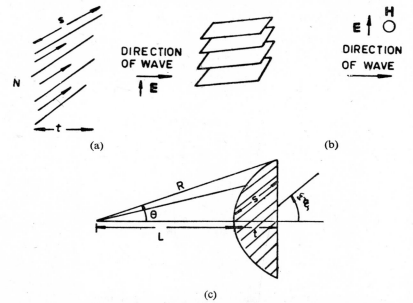

Fig. 12.8   (a) Stack of plates parallel to H-plane, (b) Stack of plates parallel to H-plane with increased path length, (c) H-plane metal plate lens.

plates as shown in Fig. 12.8(b), an increase in path length is produced which is equal to $s - t$. Using such a stack of plates, a lens can be constructed as shown in Fig. 12.8(c).

For equality of path lengths,

$$R = L + \frac{R \cos \theta - L}{\cos \xi} \tag{12.25}$$

or

$$R = \frac{(n - 1)L}{n \cos \theta - 1} \tag{12.26}$$

where $n = \dfrac{1}{\cos \xi} > 1$, and this equation represents a hyperbola. Hence an H-plane metal plate lens is of the same shape as a dielectric lens.

12.5 Luneburg Lens

If the refractive index can be varied in an artificial dielectric medium or in a mixture of natural dielectrics, the design of lenses having properties unattainable with uniform media are possible. Brown[1] has discussed this subject well in a monograph and a particularly interesting case was introduced by Luneburg[2].

The Luneburg lens is designed to provide a scanning performance which is independent of the direction of the radiated beam, and hence is of spherical symmetry. Consider the two-dimensional case shown in Fig. 12.9 in which is shown a lens of radius a and having a refractive index $n(r)$ which depends on the distance r from the centre of the circle.

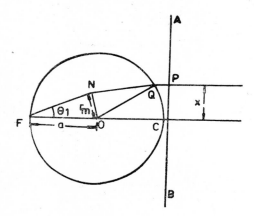

Fig. 12.9 Luneburg Lens.

[1]Brown, J., *Microwave Lenses*, Methuen and Co. Ltd., London, 1953.
[2]Luneburg, R. K., *The Mathematical Theory of Optics*, Brown University Press, Providence, R.I., 1944.

The ray paths are assumed to be as shown in the figure, and the ray pattern rotates as the feed is moved around the lens circumference. To scan the radiated beam, only the feed is moved and it is not necessary to move the lens because of its circular symmetry.

The line AB should be a phase front for beam formation, and for each of the possible rays $FNPQ$ the electrical path length must be constant. Then the rays emanating from the point source at F on the surface of the sphere are focussed into a bundle of parallel rays emerging from the opposite side of the sphere. To achieve this, the refractive index n should be a function of r given by

$$n(r) = [2 - (r/a)^2]^{1/2} \qquad (12.27)$$

Lenses of this type may be constructed of foamed dielectric containing many small glass spheres whose spacings are varied to achieve the relation $n(r)$. They may also be made of spherical shells of graded dielectric constant fitted one within the other. At least ten steps with equal increments of refractive index are desirable.

Two-dimensional lenses may be made using conducting plates with a variable spacing adjusted to provide the required refractive index variation assuming that propagation within the lens takes place as a TE_{01} waveguide mode.

PROBLEMS

1. Design a plano-convex dielectric lens for 6,000 MHz with a diameter of 12. The lens material has $\epsilon_r = 2.5$ and $F = 1$. Draw the lens cross-section. Which type of primary antenna pattern is required to produce a uniform aperture distribution?

2. Design an unzoned plano-concave E-plane metal plate lens with an aperture of 20λ square for use with a 4000 MHz line source 15λ long. The distance of the source from the lens is 20λ. Let $\epsilon_r = 3.0$. Find the spacing between the plates. Draw the shape of the lens, giving the dimensions.

3. Design an artificial dielectric to give $\epsilon_r = 2.0$ for use at 5000 MHz, using
   (i) copper spheres
   (ii) copper discs
   (iii) copper strips.

SUGGESTED READING

1. Collin, R. E. and Zucker, F. J., *Antenna Theory*, Parts 1 and 2, McGraw-Hill, New York, 1969.

2. Jasik, H., *Antenna Engineering Handbook*, McGraw-Hill, 1961.

3. Jordan, E. C. and Balmain, K. G., *Electromagnetic Waves and Radiating Systems*, Prentice-Hall of India, 1969.

4. Kraus, J. D., *Antennas*, McGraw-Hill, New York, 1950.

5. Schelkunoff, S. A. and Friis, H.T., *Antennas, Theory and Practice*, Chapman and Hall, London, 1952.

6. Fradin, A. Z., *Microwave Antennas*, Pergamon Press, Oxford, 1961.

7. Schelkunoff, S. A., *Electromagnetic Waves*, D. Van Nostrand and Co., Inc., Princeton, N.J., 1943.

8. Silver, S., *Microwave Antenna Theory. and Design*, McGraw-Hill Book Co., New York, 1949.

9. Fry, D. W. and Goward, F. K., *Aerials for Centimeter Wavelengths*, Cambridge University, 1950.

10. Love, A. W., *Electromagnetic Horn Antennas*, IEEE Press, The Institution of Electrical and Electronics Engineers, Inc., New York, 1976.

13

LEAKY-WAVE AND SURFACE-WAVE ANTENNAS—DIELECTRIC AND DIELECTRIC-LOADED METAL ANTENNAS

Leaky-wave and surface-wave antennas form sub-classes of the more general class of travelling-wave antennas. In any travelling-wave antenna, the illumination is generated by a wave, which launched from a single source of feed, propagates along the guiding structure, thus illuminating the radiating structure. The amplitude and phase distribution in a travelling-wave antenna can be controlled by having a detailed knowledge of the propagation characteristics of the travelling-wave structure. Attractive features of travelling-wave antennas are simplicity of feed arrangement, suitability for flush mounted or low silhouette design, and special pattern characteristics.

Some dielectric and dielectric-coated metal antennas belong to the class of surface-wave and leaky-wave antennas, while others may belong to the class of horn antennas, while some others do not belong to any of the classes of antennas considered so far.

13.1 Leaky-wave Antennas

The properties of a leaky-wave have been dealt with in section 2. The propagation constant of a leaky wave is given by

$$k_z = \beta_z + j\alpha_z \tag{13.1}$$

where z is the direction of propagation, and the wave is of the form $\exp[j(\omega t - k_z z)]$, β_z is the phase constant and α_z the attenuation constant. For a leaky wave, $\beta_z < k_0$ where $k_0 = \omega(\mu_1\epsilon_0)^{1/2} = 2\pi/\lambda_0 = \beta_0$, so that the wave travels faster than the velocity of light, or in other words,

$$v_p = \frac{\omega}{\beta_z} > \frac{\omega}{\beta_0} > v_0$$

where $v_0 = 1/(\mu_0\epsilon_0)^{1/2}$. The leaky-wave antenna radiates continuously along its path, and the main beam in the radiation pattern points at an angle θ with respect to the broadside given by

$$\sin \theta = \beta_z/\beta_0 = \beta_z/k_0 \tag{13.2}$$

as shown in Fig. 13.1. If β_x is the phase constant in a transverse direction

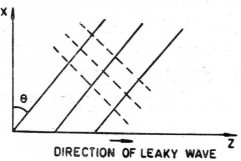

DIRECTION OF LEAKY WAVE

Fig. 13.1 Leaky-wave.

and if the field is uniform in the y-direction, then $\partial/\partial y = 0$, and

$$k_0^2 = k_x^2 + k_z^2 \tag{13.3}$$

where

$$k_x = \beta_x + j\alpha_x \tag{13.4a}$$

$$k_z = \beta_z + j\alpha_z \tag{13.4b}$$

For a leaky wave, $\beta_z < k_0$, and if $\alpha_x \ll \beta_x$ and $\alpha_z \ll \beta_z$, we have from Eqs. 13.3 and 13.4,

$$\beta_x^2 > 0 \quad \text{or} \quad \beta_x \text{ is real} \tag{13.5}$$

which means that the phase constant in the x-direction is real. Therefore, the wave travels in a direction making an angle θ with the x-direction such that

$$\tan \theta = \beta_z/\beta_x \tag{13.6a}$$

or

$$\sin \theta = \beta_z/k_0 \tag{13.6b}$$

This means that the leaky-wave structure radiates its maximum energy in the direction θ. This main beam can be frequency scanned from 10° to 80° off broadside, because β_z changes when frequency changes.

Examples of leaky-wave antennas are:
  (i) Slotted rectangular waveguide,
  (ii) Inductive grid,
  (iii) Slotted circular waveguide,
  (iv) Holey waveguide,
  (v) Dielectric-filled waveguide.

Figure 13.2 shows some of these structures. For the slotted rectangular and cylindrical waveguides, $\beta_z \simeq \beta_{z0}$, where β_{z0} is the phase constant inside the waveguide.

(a)   (b)

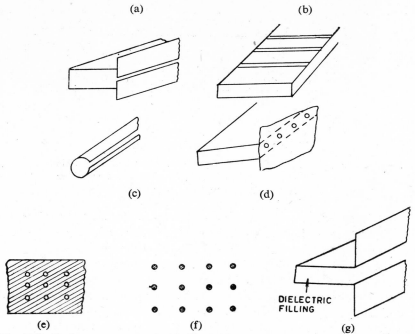

(c)   (d)

(e)   (f)   DIELECTRIC
FILLING   (g)

Fig. 13.2 Leaky-wave antennas: (a) Slotted rectangular waveguide, (b) Inductive
grid, (c) Slotted circular waveguide, (d) Holey waveguide, (e) Holey
plate, (f) Array of discs, (g) Dielectric-filled channel waveguide.

13.2 Surface-wave Antennas

The mechanism of surface-waves has been treated in section 2. Surface-wave antennas radiate endfire or off-broadside depending upon the mode of excitation, while leaky-wave antennas usually radiate only off-broadside.

A surface-wave travelling in the z-direction travels with a velocity less than the free-space velocity so that $v_z < v_0$ and $k_z = \beta_z > k_0$. Hence from the relation

$$k_0^2 = k_x^2 + k_z^2 \tag{13.7}$$

$\left(\text{assuming } \dfrac{\partial}{\partial y} = 0\right)$, we have

$$k_x^2 = k_0^2 - k_z^2 < 0 \tag{13.7a}$$

or k_x is an imaginary quantity equal to $j\alpha_x$. Therefore, the wave is of the form

$$E = E_0 \exp\left[j(\omega t - k_z z)\right] \exp\left(-\alpha_x x\right) \tag{13.8}$$

which represents a wave travelling in the z-direction with a velocity $v_z = \omega/k_z < v_0$, and decaying exponentially as $\exp\left(-k_x x\right)$ in the x-direction (or evanescent in the x-direction).

Typical surface-wave structures are (a) dielectric slab, (b) dielectric rod, (c) corrugated metal or dielectric structures, and (d) long Yagi, as shown in Fig. 13.3.

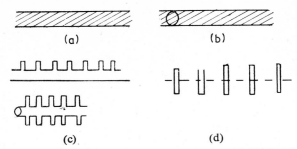

Fig. 13.3   Typical surface-wave structures: (a) Dielectric
slab, (b) Dielectric rod, (c) Corrugated metal or
dielectric structures, (d) Long Yagi.

There is a certain amount of controversy about how a surface-wave radiates. Examining Eq. 13.8, it can be seen that an infinitely long surface-wave structure with no discontinuities cannot radiate because of the decay term $\exp(-\alpha_x x)$. Hence, some workers have argued that a surface-wave structure radiates only if it has discontinuities and, further, that it radiates only at the discontinuities, namely, at the feed point and at the termination.[1,2] But some other workers have shown that a finite length surface-wave structure not only radiates at the discontinuities but also all along the structure.[3,4,5] This is probably due to the fact that the conditions that exist on an infinitely long structure cannot really exist on a finite length structure due to multiple reflections at the discontinuities. The actual current distribution $I_y(z)$ along a surface-wave structure excited at $z = 0$ does not look like Fig. 13.4(a) but like Fig. 13.4(b). The actual current distri-

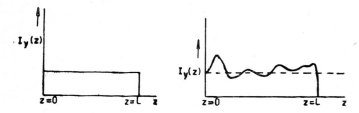

Fig. 13.4   Current distribution on surface-wave structure: (a) Ideal,
(b) Actual.

[1]Zucker, F. J., 'Theory and Application of Surface Waves', *Nuovo Cimento Suppl.*, vol. 9, p. 451, 1952.

[2]Brown J. and Spector, J. O., 'The Radiating Properties of End-fire Aerials', *Proc. Inst. Elec. Engrs.* (London), vol. 104, Part II, p. 27, 1957.

[3]Kiely, D. G., *Dielectric Aerials*, Methuen and Co. Ltd., London, 1953.

[4]Watson, R. B. and Horton, C. W., 'The Radiation Pattern of Dielectric Rods, Experiment and Theory', *J. Appl. Physics*, vol. 19, p. 661, 1948.

[5]Chatterjee, R. and Chatterjee, S. K., 'Some Investigations on Dielectric Aerials', Parts I and II, *Journal of the Indian Institute of Science*, vol. 38, No. 2, p. 93, 1956, and vol. 39, No. 3, p. 134, 1957.

bution is difficult to derive theoretically, and hence the radiation characteristics, calculated using the 'aperture' point of view where the currents on
the structure as well as the currents on the discontinuities are taken into
account, are approximate. Similarly, by considering only the radiation
from the discontinuities, only approximate results are obtained because
the radiation from the surface-wave on the structure is neglected. Hence
the results obtained theoretically by using the two approaches are almost
identical.[1]

Though surface-wave antenna theory leaves much to be desired, the
simple approximate theories using either the scalar Huyghens' Principle
or Schelkunoff's Equivalence Principle give fairly good results which check
up with experimental results.

The more rigorous problem of solving theoretically the surface-wave
antenna problem is by considering the source and the antenna together
as a boundary-value problem. Simple sources like line sources and ring
sources have been considered. Wiener-Hopf techniques are quite useful
for solving some of these problems.

13.3   Dielectric and Dielectric-Loaded Metal Antennas

Dielectric and dielectric-loaded metal antennas are primary antennas
employing a system of dielectric elements to radiate electromagnetic energy
or to receive electromagnetic energy. Dielectric lenses are not included in
this class of antennas, because they are secondary radiators. This class of
antennas can be of many shapes like cylinders, cones, spheres, tapered and
untapered rectangular rods, circular and rectangular horns, or of any
other shape. Some of them are shown in Figs. 13.5, 13.6 and 13.7.

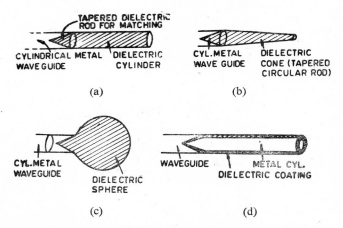

(a)

(b)

(c)

(d)

Fig. 13.5   Different types of dielectric and dielectric-loaded antennas (*contd.*).

[1]Collin, R. E. and Zucker, F. J., *Antenna Theory*, Part 2, McGraw-Hill Book Co. Inc.,
New York, Chap. 21, 1969.

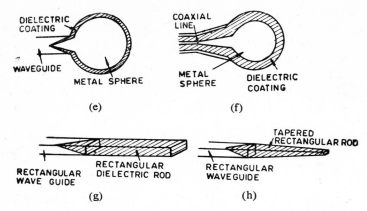

Fig. 13.5   Different types of dielectric and dielectric-loaded antennas.

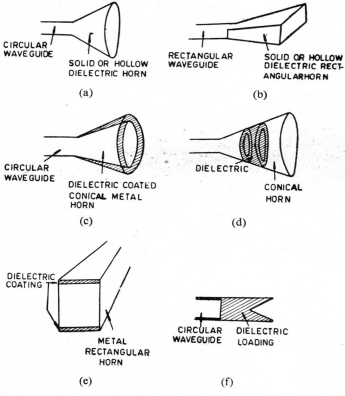

Fig. 13.6   Dielectric and Dielectric-loaded Horns: (a) Conical solid or hollow dielectric horn, (b) Rectangular solid or hollow dielectric horn, (c) Dielectric-coated metal conical horn, (d) Dielectric-loaded metal conical horn, (e) Dielectric-coated metal rectangular horn, (f) Dielectric-loaded circular waveguide horn.

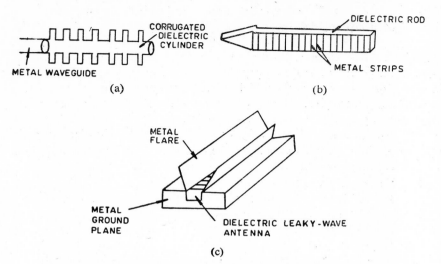

Fig. 13.7 Other types of dielectric antennas: (a) Corrugated dielectric-
rod antenna, (b) Dielectric leaky-wave antenna, (c) Horn
image-guide leaky-wave antenna.

It was Mallach in Germany who first used untapered and tapered circular and rectangular dielectric rods as directional antennas, before 1938.[1]
Since then several workers have investigated several types of dielectric
and dielectric-loaded antennas.

Circular cylindrical and untapered rectangular rod dielectric and dielectric-coated metal antennas may be considered as belonging to the general class of surface-wave antennas, because the infinitely long cylindrical
dielectric rod of circular cross-section supports a surface-wave. Hence, the
theoretical basis of studying these two types of antennas is the surface-
wave field of the infinitely long structure which the finite-length rod antenna is assumed to approximately support.

The conical dielectric and dielectric-coated metal antennas and the
tapered rectangular dielectric-rod antenna are assumed to be made up of
untapered sections and hence are also considered as approximate surface-
wave structures.

The antennas of spherical shape are treated as supporting spherical
waves which are obtained by solving the electromagnetic boundary-value
problem, or by the scattering theory.

The circular and rectangular dielectric and dielectric-loaded metal horns
are treated in a manner similar to metal horns, but with modifications.

Almost all these types have directional radiation patterns and fairly
high gains which are comparable to that of the metal horn which is one of
the traditional microwave antennas.

---

[1]Mallach, L., 'Dielektrische Richtstrahler', *Fornmeldtech Zeit*, 2, p. 33, 1949.

13.4 Applications of Dielectric and Dielectric-Loaded Antennas at Microwave and Millimetre-Wave Frequencies

Dielectric and dielectric-loaded metal antennas are light in weight, have good sealing and corrosion properties and are attracting a lot of interest in view of the ease with which dielectrics can be manufactured with present-day techniques. They are useful for the following applications at microwave and millimetre-wave frequencies.

(i) An antenna with its larger dimension in a longitudinal direction rather than in a plane transverse to the direction of maximum radiation.

(ii) An antenna with broad-band characteristics.

(iii) For sealing apertures of excitation systems.

(iv) To feed lenses, because of flat-topped radiation patterns.

(v) Used in mouth or throat of metal horns to modify radiation patterns.

(vi) In phased arrays[1].

(vii) Dielectric and dielectric-coated metal cylinders, dielectric and dielectric-coated metal spheres or dielectric-coated metal cones excited in the TM symmetric mode by a coaxial line have radiation patterns with a sharp null on the axis. These antennas can be used in direction-finding where a sharp null is better than a broad maximum. TM symmetric mode excitation is also useful for beacon antennas.

(viii) Dielectric and dielectric-coated metal spheres of dimensions greater than the free-space wavelength and excited in the unsymmetric hybrid mode by a circular metal waveguide carrying the dominant mode, have very narrow major lobes and gains as high as that of a metal horn of equal aperture. These antennas can be used to replace metal horns.

(ix) Tapered dielectric antennas of circular or rectangular cross-section have radiation patterns with suppressed minor lobes. By a proper choice of dimensions, such antennas can also be made to have high gain, and can replace metal horns.

(x) The corrugated dielectric rod antenna has suppressed minor lobes by properly adjusting the groove depth and spacing, and hence has a desirable radiation pattern. It can be also used for electronic scanning.

(xi) Dielectric and dielectric-loaded horns of circular or rectangular shape have many desirable properties. Dielguides which are dielectric guiding structures when placed between the primary feed which is a metal horn and the parabolic reflector or subreflector reduce spillover and provide a more uniform reflector illumination. They are highly efficient low noise antenna feeds[2]. Dielectric-loaded horns bring about enhancement of aperture efficiency, have rotationally symmetric beams and extremely low

[1]Shazu Tanaka, Kazuo Chiba, Teneaki Chiba and Hideo Nakasato, 'Ceramic Rod Array Scanned with Ferrite Phase Shifters', *Proc. I.E.E.E.*, vol. 56, No. 11, pp. 2000–2010.

[2]Bartlett, H. E. and Moseley, R. E., 'Dielguides, Highly Efficient Low-Noise Antenna Feeds', *Microwave J.*, vol. 9, pp. 53–58, 1966.

side lobes. A dielectric rectangular horn inserted in a metal horn raises the gain of the latter. Conical dielectric horns have highly symmetrical radiation patterns and greater frequency independence and lower cost as compared to corrugated metal horns. Cassegrain antennas employing conical dielectric horns have high efficiency, low far-out side-lobes and low mutual coupling.

Due to these desirable properties, dielectric and dielectric-loaded metal horns used as feeds for parabolic reflector antennas can be employed as satellite tracking antennas, or as terrestrial microwave or millimetre-wave relay station antennas. Dielectric and dielectric-loaded metal horns can be used in limited scanning arrays and as satellite-borne antennas.

(xii) Dielectric millimetre-wave antennas can be integrated into milli-metre-wave monolithic integrated circuits as shown by Yao *et al.*[1] They have built a monolithic silicon integrated circuit consisting of a mixer diode and an all-dielectric receiving antenna and tested it at 85 GHz. This design can be used efficiently at considerably higher frequency, and can be elaborated into more complex integrated circuits. Such integrated milli-metre-wave receivers should be suitable for short path-terrestrial communications, in applications where compactness and low cost are required.

13.5 Radiation Pattern of Circular Cylindrical Dielectric Rod Antenna

Consider a circular cylindrical dielectric rod antenna excited by a circular metal waveguide carrying the dominant TE_{11} mode as shown in Fig. 13.8(a). Figure 13.8(b) shows the electric and magnetic field lines in the circular

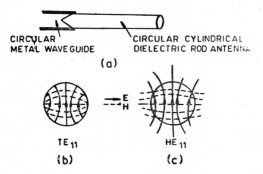

Fig. 13.8   Circular cylindrical dielectric rod antenna:
(a) Antenna excited by circular metal wave-guide, (b) Electric and magnetic field lines of TE_{11} mode in circular waveguide, (c) Electric and magnetic field lines of HE_{11} mode in dielectric cylinder.

[1]Chingehi Yao, Steven E. Schwarz and Blumenstock, B. J., 'Monolithic Integration of a Dielectric Millimetre-wave Antenna and Mixer Diode', An Embryonic Millimetre-Wave IC', *I.E.E.E. Trans.*, MTT-30, No. 8, pp. 1241–1246, 1982.

waveguide for the TE_{11} mode. The TE_{11} mode in the waveguide naturally becomes the dominant hybrid HE_{11} mode in the dielectric cylinder and the corresponding field lines are shown in Fig. 13.8(c).

A simple method of deriving the radiation pattern of such an antenna is by considering the rod as consist-ing of a large number of Huyghens' radiators, distributed all over the surface of the rod as shown in Fig. 13.9. In order to calculate the field at a distant point P due to this infinite set of radiating elements, it is necessary to calculate the field at P due to an infinite number of radiating elements distributed on the circumference of a cross-section TT',

Fig. 13.9   Distribution of Huyghens' radiators on dielectric cylinder.

as shown in Fig. 13.10, of the cylinder and then to multiply this field by the field due to one set of elements on the length AB of the cylinder.

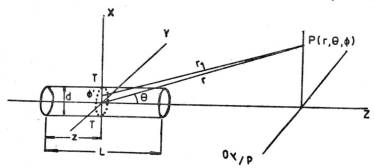

Fig. 13.10   Coordinate system for deriving the radiation pattern of a cylindrical dielectric rod antenna.

The coordinates of the distant point P are the spherical polar coordinates r, θ, ϕ, while those of a typical radiating element on the surface of the cylinder are $(d/2, \phi', z)$, where d is the diameter of the rod. When $r \gg L$, where L is the length of the dielectric cylinder, the distance r_1 of the radiat-ing element $(d/2) \, d\phi'$ from the distant point P is given by

$$r_1 \simeq r - \frac{d}{2} (\sin \phi' \sin \phi - \cos \phi' \cos \phi) \sin \theta \qquad (13.9)$$

The field dE_2 at P due to a Huyghens' point radiator in the infinitesimal element $(d/2)d\phi'$ is given by

$$dE_2 = \frac{A}{r_1} \exp \left[j(\omega t - kr_1) \right] \qquad (13.10)$$

where $k = 2\pi/\lambda_0$ and $A = $ constant involving the magnitude of excitation. Assuming that the strength of the point source on the dielectric rod varies as the electromagnetic power which varies as $\cos^2 \phi'$ or $\sin^2 \phi'$ for the HE_{11}

mode, A can be put equal to $A_0 \sin^2 \phi'$. Then the total field E_2 at P due to all the point sources on the circumference TT' is given by

$$E_2 = \int_{\phi'=0}^{2\pi} \frac{A_0}{r_1} \sin^2 \phi' \exp\left[j(\omega t - kr_1)\right] \frac{d}{2} d\phi'$$

$$\simeq \frac{A_0 d}{2r} \exp\left[j(\omega t - kr)\right] \int_{\phi'=0}^{2\pi} \sin^2 \phi' \exp\left[jk \frac{d}{2} \sin \theta\right] \cos(\phi - \phi') \frac{d}{2} d\phi'$$

$$= -\frac{A_0 d}{2r} \exp\left[j(\omega t - kr)\right]\left[\pi\left\{J_0\left(\frac{kd}{2}\sin\theta\right) + \cos(2\phi)J_2\left(\frac{kd}{2}\sin\theta\right)\right\}\right.$$

$$\left. + 4(\pi/2)^{1/2}\Gamma(2)\sin(2\phi)\frac{J_{3/2}(\tfrac{1}{2}kd\sin\theta)}{(\tfrac{1}{2}kd\sin\theta)^{1/2}}\right] \tag{13.11}$$

The field E_1 due to one set of point radiating elements on AB is given by

$$E_1 = A_0 \frac{\sin(n\Delta\phi/2)}{\sin(\Delta\phi/2)} \tag{13.12}$$

where n is the number of elements and $\Delta\phi$ is the phase difference at the distant point P between the radiated fields of two consecutive elements and hence

$$\Delta\phi = \frac{2\pi a}{\lambda} - \frac{2\pi a}{\lambda_0} \cos\theta \tag{13.13}$$

where a is the distance between the consecutive elements, λ is the wavelength in the dielectric rod and λ_0 is the free-space wavelength. If $K = \lambda_0/\lambda$, then

$$\Delta\phi = \frac{2\pi a}{\lambda_0}(K - \cos\theta) \tag{13.14}$$

If L is the length of the dielectric cylinder, then

$$L = (n-1)a \tag{13.15}$$

and if $A_0 = A'/n$, then

$$E_1 = \frac{A'}{n} \frac{\sin\left\{\dfrac{n\pi L}{(n-1)\lambda_0}(K - \cos\theta)\right\}}{\sin\left\{\dfrac{\pi L}{(n-1)\lambda_0}(K - \cos\theta)\right\}} \tag{13.16}$$

and if $n \to \infty$,

$$E_1 = A' \sin\frac{\left\{\dfrac{\pi L}{\lambda_0}(K - \cos\theta)\right\}}{\dfrac{\pi L}{\lambda_0}(K - \cos\theta)} \tag{13.17}$$

Therefore, the total field at P is given by

$$E = E_1 E_2 \tag{13.18}$$

where E_1 is given by Eq. 13.17 and E_2 is given by Eq. 13.11.

One can find $K = \lambda_0/\lambda$ from the propagation properties of the HE_{11}

mode in the dielectric rod[1, 2]. Figure 13.11 shows the dependence of K on d/λ_0 for three different values of ϵ, the relative permittivity of the dielectric cylinder for the HE_{11} mode.

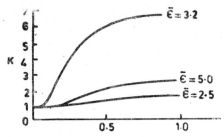

Fig. 13.11   Dependence of $K = \lambda_0/\lambda$ on d/λ_0 for the HE_{11} mode for a dielectric cylinder of diameter λ.

Figure 13.12 shows the radiation pattern in the $\phi = 0°$ plane for a polystyrene rod of length $3\lambda_0$ and diameter $0.46\lambda_0$.

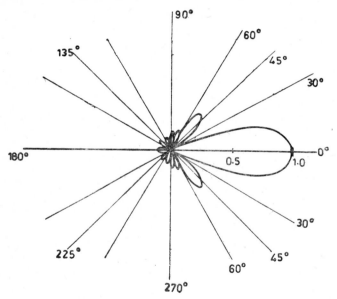

Fig. 13.12   Radiation pattern in the $\phi = 0°$ plant polystyrene cylindrical rod of length $L = 3\phi_0$ and diameter $d = 0.46\lambda_0$.

More rigorous methods of deriving the radiation pattern are the application of Schelkunoff's Equivalence Principle, the two-aperture theory,

[1]Kiely, D. G., *Dielectric Aerials*, Methuen and Co. Ltd., New York, 1953, Chaps. 2 and 3.
[2]Collin, R. E., *Field Theory of Guided Waves'*, McGraw-Hill Book Co. Inc., New York, 1960, Chap. 5.

and the scattering theory. For larger diameter rods, the higher order HE modes are also present and they have to be considered in deriving the pattern[1].

PROBLEMS

1. Calculate and plot the radiation pattern of a circular cylindrical dielectric rod of diameter 2.5 cms and length $10\lambda_0$ given that $\lambda_0 = 3$ cm and $\epsilon_r = 2.5$.

2. Derive the characteristic equation for the electromagnetic boundary-value problem of the infinitely long circular dielectric cylinder embedded in free-space for hybrid modes. Show that axially unsymmetric TE and TM modes cannot be supported on such a structure, but only axially symmetric TE and TM modes as well as hybrid modes can be supported.

3. Solve the electromagnetic boundary-value problem of a dielectric sphere embedded in free-space and derive the characteristic equation.

SUGGESTED READING

1. Collin, R. E. and Zucker, F. J., *Antenna Theory*, Parts I and II, McGraw-Hill Book Co., Inc., New York, 1969.

2. Fradin, A. Z., *Microwave Antennas*, Pergamon Press, Oxford, 1961.

3. Jasik, H., *Antenna Engineering Handbook*, McGraw-Hill Book Co., New York, 1961.

4. Kiely, D. G., *Dielectric Aerials*, Methuen and Co. Ltd., London, 1953.

5. Chatterjee, R., *Dielectric and Dielectric-loaded Antennas*, Research Studies Press, Letchworth, Hertfordshire, England, 1985; John Wiley & Sons, New York.

6. Walter, C. H., *Travelling Wave Antennas*, Dover Publications, Inc., New York, 1965.

7. Chattarjee, R., *Dielectric and Dielectric-loaded Antennas*, Research Studies, Press, U.K., 1985.

---

[1]Dilli, J., Chatterjee, R. and Chatterjee, S. K., 'Radiation Characteristics of Overmoded Dielectric Rods', *J. Ind. Inst. Sci.*, vol. 59 (11), pp. 431–73, 1977.

14

WIDE-BAND ANTENNAS

Most of the antennas considered so far with the exception of the helical, conical and horn antennas, are frequency sensitive and cannot be used over a wide-band of frequencies. Some types of antennas which are comparatively wide-band will be considered in this chapter.

14.1 General Principles

Some relatively simple but at the same time powerful ideas can be developed to design wide-band antennas.

The first idea is based on the fact that the properties of an antenna like radiation pattern, gain and impedance depend on its shape and on its dimensions relative to the wavelength.[1] If by any arbitrary scaling, the antenna is transformed into a structure which is equal to the original one, its properties will be independent of the frequency of operation. The antenna is then said to satisfy the *angle condition*, which makes its form to be specified only by angles and not by any dimensions. *Conical antennas* of any arbitrary cross-section have a common apex and infinitely long, and equiangular antennas having their surfaces generated by equiangular spirals having a common axis and the same defining parameter, belong to this class.

The second idea[2] depends on the fact that a structure which becomes equal to itself by a particular scaling of dimensions, by some ratio τ, will have the same properties at a frequency f and at the frequency τf. Due to this, the characteristics like impedance, etc. of the antenna is a periodic function, with the period log τ, of the logarithm of frequency. Antennas based on this idea are called log-periodic. The variation of the properties over the frequency band $(f, \tau f)$ is very small.

To satisfy these conditions, equiangular as well as log-periodic antennas

---

[1]Rumsay, V. H., 'Frequency Independent Antennas', *IRE National Convention Record*, Pt. 1, pp. 114–118, 1957.

[2]Duhamel, R. H. and Isbell, D. E., 'Broadband Logarithmically Periodic Antenna Structures', *IRE Convention Record*, Part I, pp. 119–128, 1957.

should extend from the origin 0 which is the centre of expansion and also the feed point, up to infinity. However, in practice, the antenna is a section of the ideal infinite structure, which is contained between two spheres of radii r_1 and r_2 and with their centre at 0. The feed region is inside the smaller sphere of radius r_1. The length r_1 determines the highest frequency f_1 of operation, and r_1 should be much smaller than the wavelength λ_1 corresponding to the frequency f_1. Thus, the feed-coupling mechanism has little influence on impedance or on the resulting current distribution over the antenna. The dimension r_1 is also determined by the size of the wave-guide or transmission line connected to the antenna.

The radius r_2 of the larger sphere determines the outer dimension of the antenna, and also the lowest frequency of operation. For many structures, the current decreases rapidly with the distance from the centre 0. Therefore, it is possible to cut-off the structure at a distance where the currents are very small as compared to the feed current I_0. This happens at a distance usually proportional to the wavelength of operation.

For conical antennas excited at the apex 0, the field is transverse to the radial direction and decreases as $1/r$ and, hence, the current also decreases as $1/r$. But for equiangular and log-periodic antennas, the current decrease is faster than $1/r$. Hence, the end-effect for equiangular and log-periodic antennas is negligible approximately at a distance of a wavelength, while for a conical antenna this is not so.

14.2   Equiangular Antennas

If an antenna has to satisfy the angle condition, then if it is expanded in an arbitrary ratio τ about the feed point 0, the new structure will coincide either with the original structure, or with a structure deduced from the original structure by a rotation about some axis passing through the point 0.

Conical structures having a common apex satisfy the first condition, namely, the expanded structure coincides with the original structure. For example, the discone antenna shown in Fig. 14.1 has good wideband properties. However, any finite length antenna of this type shows end-effects. Hence, the bandwidth cannot be increased at will by increasing the size.

The equiangular structures characterized by the following conditions are much more frequency-independent.

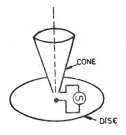

Fig. 14.1   Discone antenna.

1.   The axis of rotation is independent of τ.
2.   The angle of rotation is proportional to log τ, or

$$\tau = \exp{(a\phi)} \qquad (14.1)$$

where ϕ is the angle of rotation and 'a' is a constant. Such equiangular structures also show fast decrease of the current with distance.

If the axis of rotation is the z-axis, the equiangular property may be expressed by stating that the surfaces bounding the antenna have an equation of the form

$$F(\theta, re^{-a\phi}) = 0 \tag{14.2}$$

where r, θ, ϕ are the spherical polar coordinates. The rotation ϕ about the z-axis together with an expansion about 0 in the ratio $\tau = e^{a\phi}$, will carry the structure onto itself. Figure 14.2 shows the geometry of the equiangular spiral. The point M describes an equiangular spiral, either in the plane $z = 0$ or on a cone of revolution having the z-axis as axis. Let the cone have a semiangle θ_0. The equation of the equiangular spiral is

$$r = r_0 \exp(a\phi) \tag{14.3a}$$

or

$$\rho = \rho_0 \exp(a\phi) \tag{14.3b}$$

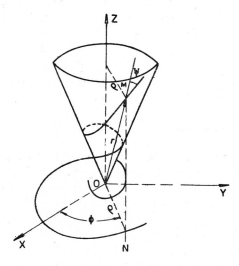

Fig. 14.2   Equiangular spiral.

The spiral makes a constant angle ψ with the radius vector r and hence it is called equiangular.

$$\tan \psi = \frac{\sin \theta_0}{a} \tag{14.4}$$

The length s of the arc from the origin 0 is proportional to the radial distance r, and is given by

$$s = \frac{r}{\cos \psi} \tag{14.5}$$

The projection of the spiral on the $z = 0$ plane is also equiangular with the same parameter a. The equiangular spirals with the parameter a'

given by

$$a' = -\frac{1}{a} \qquad (14.6)$$

are orthogonal to those with the parameter a.

The spirals of the orthogonal set on the cone $\theta = \theta_0$ have the parameter

$$a' = -\frac{\sin^2 \theta_0}{a} \qquad (14.7)$$

If the infinitely long equiangular spiral is fed at the centre 0 by a constant current source of unit magnitude, the electric field at frequency f at point M, is $E(M, f)$. If the frequency is changed to $f' = f/\tau$, then all dimensions are scaled by the factor τ, and the field becomes,

$$E\left(M, \frac{f}{\tau}\right) = \frac{1}{\tau^2} T^\phi (T^{-\phi} M, f) \qquad (14.8)$$

where the rotation T^ϕ corresponds to $\phi = (1/a) \ln \tau$. The radiation pattern is given by the field on a large sphere of radius R and is given by

$$E(\theta, \phi, R) = \frac{e^{-jkR}}{R} [\mathbf{u}_\theta P_\theta(\theta, \phi) + \mathbf{u}_\phi P_\phi(\theta, \phi)] \qquad (14.9)$$

Both P_θ and P_ϕ satisfy the relation

$$P\left(\theta, \phi, \frac{f}{\tau}\right) = P\left(\theta, \phi - \frac{1}{a} \ln \tau, f\right) \qquad (14.10)$$

When the frequency is changed from f to f/τ, the pattern is rotated about the z-axis through the angle $(1/a \ln \tau)$.

Practical equiangular spiral structures are made of conducting wires or sheets. The wires or sheets should be tapered with increasing distance to satisfy the equiangular condition, but in practice, they are used over fairly wide frequency bands even if they are not tapered.

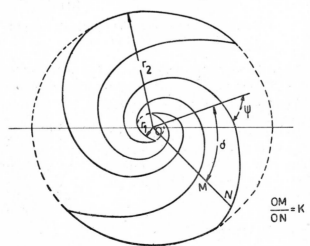

Fig. 14.3 Two-arm planar equiangular spiral antenna.

The planar structures are made of two symmetrical equiangular arms as shown in Fig. 14.3. The expansion ratio is $e^{a\phi}$ corresponding to the rotation ϕ. The angle ψ is given by

$$\psi = \tan^{-1}(1/a)$$

and is the angle at which the radial lines cut the spirals. The arm width is specified by the angle δ or by the ratio $K = e^{-a\delta}$. When $\delta < 90°$, the structure may be considered as a slot of angular width $180° - \delta$ rather than as a dipole. The cut-off radii of the finite antenna are r_1 and r_2. The exact shape of the termination is not very important if the antenna is large enough to accommodate the given frequency band.

Typical radiation patterns as functions of the cone angle of equiangular

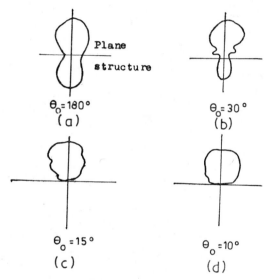

$\theta_0 = 180°$
(a)

$\theta_0 = 30°$
(b)

$\theta_0 = 15°$
(c)

$\theta_0 = 10°$
(d)

Fig. 14.4   Radiation patterns of equiangular spiral as functions of cone half angle.

spirals are shown in Fig. 14.4. The antenna is usually fed by a balanced line or by coaxial cable embedded into one arm of the spiral.

Though the impedance of the infinitely long equiangular spiral is real and independent of frequency, the impedance of a finite length antenna differs from this ideal, but the VSWR is nearly unity over the band of operation. Figure 14.5 shows an equiangular spiral on a cone.

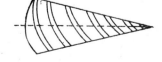

Fig. 14.5   Equiangular spiral antenna on cone.

14.3   Log-Periodic Antennas

Log-periodic antennas are designed so that the electrical properties

repeat periodically with the logarithm of the frequency. Frequency independence is obtained if the variation of the properties over a period is
small. Figure 14.6(a) shows a log-periodic antenna with trapezoidal tooth.

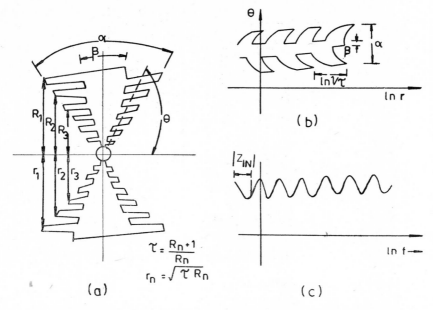

Fig. 14.6   Log periodic antenna and characteristics: (a) Trapezoidal tooth log-
periodic structure, (b) Plot θ vs. ln r, (c) Plot of $|Z_{1N}|$ vs. Inf.

θ is a function $f(\ln r)$ of r, which is a periodic function as shown in
Fig. 14.6(b). The impedance Z_{1N} also varies periodically as shown in
Fig. 14.6(c). Other shapes of teeth like triangular or curved teeth are also
used either in wire or sheet form as shown in Fig. 14.7. Theoretically, the
thickness of the sheets and diameter of the wires should increase linearly
with distance from the vertex of the half structures, though this is not very
necessary for bandwidths less than 5 : 1. The greater the bandwidth, the
greater should be the tapering. Bandwidths of 20 : 1 are easily achieved.

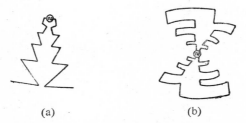

(a)   (b)

Fig. 14.7   Other types of log-periodic antennas:
(a) Sheet circular tooth, (b) Wire
triangular tooth.

Log-periodic dipole arrays as shown in Fig. 14.8 are also used. The dipoles are excited by a uniform two-wire line with the line transposed between dipoles.

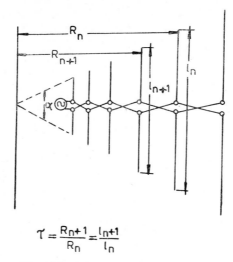

$$\tau = \frac{R_{n+1}}{R_n} = \frac{l_{n+1}}{l_n}$$

Fig. 14.8   A log-periodic dipole array.

14.4   Self-Complementary Antennas

A self-complementary structure is a planar structure in which the metal area is congruent to the open area and has a constant input impedance. This follows from Babnet's principle which shows that the product of the impedance of two complementary planar structures is equal to $\eta_0^2/4$, where η_0 is the free-space intrinsic impedance of 377 ohms. Therefore, a self-complementary structure has an impedance of $\eta_0/2 = 189$ ohms, which is independent of frequency. This does not necessarily mean that the radiation pattern is also independent of frequency. However, log-periodic structures which are also self-complementary have a radiation pattern which remains more or less constant with frequency. Two such structures are shown in Figs. 14.9(a) and (b).

Fig. 14.9   Self-complementary log-periodic antennas: (a) With four-fold symmetry, (b) With two-fold symmetry.

14.5   Applications of Wide-band Antennas

Wide-band antennas used as feeds for lens or reflector-type antennas make the system work over wide frequency ranges. Arrays of log-periodic antennas have frequency independent operation. If the log-periodic elements are inclined with respect to the ground and with their feed points at ground level, then the resultant radiation pattern will be frequency independent. The elements can be so placed that with their images they form either H-plane or E-plane arrays or a combined E- and H-plane array.

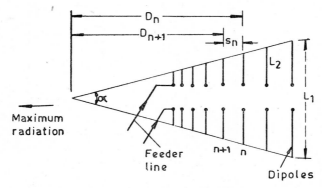

Fig. 14.10   Geometry of log-periodic dipole antenna.

A wire log-periodic antenna with a trapezoidal tooth structure is a useful H-F communications antenna. Two log-periodic structures placed at right angles to each other give a radiation pattern which is omnidirectional in the plane transverse to the axis of the antennas.

Two opposite log-periodic half-structures are fed against each other to provide a linearly polarized unidirectional beam and two other similar structures are fed against each other to provide a linearly polarized unidirectional beam with orthogonal polarization. Such an arrangement can give circular polarization over wide bandwidths.

Figure 14.10 shows a log-periodic dipole antenna.

SUGGESTED READING

1. Collin, R. E. and Zucker, F. J., *Antenna Theory*, Parts 1 and 2, McGraw-Hill, New York, 1969.

2. Jasik, H., *Antenna Engineering Handbook*, McGraw-Hill, 1961.

3. Jordan, E. C. and Balmain, K. G., *Electromagnetic Waves and Radiating Systems*, Prentice-Hall of India, 1969.

4. Kraus, J. D., *Antennas*, McGraw-Hill, New York, 1950.

5. Schelkunoff, S. A. and Friis, H. T., *Antennas, Theory and Practice*, Chapman and Hall, London, 1952.

6. Fradin, A. Z., *Microwave Antennas*, Pergamon Press, Oxford, 1961.

7. Schelkunoff, S. A., *Electromagnetic Waves*, D. Van Nostrand and Co., Inc., Princeton, N.J., 1943.

8. Silver, S., *Microwave Antenna Theory and Design*, McGraw-Hill Book Co., New York, 1949.

9. Fry, D. W. and Goward, F. K., *Aerials for Centimeter Wavelengths*, Cambridge University, 1950.

10. Love, A. W., *Electromagnetic Horn Antennas*, IEEE Press, The Institution of Electrical and Electronics Engineers, Inc., New York, 1976.

11. Rumsey, V. H., *Frequency Independent Antennas*, Academic Press, 1966.

15

ANTENNA SYNTHESIS

In antenna analysis, the fields produced by a given source distribution are derived. Antenna synthesis is the inverse process which determines the source distribution for producing a given radiation pattern. Although the general synthesis problem is difficult to solve, a number of methods have been developed for certain restricted classes of patterns and sizes of sources.

The earlier work on the synthesis problem has been the determination of an array of discrete sources to produce a given far-field pattern. Any practical antenna may consist of an array of discrete sources, a continuous source distribution, or a combination of the two types. In many practical applications a continuous source may be approximated by an array of discrete sources having separations of $\lambda/2$ or less. Hence, synthesis techniques for continuous sources may be sometimes used for discrete sources and vice versa.

There is a second problem in synthesis, namely, the attempt to increase the gain of an antenna beyond a certain value, i.e., the attempt to obtain a supergain antenna. Any attempt for a supergain antenna results in a rapid rate of change of phase and amplitude along the source which produces strong destructive interference in the field of the source. It is also evident that a radiation pattern having a fixed maximum strength but an increased gain must show a decrease in radiated power. Large reactive currents result in large ohmic losses, while rapid phase and amplitude variations along the source are difficult to achieve in practice. These are some of the practical limitations of supergain antennas.

In practice, the synthesis problem is one of finding a physically realizable source distribution which gives approximately the specified pattern.

15.1 Approximate Far-field Pattern of a Line-Source

Let there be a line source of length L along the z'-axis, whose amplitude and phase vary as $A(z')$ and $\psi(z')$ respectively, as shown in Fig. 15.1. The

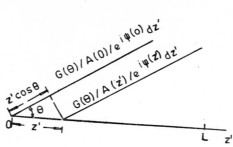

Fig. 15.1   Line source.

far-field $E(\theta)$ of such a line source is given by

$$E(\theta) = G(\theta) \int_{z'=0}^{L} |A(z')| \exp\left[j\{\psi(z') + kz' \cos \theta\}\right] dz' \qquad (15.1)$$

where $G(\theta)$ is the pattern of an element dz' of the line source, and $k = 2\pi/\lambda_0$. Equation 15.1 can be either integrated or evaluated approximately by numerical methods.

An approximate method of evaluating $E(\theta)$ is by representing the continuous source by an array of discrete elements which have the amplitude and phase of the continuous distribution at the points where the elements are located. An element of the array will have the same radiation characteristics as the differential element dz', and the elements of the array need not be equally spaced. Experience with dipole arrays and continuous distributions indicates that an array with spacing of $\lambda/2$ between consecutive elements would, in most cases, give essentially the same pattern as a continuous distribution with the same amplitude and phase at corresponding points. For a spacing of $\lambda/2$, the integral of Eq. 15.1 can be approximated by

$$E(\theta) = \int_{0}^{L} |A(z')| \exp\left[j\{\psi(z') + kz' \cos \theta\}\right] dz'$$

$$\simeq \sum_{m=0}^{2L/\lambda} A_m \exp\left[j(\psi_m + m\pi \cos \theta)\right] \qquad (15.2)$$

where A_m and ψ_m are the amplitude and phase, respectively, at distance $m\lambda/2$ along the source. Equation 15.2 is actually the form of the integral (Eq. 15.1) if the trapezoidal rule is applied, with the difference that the end terms of the series are different by a factor of 2.

As an example, if a line source of uniform distribution is considered, then Eq. 15.1 gives

$$E(\theta) = \sin X/X \qquad (15.3)$$

where $X = (\pi L/\lambda) \cos \theta$, and Eq. 15.2 gives

$$E(\theta) = \frac{\sin (NY)}{N \sin Y} \qquad (15.4)$$

where $Y = (\pi S/\lambda) \cos \theta$, and $N = L/S$, and S is the spacing between elements. Figure 15.2 shows $E(\cos \theta)$ vs. $\cos \theta$ as calculated by using Eqs. 15.3 and 15.4.

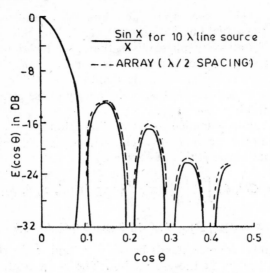

Fig. 15.2   Far-field pattern of line source 10λ long with uniform source distribution $A(z')=1$ and $\psi(z') = \psi_m = 0$.

15.2   The Angle of Maximum Radiation by the Principle of Stationary Phase

The principle of stationary phase states that for an integral of the form,

$$P(\cos \theta) = \int_{z_1}^{z_2} U(z', \cos \theta) \exp\left[jV(z', \cos \theta)\right] dz' \qquad (15.5)$$

where U varies slowly with z', there is general cancellation of positive and negative portions of the integral except for ranges of stationary phase, i.e., where $\partial V/\partial z' = 0$. This means that the main contribution of the integral is for those values of $\cos \theta$ for which $V(z', \cos \theta)$ has stationary values.

As an example, for a line source of length L in free space, with uniform amplitude of unity and a constant phase velocity v, the integral in the far-field expression is

$$P(\cos \theta) = \int_0^L \exp\left[-jkz'\left(\frac{c}{v} - \cos \theta\right)\right] dz' \qquad (15.6)$$

where c is the velocity of light in free space. By the method of stationary phase, the far-field maximum should occur for the value of $\cos \theta$ for which

$$\frac{\partial}{\partial z'}\left[\left(\frac{c}{v} - \cos \theta\right) kz'\right] = 0 \qquad (15.7a)$$

or

$$c/v = \cos \theta_m \qquad (15.7b)$$

where θ_m is the angle of maximum radiation. This result can be verified by performing the actual integration of Eq. 15.6. For this particular case the principle of stationary phase gives the exact location of maximum. However, if the amplitude of the source is not constant, this is not true because of the modifying effect of the amplitude function $U(z', \cos \theta)$ in Eq. 15.5. But even then the method of stationary phase gives the location of the maximum fairly accurately.

The method can be extended to a two-dimensional source. For a rectangular source in the x-z plane, the relative far-field of a source of amplitude $A(x', z')$ and phase $\psi(x', z')$ is given by

$$E(\theta, \phi) = G(\theta, \phi) \int_{x_1}^{x_2} \int_{z_1}^{z_2} |A(x', z')| \exp [j\xi(x', z', \theta, \phi)] \, dz' \, dx' \quad (15.8)$$

where

$$\xi = \psi(x', z') + kx' \sin \theta \cos \phi + kz' \cos \theta \qquad (15.9)$$

and the rectangular source extends between x_1 and x_2 in the x-direction and between z_1 and z_2 in the z-direction. The main contribution of the integral (Eq. 15.8) is in the region where

$$\frac{\partial \xi}{\partial x'} = \frac{\partial \xi}{\partial z'} = 0 \qquad (15.9a)$$

15.3   Synthesis of Line Sources

Let a line source $|A(z')| \exp [j\psi(z')]$ of length L along the z-axis have an element pattern $G(k \cos \theta)$. Then its relative field pattern $E(k \cos \theta)$ is

$$E(k \cos \theta) = G(k \cos \theta) \int_0^L |A(z')| \exp [j\{\psi(z') + kz' \cos \theta\}] \, dz'$$

$$(15.10)$$

Once the source distribution $|A(z')| \exp [j\psi(z')]$ is specified, the far-field pattern is determined uniquely in both magnitude and phase. The synthesis problem of finding a source function, which will produce a given radiation pattern function, is much more difficult. An exact solution exists only for certain special cases.

15.4   Fourier-Transform Method for Line Sources

Equation 15.10 can be rewritten as

$$F(k \cos \theta) = \frac{E(k \cos \theta)}{G(k \cos \theta)}$$

$$= \int_{-\infty}^{\infty} |A(z')| \exp [j\{\psi(z') + kz' \cos \theta\}] \, dz' \qquad (15.11)$$

It is possible to use the infinite integral by making $A(z')$ zero outside the finite interval $0 \leqslant z' \leqslant L$, where L is the length of the source.

Equation 15.11 is one integral of a Fourier transform pair. The other integral of the pair is given by

$$A(z') = \frac{1}{2\pi} \int_{-\infty}^{\infty} F(k \cos \theta) \exp [-jz'k \cos \theta] d(k \cos \theta) \quad (15.12)$$

The factor $1/2\pi$ may be omitted if only relative distributions are to be considered. Equation 15.12 may be called the basic equation of synthesis of a linear antenna source. If the far-field function $F(k \cos \theta)$ is given, the corresponding source function $A(z')$ can be calculated using this equation. The integral (Eq. 15.12) exists if $F(k \cos \theta)$ is at least piecewise-continuous.

In a practical antenna, the magnitude of $F(k \cos \theta)$ is specified over only a portion of the angular spectrum. Therefore, depending on the form that is assigned to the field pattern $F(k \cos \theta)$ outside the specified interval and the form assigned to the phase of the field function over the entire interval, there will be an infinite number of possibilities for $A(z')$. If, however, $A(z')$ is made zero outside the finite interval L, then $A(z')$ is unique for a given $F(k \cos \theta)$, if $F(k \cos \theta)$ is known over a continuous but finite range of $k \cos \theta$. If $F(k \cos \theta)$ is zero except over a finite interval, then $A(z')$ will exist over an infinite interval, and the source will be of infinite extent.

Regarding the phase of $F(k \cos \theta)$, if it is assumed to be constant (usually zero), then the mathematical problem of finding $A(z')$ is simplified.

The Fourier integral given by Eq. 15.12 is evaluated in an easy manner by making use of the convolution integral of Fourier transform theory. If the field function $F(k \cos \theta)$ can be expressed as the product of two functions $F_1(k \cos \theta)$ and $F_2(k \cos \theta)$, or

$$F(k \cos \theta) = F_1 (k \cos \theta)F_2(k \cos \theta) \quad (15.13)$$

and if

$$
\begin{aligned}
F(k \cos \theta) &= \text{FT } A(z') \\
F_1(k \cos \theta) &= \text{FT } A_1(z') \quad (15.14) \\
F_2(k \cos \theta) &= \text{FT } A_2(z')
\end{aligned}
$$

where FT stands for 'Fourier Transform'. Then by the Convolution Theorem,

$$A(z') = \int_{-\infty}^{\infty} A_1(z' - w)A_2(w)dw = \int_{-\infty}^{\infty} A_1(w)A_2(z' - w)dw \quad (15.15)$$

Some of the simple Fourier Transforms are shown in Fig. 15.3.

The source function $A(z')$ which is determined by using Eq. 15.12 may not be restricted in length. Using a finite length L of the source function, the field function $F_a (k \cos \theta)$ is given approximately by

$$F_a(k \cos \theta) = \int_{-L/2}^{L/2} A(z') \exp \{jz'k \cos \theta)\} \, dz' \quad (15.16)$$

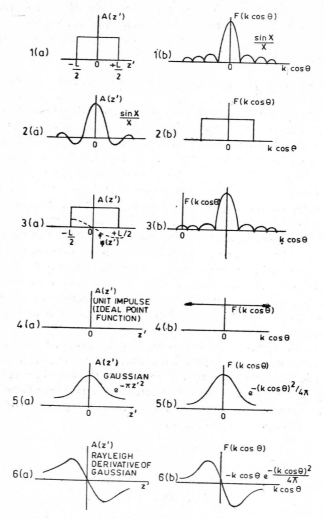

Fig. 15.3 Simple Fourier Transforms.

which becomes,

$$F_a (k \cos \theta) = \int_{-L/2}^{L/2} \int_{-\infty}^{\infty} F(\xi) \exp [jz'\{k \cos \theta - \xi\}] \, d\xi \, dz' \quad (15.17)$$

where $\xi = k \cos \theta$ in Eq. 15.12. Equation 15.17 can be reduced to

$$F_a (k \cos \theta) = L \int_{-\infty}^{\infty} F(\xi) \frac{\sin \left\{ \dfrac{L(k \cos \theta - \xi)}{2} \right\}}{\dfrac{L(k \cos \theta)}{2}} \, d\xi \quad (15.18)$$

Though Eq. 15.18 is an approximate expression, it is quite good for increasing source length L. This equation is called the Dirichlet formulation of the approximation, and can be easily evaluated using the convolution integral.

15.5 The Antenna as a Filter

The antenna can also be considered as a filter which is an useful concept in radio astronomy. Consider the Fourier transform pair

$$F(k \cos \theta) = \frac{1}{2\pi} \int_{-\infty}^{\infty} A(kz') \exp (jkz' \cos \theta) \, d(kz') \qquad (15.19)$$

and

$$A(kz') = \int_{-\infty}^{\infty} F(\cos \theta) \exp (-jkz' \cos \theta) d(\cos \theta) \qquad (15.20)$$

In radio astronomy, the power density $U(\cos \theta)$ in the far-field is important. The power density $U(\cos \theta)$ is given by

$$U(\cos \theta) \simeq F(\cos \theta) F^*(\cos \theta) \qquad (15.21)$$

or

$$U(\cos \theta) \simeq \text{FTA } (kz') \text{ FTA}^*(-kz') \qquad (15.22)$$

where * denotes the complex conjugate. $U(\cos \theta)$ gives the power pattern of the antenna. If the inverse transform of $U(\cos \theta)$ is $P(kz')$, then by the convolution integral,

$$P(kz') = \int_{-\infty}^{\infty} A^*(w) A(kz' + w) \, dw \qquad (15.23)$$

Equation 15.23 is called the source correlation function. The power density functions $U(\cos \theta)$ and $P(kz')$ form the Fourier transform pair,

$$U(\cos \theta) = \frac{1}{2\pi} \int_{-\infty}^{\infty} P(kz') \exp (jkz' \cos \theta) \, d(kz') \qquad (15.24)$$

and

$$P(kz') = \int_{-\infty}^{\infty} U(\cos \theta) \exp (-jkz' \cos \theta) \, d(\cos \theta) \qquad (15.25)$$

If $\cos \theta$ and kz' correspond to t and w, respectively, then z'/λ corresponds to frequency f. The quantity z'/λ is called the spatial frequency and is expressed in cycles per radian.

The function $P(kz')$ as obtained from Eq. 15.25 may be thought of as the filter characteristic of the antenna. The antenna filters spatial frequencies just as a network filters conventional time frequencies.

For example, if there is a uniformly illuminated line source of length L, as shown in Fig. 15.4(a), then the far-field pattern $F(\cos \theta)$ is the $\sin X/X$ function as shown in Fig. 15.4(b). The far-field power pattern is shown in Fig. 15.4(c), and the source auto-correlation function $p(kz')$ is shown in Fig. 15.4(d). This auto-correlation function is the filter characteristic. It can be seen from Fig. 15.4(d) that the antenna will not respond to spatial frequencies greater than L/λ.

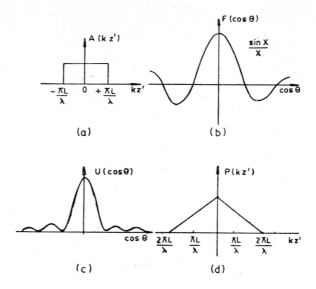

Fig. 15.4 Uniformly illuminated line source: (a) Source,
(b) Far-field pattern, (c) Far-field power
pattern, (d) Source autocorrelation function or
filter characteristic.

The antenna aperture is viewed as a filter with a certain 'spatial frequency response' $P(kz')$ and a certain 'waveform response' $U(\cos\theta)$ to a unit impulse $\delta(\cos\theta)$. The unit impulse $\delta(\cos\theta)$ is interpreted as a point-source transmitter located in direction θ in the far-field of the receiving antenna. The quantity L/λ is the cut-off frequency of an aperture extending from $-L/2$ to $+L/2$ along the z' axis considered as a filter. Such an antenna will not show any response to spatial frequency components greater than L/λ of an incoming signal. To detect fine details of a far-field distribution of emitters, the filter bandwidth has to be increased, which means that the aperture has to be widened in order to cover all spatial frequencies in the spectrum of the incoming signal. The entire spatial frequency spectrum of the far-field distribution of emitters can be constructed, however, by knowing the measured finite spectrum detected by the antenna of length L. This is possible in principle because the spatial frequency spectrum will be an analytic function and, if any part of it is known exactly, it can be mathematically extended to construct the entire spectrum. However, there are difficulties in practice, because of the precision with which the output of the antenna should be measured and also because of the deteriorating effects of noise.

It must be noted that when the antenna is considered as a filter, $\cos\theta$ and kz' correspond to t and ω respectively, while in the Fourier transform network synthesis method, $-k\cos\theta$ and z' correspond to ω and t respectively. In considering antenna as a filter, the concept of spatial frequency is introduced, while in network synthesis phase constant and angular frequency are compared and the latter also enables complex frequency and complex propagation constant to be compared. This enables other network synthesis methods like the Laplace Transform method to be applied to antenna synthesis.

15.6 Laplace Transform Method

The Laplace Transform pair may also be used to represent the source and the far-field functions. The poles of the pattern function can be interpreted as corresponding to damped travelling-waves at the source, just as the poles of the complex frequency response of a network correspond to damped oscillations in the time domain.

In network theory, the Laplace transform pair is

$$f(t) = \frac{1}{2\pi j} \int_{c-j\infty}^{c+j\infty} g(s) \exp(st)\, ds \tag{15.26}$$

and

$$g(s) = \int_0^\infty f(t) \exp(-st)\, dt \tag{15.27}$$

where s is the complex frequency $(\sigma + j\omega)$. The function $f(t)$ must be single-valued almost everywhere in the range $t \geqslant 0$, and must not increase so rapidly as $t \to \infty$ that $\exp(-\sigma t)$ will not predominate. The function $g(s)$ is defined only in the region of absolute convergence of the integral in Eq. 15.27.

The Laplace transform pair can be formed in antenna theory by putting

$$\gamma = \alpha + j\beta \tag{15.28}$$

where

$$\beta = k \cos \theta \tag{15.29}$$

The quantity γ is the complex propagation constant and it is analogous to the complex frequency s in network theory. While s has the dimension of inverse time, γ has the dimension of inverse length. While the Laplace transform of a function of s gives a function of time, the Laplace function of a function of γ gives a function of length. The field function $F'(\gamma)$ and the source function $A(z')$ are related to each other by the Laplace transform pair

$$F(\gamma) = \int_0^\infty A(z') \exp(\gamma z')\, dz' \tag{15.30}$$

and

$$A(z') = \frac{1}{2\pi j} \int_{-c-j\infty}^{-c+j\infty} F(\gamma) \exp(-\gamma z')\, d\gamma \tag{15.31}$$

where $-\gamma$ and z' are analogous to s and t, respectively. Equations 15.30 and 15.31 reduce to the Fourier transform pair Eqs. 15.11 and 15.12 for $\alpha = 0$.

The profile of $F(\gamma)$ along the $j\beta$ axis, in the interval $-k \leqslant \beta \leqslant k$ is the far-field pattern. This is what is represented in a measured radiation pattern of an antenna. The region $-k \leqslant \beta \leqslant k$ is called the visible region. The region $|\beta| > k$ is the region where the energy is stored in the field of the antenna.

The synthesis problem of a line source is to find $A(z')$ when $F(\gamma)$ is specified. The analogous problem in network theory is that of finding the time function corresponding to a given function of complex frequency. In network theory, a pole at $s_1 = -\sigma_1 + j\omega$, as shown in Fig. 15.5(a) corresponds to a damped time function of the form $\exp(-\sigma_1 t + j\omega_1 t)$. In antenna synthesis, a pole at $\gamma_1 = \alpha_1 + j\beta_1$, as shown in Fig. 15.5(b), corresponds to an attenuated travelling-wave of the form $\exp(-\alpha_1 z')\exp(-j\beta_1 z')$ along the z' axis.

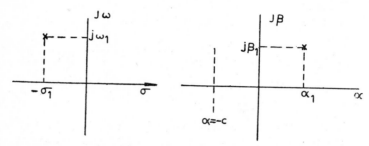

Fig. 15.5   Complex s and r planes: (a) Complex s plane with pole at $(-\sigma_1 + j\omega_1)$, (b) Complex r plane with pole at $(\sigma_1 + j\omega_1)$.

15.7   Woodward's Synthesis Method[1]

This is one of the simplest methods which can be applied when the field pattern is given graphically or analytically.

The far-field pattern of a line source of length L is

$$F(k \cos \theta) = \int_{-L/2}^{L/2} A(z') \exp(jz'k \cos \theta)\, dz' \tag{15.32}$$

If $A(z')$ is represented by a Fourier series as given below:

$$A(z') = A_N(z') = \sum_{-N}^{N} a_n \exp[-j(2\pi nz'/L)] \tag{15.33}$$

then the far-field pattern is given by

$$F_N(k \cos \theta) = \sum_{-N}^{N} a_n \int_{-L/2}^{L/2} \exp\left\{-jz'\left[\frac{2n\pi}{L} - k \cos \theta\right]\right\} dz' \tag{15.34}$$

After integrating Eq. 15.34, we obtain

$$F_N(k \cos \theta) = \sum_{-N}^{N} a_n \frac{\sin\left[n\pi - \dfrac{Lk \cos \theta}{2}\right]}{\left(n\pi - \dfrac{Lk}{2} \cos \theta\right)} \tag{15.35}$$

[1]P. M. Woodward, 'A method of calculating the field over a plane aperture required to produce a given polar pattern', *J.I.E.E.*, Part III (a), p. 1554, 1946.

The synthesis problem reduces to a problem of determining the constants a_n in such a way that $F_N(\cos \theta)$ is a satisfactory approximation to the given pattern $F(k \cos \theta)$ over a specified interval. The coefficients a_n can be chosen such that the approximate pattern $F_N(k \cos \theta)$ will fit the given pattern $F(k \cos \theta)$ at the points

$$k \cos \theta = \frac{2\pi n}{L}, \ n = 0, \pm 1, \pm 2, \ldots, \pm N \qquad (15.36)$$

or

$$\frac{\omega}{c} \cos \theta = \frac{2n\pi}{L} = \frac{\omega}{v} = \beta$$

or

$$\cos \theta = \frac{c}{v} \qquad (15.37)$$

where $N = L/\lambda$.

The coefficients a_n are determined from Eq. 15.35 and are given by

$$a_n = F(2n\pi/L) \qquad (15.38)$$

If a_n is changed to

$$a_n = F(2n\pi/L \pm k \cos \theta) \qquad (15.39)$$

then Eq. 15.33 becomes,

$$A_N(z') = \sum_{-N}^{N} a_n \exp \left[-j \left\{ \frac{2n\pi}{L} \pm k \cos \delta \right\} z' \right] \qquad (15.40)$$

The quantity $k \cos \delta$ can be chosen so that the field pattern is fitted exactly at the most desirable points.

In Woodward's method, each term in the Fourier series expansion represents a travelling-wave of constant amplitude a_n and phase constant $2n\pi/L = \beta = \omega/v$ or $[(2n\pi/L) \pm k \cos \delta]$. The far-field of each of these waves is of the form $a_n(\sin X_n/X_n)$ and consists of a main lobe in a direction determined by the phase constant of the wave, together with a number of equally spaced side-lobes. The direction of each main lobe coincides with a null in the pattern of each of the other travelling-waves. Therefore, the field in the direction of each main lobe is determined entirely by the magnitude of that lobe.

For large apertures, the desired pattern may be fitted exactly at a large number of points. For short apertures, since the number of points are fewer, the accuracy will be less.

Figure 15.6 shows the desired pattern of a source 10λ long and the actual pattern obtained by Woodward's synthesis. From Eq. 3.36,

$$\cos \theta = \frac{2n\pi}{kL} = \frac{n}{L/\lambda} = \frac{n}{10} \qquad (15.41)$$

where $n = 0, \pm 1, \pm 2, \ldots, \pm 10$.

The desired pattern and the approximate pattern will be made to coincide at the values of $\cos \theta = 0.5, 0.6, \ldots, 0.9$. In between these values, the actual pattern will oscillate about the desired pattern. The actual

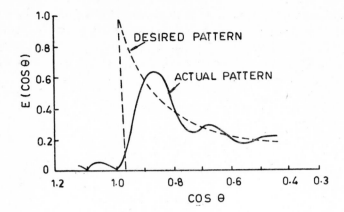

Fig. 15.6   Field pattern of 10λ line source, showing the desired pattern and the actual pattern obtained by Woodward's synthesis.

pattern is given by

$$E(\cos\theta) = \frac{0.22 \sin(5\pi - 10\pi \cos\theta)}{(5 - 10\pi \cos\theta)}$$

$$+ \frac{0.25 \sin(6\pi - 10\pi \cos\theta)}{(6\pi - 10\pi \cos\theta)} + \frac{0.32 \sin(7\pi - 10\pi \cos\theta)}{(7\pi - 10\pi \cos\theta)}$$

$$+ \frac{0.40 \sin(8\pi - 10\pi \cos\theta)}{(8\pi - 10\pi \cos\theta)} + \frac{0.58 \sin(9\pi - 10\pi \cos\theta)}{(9\pi - 10\pi \cos\theta)}$$

$$(15.42)$$

15.8   Optimization Methods

An antenna can be optimized in a number of ways. An usual problem is the minimization of the beamwidth for a given size-lobe level, and another problem is maximization of the gain.

The *Dolph-Chebyshev method* of optimizing an array to give the minimum beamwidth for a given side-lobe level has been considered in Section 6.11. The same method can be extended to a continuous line source, by obtaining a line source distribution whose pattern approximates the optimum pattern by converting the source distribution of the discrete elements of the array into a continuous distribution. For example, for a Dolph-Chebyshev linear broadside array of eight-elements with λ/2 spacing, the dots in Fig. 15.7 represent the excitation coefficients of the discrete elements. If a smooth curve is drawn through these dots, this curve can represent the continuous source. Figure 15.7 shows the field patterns for the array with discrete elements as well as for the continuous source distribution. This shows that the continuous source gives a fair approximation to the Dolph-Chebyshev pattern.

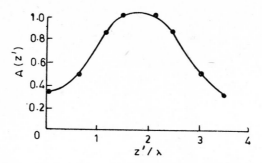

Fig. 15.7 Conversion of the Dolph-Chebyshev excitation coefficients into a continuous source.

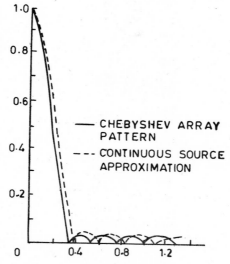

Fig. 15.8 Dolph-Chebyshev array pattern and approximate continuous source pattern.

The *gain* of a continuous linear source can be *optimized* as follows.

Consider a line source of length L on the z-axis excited by a finite sum of travelling-waves so that the source distribution is

$$A(z) = \sum_{i=1}^{N} a_i A_i(z) \tag{15.43}$$

Then the pattern $F(k \cos \theta)$ of this source is

$$F(k \cos \theta) = \int_{-L/2}^{L/2} A(z) \exp(jkz \cos \theta)\, dz = \sum_{i=1}^{N} a_i F_i(k \cos \theta) \tag{15.44}$$

where

$$F_i(k \cos \theta) = \int_{-L/2}^{L/2} A_i(z) \exp(jkz \cos \theta)\, dz \tag{15.45}$$

It may be assumed that a_i and $F_i(k \cos \theta)$ are real and that the element dz has an isotropic pattern.

The directivity (or gain if the efficiency is 100 per cent) of the line source is given by

$$D = \frac{4\pi \left[\sum\limits_{i=1}^{N} a_i F_i(k \cos \theta_m) \right]^2}{\int_0^{2\pi} \int_0^{\pi} \left[\sum\limits_{i=1}^{N} a_i F_i(k \cos \theta) \right]^2 \sin \theta \, d\theta \, d\phi} \tag{15.46}$$

$$= \frac{2 \sum\limits_{i=1}^{N} \sum\limits_{j=1}^{N} f_{ij} a_i a_j}{\sum\limits_{i=1}^{N} \sum\limits_{j=1}^{N} g_{ij} a_i a_j} \tag{15.47}$$

where

$$f_{ij} = F_i(k \cos \theta_m) F_j(k \cos \theta_m) = f_{ji} \tag{15.48}$$

and

$$g_{ij} = \int_0^{\pi} F_i(k \cos \theta) F_j(k \cos \theta) \sin \theta \, d\theta = g_{ji} \tag{15.49}$$

and θ_m = the angle of maximum field.

If the coefficients a_i have to be selected for optimum gain but the function $A_i(z)$ are fixed,

$$\frac{\partial D}{\partial a_n} = 0, n = 1, 2, \ldots, N \tag{15.50}$$

$$\frac{\partial D}{\partial a_n} = \frac{2 \sum\limits_{i=1}^{N} \sum\limits_{j=1}^{N} g_{ij} a_i a_j \left[\sum\limits_{j=1}^{N} f_{nj} a_j + \sum\limits_{i=1}^{N} f_{in} a_i \right]}{\left[\sum\limits_{i=1}^{N} \sum\limits_{j=1}^{N} g_{ij} a_i a_j \right]^2}$$

$$- \frac{2 \sum\limits_{i=1}^{N} \sum\limits_{j=1}^{N} f_{ij} a_i a_j \left[\sum\limits_{j=1}^{N} g_{nj} a_j + \sum\limits_{i=1}^{N} g_{in} a_i \right]}{\left[\sum\limits_{i=1}^{N} \sum\limits_{j=1}^{N} g_{ij} a_i a_j \right]^2} \tag{15.51}$$

Putting the numerator of Eq. 15.51 equal to zero, we obtain

$$\sum\limits_{i=1}^{N} \sum\limits_{j=1}^{N} g_{ij} a_i a_j \sum\limits_{m=1}^{N} f_{nm} a_m - \sum\limits_{i=1}^{N} \sum\limits_{j=1}^{N} f_{ij} a_i a_j \sum\limits_{m=1}^{N} g_{nm} a_m = 0 \tag{15.52}$$

where $n = 1, 2, \ldots, N$ and remembering $f_{ij} = f_{ji}$ and $g_{ij} = g_{ji}$. If the amplitude of $A_i(z)$ is so chosen that $F_i(k \cos \theta_m) = 1$ for all i, then $f_{ij} = 1$ for all i, j. Then Eq. 15.52 becomes

$$\sum\limits_{i=1}^{N} \sum\limits_{j=1}^{N} g_{ij} a_i a_j \sum\limits_{m=1}^{N} a_m - \sum\limits_{i=1}^{N} \sum\limits_{j=1}^{N} a_i a_j \sum\limits_{m=1}^{N} g_{nm} a_m = 0 \tag{15.53}$$

which can be written as

$$\sum\limits_{m=1}^{N} g_{nm} a_m = \frac{\sum\limits_{i=1}^{N} \sum\limits_{j=1}^{N} g_{ij} a_i a_j}{\sum\limits_{i=1}^{N} a_i}, n = 1, 2, \ldots, N \tag{15.54}$$

Equation 15.54 can be written as

$$(g)(a) = (K) \tag{15.55}$$

where

$$(g) = \begin{pmatrix} g_{11} & \cdots & g_{1N} \\ \vdots & & \\ g_{N1} & \cdots & g_{NN} \end{pmatrix}, \quad (a) = \begin{pmatrix} a_1 \\ \vdots \\ a_N \end{pmatrix},$$

$$(K) = \begin{pmatrix} K \\ K \\ \vdots \\ K \end{pmatrix}, \; K \text{ being a constant.}$$

Therefore,

$$(a) = (g)^{-1}(K) = \frac{(b)}{|g|} (K) \tag{15.55a}$$

where

$$(b) = \begin{pmatrix} b_{11} & \cdots & b_{1N} \\ \vdots & & \\ b_{N1} & \cdots & b_{NN} \end{pmatrix} \tag{15.56}$$

where $|g|$ is the determinant of (g), and (b) is the adjoint of (g), where b_{ij} is the cofactor of g_{ij}. Then the ratio of the unknown coefficients a_i are given by

$$a_1 : a_2 : \ldots : a_N = \sum_{j=1}^{N} b_{ij} : \sum_{j=1}^{N} b_{2j} \ldots \sum_{j=1}^{N} b_{Nj} \tag{15.57}$$

The method works for only a finite number of waves.

15.9 Hansen-Woodyard Increased Directivity

Hansen and Woodyard have considered the directivity of a line source with constant amplitude and phase velocity, and have adjusted the phase velocity to give optimum directivity.[1]

The far-field pattern of a uniform line source of length L and with a travelling-wave with phase constant β is given by

$$F(k \cos \theta) = \frac{\sin \{(k \cos \theta - \beta)L/2\}}{\{(k \cos \theta - \beta)L/2\}} \tag{15.58}$$

where θ is measured from the axis of the source and $\theta = 0°$ corresponds to the end-fire direction. The directivity of this source is

$$D = \frac{2 \left[\dfrac{\sin \{(k - \beta)L/2\}}{\{(k - \beta)L/2\}} \right]^2}{\displaystyle\int_{-1}^{+1} \left[\dfrac{\sin \{(k \cos \theta - \beta)L/2\}}{\{(k \cos \theta - \beta)L/2\}} \right]^2 d(\cos \theta)} \tag{15.59}$$

---

[1] W. W. Hansen and and J. R. Woodyard, 'A New Principle in Antenna Design', *Proc. I.R.E.*, 26(3), pp. 333–345, March 1938.

Hansen and Woodyard minimized $1/D$ to obtain the maximum directivity. The integral in the denominator of Eq. 15.59 is given by

$$f(u) = \frac{2}{kL}\left(\frac{u}{\sin u}\right)^2\left[\frac{\pi}{2} + \frac{\cos (2u - 1)}{2u} + Si\,(2u)\right] \qquad (15.60)$$

where $u = (k - \beta)\dfrac{L}{2}$.

If Eq. 15.60 is plotted against u, the curve shown in Fig. 15.9 is obtained. The minimum of $f(u)$ occurs for $u = -1.47$, or when

$$\beta = k + \frac{2.94}{L} \qquad (15.61)$$

or

$$\frac{c}{v} = 1 + \frac{1.47}{\pi L/\lambda} \qquad (15.62)$$

which says that the phase delay along the source with increased directivity is 2.94 radians greater than along an ordinary end-fire source ($\beta = k$) of the same length.

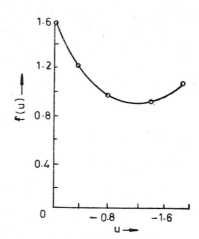

Fig. 15.9 $f(u)$ vs. u.

The region of minimum of $f(u)$ is quite broad, and hence in practice the value of π is used rather than 2.94, and then

$$\frac{c}{v} = 1 + \frac{1}{2L/\lambda} \qquad (15.63)$$

15.10   Synthesis of Planar Rectangular Source

It is convenient to assume that the source distribution $A(x', z')$ is separable and is equal to $A_1(x')A_2(z')$ over a rectangular source of dimensions L and W in the x'-z' plane as shown in Fig. 15.10. A two-dimensional Fourier transform of the far-field can be used to synthesise such a source. The two-dimensional Fourier transform pair is given by

$$F(u,\,v) = \int_{-\infty}^{\infty}\int_{-\infty}^{\infty} A(x',\,z')\,\exp\,[j(ux' + vz')]\,dz'\,dx' \qquad (15.63a)$$

and

$$A(x',\,z') = \int_{-\infty}^{\infty}\int_{-\infty}^{\infty} F(u,\,v)\,\exp\,[-j(ux' + vy')]\,du\,dv \qquad (15.64)$$

where $u = k \sin \theta \cos \phi$, and $v = k \cos \theta$. Equations 15.63 and 15.64 form a Fourier transform pair if $\displaystyle\int_{-\infty}^{\infty}\int_{-\infty}^{\infty} |F(u,\,v)|^2\,du\,dv$ exists. Woodward's method can be applied to such a source. If the field is of the form

$$F(u,\,v) = \underset{n\ \ m}{\Sigma\ \Sigma}\ a_{mn}\ \frac{\sin X_m}{X_m}\ \frac{\sin Z_n}{Z_n} \qquad (15.65)$$

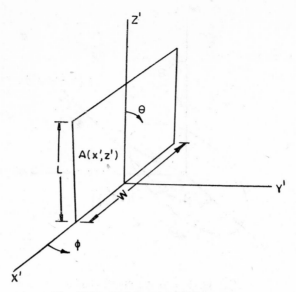

Fig. 15.10   Rectangular planar source.

where

$$X_m = \left(u - \frac{2\pi m}{W}\right)\frac{W}{2} \quad \text{and} \quad Z_m = \left(v - \frac{2\pi n}{L}\right)\frac{L}{2},$$

then the corresponding source distribution is

$$A(x', z') = \sum_n \sum_m a_{mn} \exp\left[-j(2\pi m x'/W + 2\pi n z'/L)\right] \qquad (15.66)$$

If $u = 2\pi p/W$ and $v = 2\pi q/L$, where p and q are integers, all the terms in Eq. 15.65 are zero except the one where $p = m$ and $q = n$. This makes it convenient to select the coefficients a_{mn} such that

$$a_{pq} = F\left(\frac{2\pi p}{W}, \frac{2\pi q}{L}\right) \qquad (15.67)$$

15.11   Synthesis of Planar Circular Source

Just as the Fourier transform pair is used to synthesise linear and rectangular planar sources, the Hankel transform pair can be used to synthesise a planar circular source. If polarization is neglected, the far-field $F(k \sin \theta, \phi)$ of a source function $A(\rho', \phi')$ on a circular planar source of radius 'a' as shown in Fig. 15.11, then

$$F(k \sin \theta, \phi) = \int_0^{2\pi} \int_0^a A(\rho', \phi') \exp\left[j\rho' k \sin \theta \cos(\phi - \phi')\right]\rho' \, d\rho' \, d\phi'$$

$$(15.68)$$

If the source function is independent of ϕ', then $A(\rho', \phi') = A(\rho')$, and

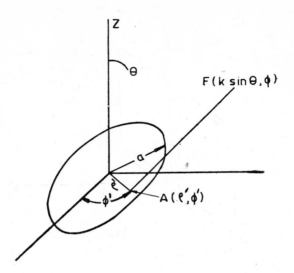

Fig. 15.11   Circular planar source.

then the field function is

$$F(k \sin \theta) = \int_0^\infty A(\rho') \int_0^{2\pi} \exp\left[j\rho'k \sin \theta \cos (\phi - \phi')\right] d\phi' \rho' \, d\rho'$$

(15.69)

or

$$F(k \sin \theta) = \int_0^\infty \rho' A(\rho') J_0(\rho'k \sin \theta) \, d\rho'$$

(15.70)

Then the source function $A(\rho')$ is given by the Hankel transform

$$A(\rho') = \int_0^\infty k \sin \theta F(k \sin \theta) J_0(\rho'k \sin \theta) \, d(k \sin \theta)$$

(15.71)

The visible region is $(-1 \leqslant \sin \theta \leqslant 1)$ and the invisible region is $(|\sin \theta| > 1)$.

Figure 15.12 shows some useful Hankel transform pairs. Equation 15.71 does not specify the radius 'a' of the source. If only the portion of the source in the region $0 \leqslant \rho' \leqslant a$ is to be used,

$$F_a(k \sin \theta) = \int_0^a \rho' A(\rho') J_0(\rho'k \sin \theta) \, d\rho'$$

(15.72)

Using Eq. 15.71 in Eq. 15.72, we obtain,

$$F_a(k \sin \theta) = \int_0^a \rho' \, d\rho' \int_0^\infty \xi F(\xi) J_0(\rho'\xi) J_0(\rho'k \sin \theta) \, d\xi$$

(15.73)

If the inner integral of Eq. 15.73 is uniformly convergent, the order of

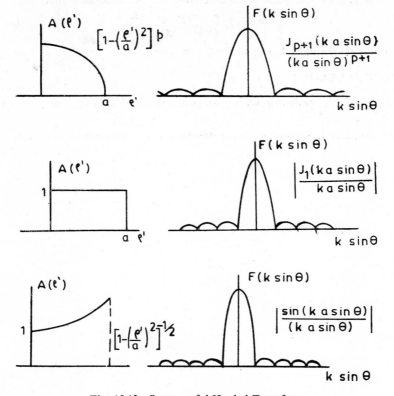

Fig. 15.12   Some useful Hankel Transforms.

integration may be interchanged, and then

$$F_a(k \sin \theta) = \int_0^\infty \xi F(\xi) \, d\xi \int_0^a \rho' J_0(\rho'\xi) J_0(\rho'k \sin \theta) \, d\rho'$$

$$= \int_0^\infty \frac{a\xi F(\xi)}{\{\xi^2 - (k \sin \theta)^2\}} [\xi J_1(a\xi) J_0(ak \sin \theta)$$

$$-k \sin \theta J_1(ak \sin \theta) J_0(a\xi)] \, d\xi \qquad (15.74)$$

The approximate pattern F_a approaches the desired pattern as $a \to \infty$.

PROBLEMS

1. Show that the visible region of an antenna pattern ($|\cos \theta| \leqslant 1$) corresponds to the power radiated by the source, and that the invisible region ($|\cos \theta| > 1$) corresponds to energy stored in the near-field of the source.

2. An equal ripple beam is given by

$$|F(\beta)| = \frac{1}{[1 + \epsilon T_n^2(\beta)]^{1/2}},$$

where T_n is the Chebyshev polynomial and ϵ is a constant. Show that

$$F(\beta) = \left[\frac{\epsilon^2 T_n^2(1/\beta)}{1 + \epsilon^2 T_n^2(1/\beta)}\right]^{1/2}$$

gives a pattern which has equal ripple side lobes.

SUGGESTED READING

1. Walter, C. H., *Travelling Wave Antennas*, Dover Publications, Inc., New York, 1965.

2. Rhodes, D. R., *Synthesis of Planar Antenna Sources*, Clarendon Press, 1974.

16

ANTENNA PRACTICE

16.1 Antenna for Low Frequencies (15 to 500 KHz)

The propagation characteristics of low-frequency (LF) waves make the region useful for certain special services. At these frequencies, the ground-waves have less attenuation and the sky-waves are less affected by ionospheric conditions. Hence, this region is useful for communication and navigation aids to ships and aircraft which are far away. At very low frequencies (VLF) the waves can penetrate into sea water and, hence, this region is useful for submarine communication.

However, atmospheric noise affects propagation in the LF and VLF regions. In the polar regions ionospheric disturbances are very severe while atmospheric noise level is low, so that low frequencies are very suitable in these regions.

At these frequencies, the antenna becomes very large if its dimension is comparable with the wavelength, its cost becomes prohibitive and there are practical difficulties in constructing it. Due to these reasons only small antennas are used whose dimensions are not greater than $\lambda/8$. These small antennas do not have high efficiency, power capacity nor bandwidth.

The radiators used at these frequencies are usually vertical radiators, umbrella-loaded vertical radiators, and flat-top radiators fed by down leads. These three different types of radiators are shown in Fig. 16.1. The adding of a top loading in the form of an umbrella increases the antenna effective height, increases the bandwidth, and decreases the voltage, and hence, the power loss. However, the umbrella cables, place an extra mechanical load on the tower and hence increase the cost. Figure 16.2 shows a typical low-frequency antenna with flat-top umbrella and supporting towers.

In a multiple tuned antenna as shown in Fig. 16.1(c), the ground loss is reduced. But the several down leads subtract from the increased effective height brought about by the flat-top, and the efficiency may be decreased. There is an increase in cost as well.

For these types of low-frequency antennas, proper ground systems have to be used. Most of them have radial ground systems. Figure 16.3 shows

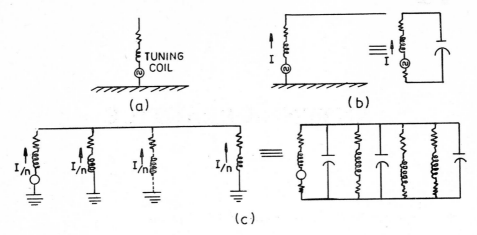

Fig. 16.1   Types of low-frequency antennas: (a) Vertical radiator, (b) Loaded vertical radiator, (c) Flat-topped vertical radiator with *n* down-leads.

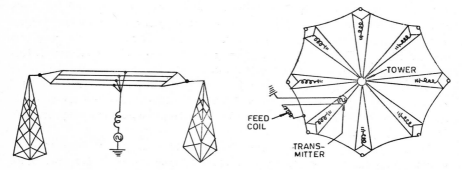

Fig. 16.2   Typical low-frequency antenna.   Fig. 16.3   VLF umbrella antenna.

a typical VLF umbrella antenna which is multiple-tuned. A 'multiple-star' ground system can be used for this type of antenna. A number of relatively small radial wire systems are scattered over the ground area in which all significant ground currents occur. The radial wires are laid or buried in the ground, extending out from the points of feed. Copper mesh is commonly included near the feed-points to shield the ground from the high field existing there and to serve as a convenient termination for the inner ends of the radius.

Receiving antennas at low frequencies are usually loop or wave antennas.

16.2   Antennas for Medium Frequencies (300 to 3000 KHz)

Medium frequencies are usually used for broadcasting purposes. Medium-frequency broadcast transmitting antennas are usually vertical radiators ranging in height from $\lambda/6$ to $5\lambda/8$ or higher, depending upon the operating characteristics, and on the considerations of cost. The physical heights vary from 150 ft. to 900 ft., and hence, towers are used as radiators. The towers may be supported by guys or may be self-supporting. They are usually insulated from the ground, though not always.

The maximum radiation of such vertical radiators is in the horizontal plane and it is uniform in this plane, but decreases with angle about the horizon and is zero towards the zenith. In short, they have the characteristics of a monopole. The ground currents are high because the current return takes place through the ground. Metallic ground systems are used to minimize losses. Vertical polarization is usually used because of better ground-wave and sky-wave propagation properties. A typical ground system is a radial system.

For directional antennas, arrays of two or more radiators are used to give specified radiation patterns in the horizontal and vertical planes. A typical ground system for a two-element array is shown in Fig. 16.4. Figure 16.5 shows a block diagram of an arrangement to supply radio frequency power to the individual radiators of a directional system of antennas. If suitable precautions are not taken, static charges accu-

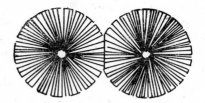

Fig. 16.4 Typical ground system for a two-element directional antenna.

mulate on towers and guy wires and they may discharge to ground. A dc path from the tower to the ground will minimize static accumulation due to lightning hits. A lightning rod or rods extending above the beacon on the tower-top will provide some protection.

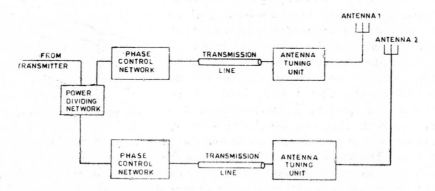

Fig. 16.5 Block diagram of arrangement for feeding radio-frequency power to a directional system consisting of two antennas.

Reradiation from a tall tower may be controlled by insulating the tower from the ground and by installing sectional insulators at one or more levels. High tension ac power lines, large tall buildings, etc. may produce unwanted reradiation.

16.3   Antennas for High Frequencies (3 to 30 MHz)

High frequency antenna arrays are usually used in this range of frequency for medium and long-distance communications and broadcasting by means of ionospheric propagation. High-power expensive transmitters are used and so high-gain transmitting and receiving antennas are desirable. Also wherever international broadcasting to specific areas are required, high-gain directional antennas are desirable.

While designing high frequency transmitting or receiving arrays, the Optimum Working Frequency (OWF) for the path involved has to be determined. This OWF varies according to the distance and location, time of day, season of the year, and also on sunspot activity. The range of vertical angles pertinent to ionospheric propagation depends on the distance, effective layer height, and mode of propagation.

Multi-element horizontal dipole arrays are usually used, which are supported at a suitable height above the ground. The vertical directivity is increased by stacking elements one above the other to form a bay of two or more radiators. Horizontal, or more correctly, conical directivity may be increased by adding additional bays with the elements of adjacent bays collinear. A group of two or more such bays is called a curtain. The antenna array may be made unidirectional by adding a reflector curtain behind the radiator curtain, and this gives an increase in gain of about 3 dB, with very little change in pattern shape in the forward direction. The reflector curtain may be formed of tuned elements operated as parasitic elements, or may be a vertical reflector screen composed of closely spaced horizontal wires. Figure 16.6 shows a curtain antenna with the typical transmission-line arrangement.

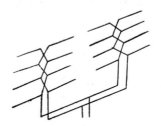

Fig. 16.6   Curtain antenna.

In selecting a site for a transmitter or receiver station, several factors have to be considered. If long-distance circuits are involved, an area should be selected with no obstructions projecting more than a few degrees in elevation angle in the direction of transmission or reception. A reasonably flat area of good conductivity is suitable. The area should be flat extending from the antennas to at least a distance where ground reflections of signals will occur in the directions of transmission or reception. If a transmitting station and receiving station are near to each other, the minimum distance between them should be such that there is no interference. Care should also be taken that there is no interference from other sources like power lines or any other man-made interference.

In addition to curtain antennas, rhombic antennas also are used at high frequencies, and they offer very good impedance characteristics over fairly wide frequency bands.

16.3.1 Horizontal Dipoles Above Ground

The simplest type of antenna used for short-wave applications is a single horizontal-dipole supported at a suitable height above the ground as shown in Fig. 16.7. If a perfect conducting ground and sinusoidal distri-

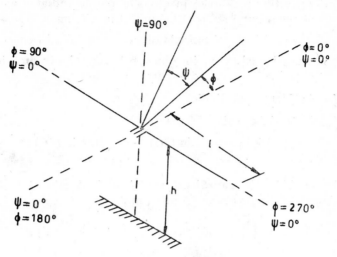

Fig. 16.7   Horizontal dipole above ground.

bution of the current in the dipole are assumed, the field intensity at unit distance (1 mile) is given by

$$F = 2E_1\sqrt{P}f(\phi') \sin (h \sin \psi) \qquad (16.1)$$

where

$$E_1 = E_0(R_r/R_1)^{1/2} \qquad (16.2)$$

$$R_1 = R_r + R_a - R_m(2h) \qquad (16.3)$$

$$f(\phi') = \frac{\cos (l \sin \phi') - \cos l}{(1 - \cos l) \cos \phi'} \qquad (16.4)$$

$$\sin \phi' = \sin \phi \cos \psi$$

$F =$ field intensity in mv/meter at unit distance (1 mile) at horizontal angle ϕ from the equatorial plane and vertical angle ψ above horizon.

$E_1 =$ field intensity in mv/meter at unit distance (1 mile) in the equatorial plane radiated by dipole element for 1 kW power.

$P =$ power input in kW.

$f(\phi') =$ directivity factor of dipole element.

$\phi' =$ angle between equatorial plane and line of propagation.

$h =$ height of radiator above ground.

E_0 = field intensity at 1 mile in mv/meter in equatorial plane radiated by an isolated dipole for a power of 1 kW.

R_r = radiation resistance in ohms of isolated dipole referred to current maximum.

R_1 = operating resistance in ohms of dipole radiator at height h above ground referred to current maximum.

R_a = assumed loss resistance in ohms referred to current maximum.

$R_m(2h)$ = mutual resistance in ohms between dipole and its image (spacing $2h$) referred to current maximum.

l = half length of dipole.

R_r and $R_m(2h)$ can be calculated from

$$R_r = 15\{4 \cos^2 G \operatorname{Cin} (2G) - \cos 2G \operatorname{Cin} (4G)$$
$$- \sin 2G[2 \operatorname{Si} (2G) - \operatorname{Si} (4G)]\} \tag{16.5}$$

(where the symbols Cin and Si are explained in Section 3.2, and G = electrical length of dipole above ground), and

$$R_m = 60 \int_0^{\pi/2} K_1 f_1(\phi')K_2 f_2(\phi') \cos \psi J_0(s \cos \phi') \, d\phi' \tag{16.6}$$

where $K_1 = K_2 = (1 - \cos G) = $ form factor, $f_1(\phi')$ and $f_2(\phi')$ are the expressions $f(\phi')$ for the dipole and its image, and $s = 2h = $ the separation between the dipole and its image, and $J_0(s \cos \phi')$ is the Bessel function of zeroth order.

16.3.2   Horizontal Rhombic Antenna

Rhombic antennas are usually used for short-wave transmitting and receiving applications, especially for point-to-point communication, because they have relatively high gain, are cheap and have broad band input impedance characteristics. However, their disadvantages are loss of power in the terminating load, relatively large secondary lobes, and lack of freedom in controlling vertical and horizontal beamwidths.

Figure 16.8 shows a horizontal rhombic antenna above ground. The radiation pattern of such an antenna is given by

$$F = 468.6 I_0 l \sin A \frac{\sin m_1 \sin m_2}{\sqrt{m_1}\sqrt{m_2}} \tag{16.7}$$

where

$$m_1 = \pi l[1 - \cos \psi \cos (\phi + A)] \tag{16.8}$$

$$m_2 = \pi l[1 - \cos \psi \cos (\phi - A)] \tag{16.9}$$

F = field intensity in mv/meter at 1 mile in direction at elevation angle ϕ from principal axis ($\phi = 0$ for direction of wave travel).

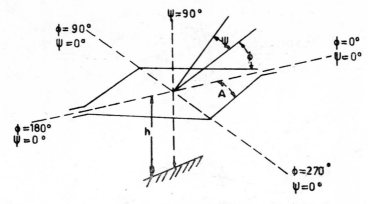

Fig. 16.8   Rhombic antenna.

l = side of rhombic in wavelengths.

A = angle in degrees between principal axis and side.

I_0 = current in amperes in rhombic wires.

For $\phi = 0$ (the principal vertical radiation plane), Eq. 16.7 becomes,

$$F_{(\phi=0)} = 74.62 I_0 \sin A \frac{1 - \cos\left[2\pi l(1 - \cos\psi \cos A)\right]}{1 - \cos\psi \cos A} \qquad (16.10)$$

The radiation in $\phi = 0$ plane given by Eq. 16.10 is horizontally polarized, but in any other plane it has both horizontal and vertical in-phase components and Eq. 16.7 gives the vector sum of these two components. Usually the separate vertical and horizontal polarized components need not be computed. However, the equations for these are found in literature.[1]

An approximate expression for the radiation resistance of a rhombic antenna is[2]

$$R_r \simeq 240(\log_e 4\pi l \sin^2 A + 0.577) \qquad (16.11)$$

16.3.3   Other Types of High-Frequency Antennas

Horizontal resonant and terminated V antennas as shown in Fig. 16.9(a) and (b) and fishbone antenna as shown in Fig. 16.10 are used at these frequencies. The terminated V gives a unidirectional pattern and gives maximum gain when the radiation lobes for both conductors are aligned, while the resonant V antenna has bidirectional characteristics, and hence, less gain. A fishbone antenna is an array with a two-wire balanced central feeder terminated at the far end, and the half-wave dipole elements are coupled to the transmission line by high reactance capacitors. Two

[1]A. E. Harper, *Rhombic Antenna Design*, D. Van Nostrand Co., Inc., New York, 1941.
[2]L. Lewin, Discussion on 'Radiation from Rhombic Antennas', by D. Foster, *Proc. I.R.E.*. vol. 29, p. 523, Sept. 1941.

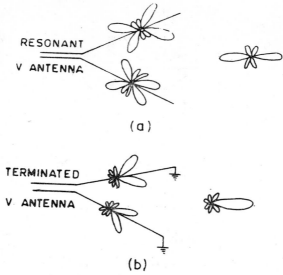

Fig. 16.9   *V*-Antennas: (a) Resonant, (b) Terminated.

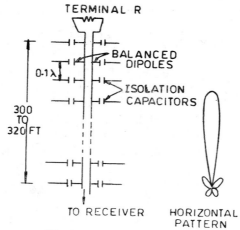

Fig. 16.10   Fishbone Antenna.

such antennas may be used in broadside to give a directional radiation pattern.

The corner reflector antenna is also used sometimes.

16.4   VHF and UHF Antennas (25–470 MHz)

VHF and UHF communication systems are used in land mobile service and coastwise and inland maritime service. Fire, police, industry, radio and telephone are some of the users. The vertical radiator as a quarter-wave 'whip' or 'monopole' as shown in Fig. 16.11 is the simplest form of antenna used for these purposes. Since base-station monopole antennas are usually mounted on masts on top of buildings, a perfect ground plane is not present. Hence the ground plane is simulated by lines extending horizontally from the base of the monopole. Figure 16.12 shows the usual form of such an antenna with four ground plane wires which are about 0.28 to 0.30 in length. By adding a parallel ground section to the monopole to form a folded monopole, a dc path formed to ground gives protection against lightning charges, and the folded configuration is more broad-banded than the single monopole. This is shown in Fig. 16.12(a). An alternative feed is shown in Fig. 16.12(b), in which a quarter-wave matching section is placed in the transmission line at the feed-point to improve the impedance match between the antenna and the transmission line.

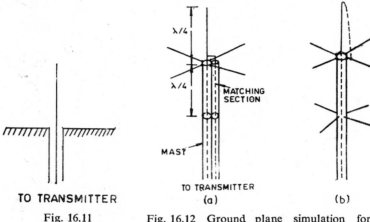

TO TRANSMITTER

Fig. 16.11

Fig. 16.12 Ground plane simulation for mast-mounted vertical antenna: (a) Folded monopole with two parallel ground sections, (b) With matching section.

In the quarter-wave monopole over a ground plane, the radiation projection of the ground rods are sometimes objectionable physically. The coaxial-skirted antenna using a choke to isolate the antenna from the mast as shown in Fig. 16.13, behaves as a half-wave dipole in free-space, and is quite useful.

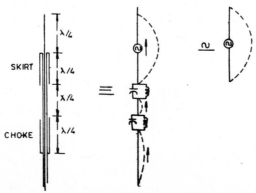

Fig. 16.13 Coaxial, centre-fed half-wave dipole with choke.

In the Franklin array, short-circuited quarter-wave lines are used to perform the function of the skirts. A modified Franklin array is shown in Fig. 16.14, which improves the overall excitation and the gain is about 2 to 3 db over that of a half-wave dipole.

A series-fed collinear array is shown in Fig. 16.15, which employs annular slots in the outer conductor of a coaxial line to excite half-wave

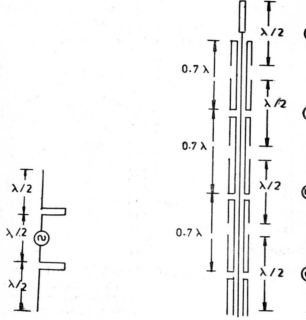

Fig. 16.14 Franklin array.   Fig. 16.15 Series-fed collinear array.

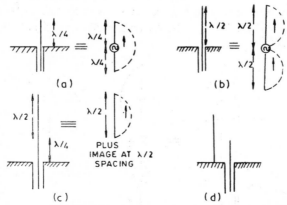

Fig. 16.16 Roof-top antennas: (a) λ/4 monopole, (b) λ/2 monopole, (c) Two-wire quarter-wave matching section, (d) Grounded wire as radiating element to provide lightning protection.

dipoles formed by adding skirts symmetrically about the slots. The slots are spaced 0.7 apart and dielectric loading of the feed-line makes the internal spacing a full wavelength to provide proper phasing. The entire antenna is placed in a fibreglass radome for protection and rigidity. The measured gain is about 4 dB above that of a half-wave dipole. A quarter-wave or half-wave monopole mounted on a flat surface of a vehicle such as the roof of an automobile, train, deck of a ship, a trailer or a truck, is usually called a 'rooftop antenna'. Figure 16.16 shows such antennas. When directional antennas are required, usually an array of two mono-poles are used.

16.5   Television and FM (Frequency Modulation) Transmitting Antennas

Antennas for TV and FM signals should be broadband. The frequency band for FM broadcast service is 88 to 108 MC, and for TV there are four bands : 54 to 72, 76 to 88, 174 to 216, and 470 to 890 MC. The input impedance of the antennas used for this purpose should remain constant over a frequency range which is 10 per cent of the operating frequency for the lower TV channels, and 0.2 per cent of the operating frequency for the FM channels because the individual channels in the FM band are 200 kc wide and in the TV band the channels are 6 MC wide. The power gains of the antennas should be high. It is usual to use a transmitting antenna having a horizontal radiation pattern which is almost circular. The antennas are usually mounted on a tower to obtain as large a coverage area as possible within radio line of sight. Care must be taken in locating the tower and antennas structure so that it is not a hazard to air navigation.

For FM transmitting antennas, square-loop antennas, clover-leaf antennas, circular or ring antennas, vertical array of V antennas and

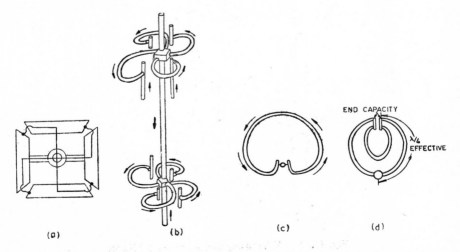

Fig. 16.17   F. M. Antennas: (a) Square-loop, (b) Cloverleaf, (c) Slotted cylinder, (d) Circular loop.

slotted-cylinder antennas are suitable. Some of them are shown in Fig. 16.17. In the clover-leaf antenna, there are two or more vertically stacked radiating units, each one being made up of a cluster of four curved elements, the plan view of which resembles a four-leaf clover. The slotted-cylinder consists of an infinite number of loops carrying an electric current stacked on top of each other and connected in parallel. The termination of these loops takes the form of a slot running the full length of the cylinder. Circular and ring antennas are different adaptations of half-wave dipoles formed in a circular arrangement, and is essentially a folded half-wave dipole bent in a circular form with capacitor plates attached to each end.

For TV, the transmitting antennas used are usually super turnstile, 'super-gain', helical, slotted ring, cylinder with multiple slots, etc. some of which are shown in Fig. 16.18. In the super-turnstile antenna, the arms of the dipole are enlarged and made cone-shaped to give broader band characteristics. Four of these are mounted around a tubular steel pole at 90° intervals. The 'super-gain' VHF antenna consists of dipoles mounted in front of reflecting screens on the faces of a four-sided supporting structure. This antenna gives better broadband impedance characteristics. Both

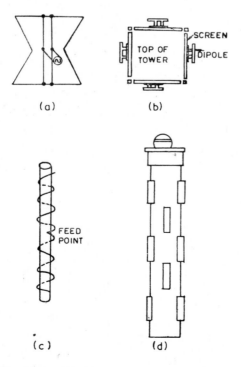

Fig. 16.18   TV Transmitting antennas (a) Super turnstile, (b) Supergain, (c) Helical, (d) Slotted cylinder.

these antennas have omnidirectional radiation patterns in the horizontal plane. The helical antenna gives high gain, right- and left-hand helices are used in each bay, and they are fed at a common point. In the slotted-cylinder antenna, each slot is about 1.3λ and are parallel to the axis of the cylinder. The slots are fed by a coaxial line feeder system within the cylinder, the slotted-cylinder serving as the outer conductor of a coaxial line, the inner conductor being a copper tube within the cylinder. Another coaxial line is installed within the inner conductor to obtain a method of centre-feeding the antenna array. Each slot is excited by a tuned loop placed across it's face.

16.6   TV-Receiving Antennas

Figure 16.19 shows several TV-receiving natennas like Vee, twin-driven Vee, tilted-fan dipole with reflector, twin-driven tilted fan dipole of two

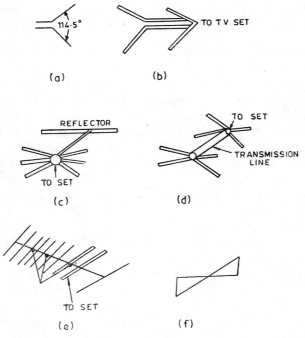

Fig. 16.19   TV Receiving Antennas: (a) Vee, (b) Twin-driven Vee. (c) Tilted fan-dipole with reflector, (d) Twin-driven tilted fan dipole of two elements, (e) Twin-driven Yagi, ( ) Triangular dipole.

elements, twin-driven Yagi and triangular dipole. The antennas with the reflector are more directional. In the Yagi antenna only one dipole is driven while the others are parasitic antennas of reflectors. Only one of the dipoles is a reflector and the others are parasitic and are called directors. The triangular flat antenna is quite broad-band while the Yagi is not broad-band. Corner-reflector antennas are also used sometimes as TV-receiving antennas.

16.7   Radar and Microwave Communication Antennas

For microwave communication, usually centre-fed paraboloid antennas are used. For radar, centre-fed or off-set paraboloid and Cassegrain antennas are usually used to give pencil beams. To give fan-shaped beams, parabolic cylinders with line sources and hog-horns are used. The cosecant antenna is used in air-to-ground or ground-to-air radar to give the field strength or gain in the vertical plane proportional to the cosecant or cosecant squared, respectively, of the angle measured from the horizon. Such special beam shapes can be produced by almost any type of antenna, including lenses, dielectric rods, cylindrical reflector, and doubly-curved reflector antennas both with various feed arrangements. It can also be done using two-dimensional arrays. Earlier, shaped-beam antennas were modified paraboloids. Phased arrays are used for scanning purposes.

16.8   Microwave Beacon Antennas

In many radars, it is necessary to reinforce the radar response by a beacon system. The beacon transponder has a response which is frequently coded for navigation or homing purposes, and it operates in conjunction with an antenna system having an omnidirectional pattern in the azimuth plane. The gain of the antenna is obtained by narrowing the vertical plane pattern. For ground- and ship-borne antennas, the vertical gain is obtained by stacking a number of elements like dipoles or slots in a linear vertical array and feeding the elements so that their contributions combine in phase in the horizontal plane. Depending on the type of radar, the polarization may be vertical or horizontal. A vertically polarized field can be obtained by cutting horizontal slots in a vertical waveguide in such a way as to interrupt vertical currents flowing in the waveguide. Slots placed symmetrically on a circular cross-section of the waveguide and being excited by equal currents in phase, give a suitable radiating element. This situation is possible for a circular waveguide excited in the TM_{01} mode and in a coaxial line carrying the TEM mode. This is best done on the coaxial line carrying the TEM mode, because in a waveguide, if the horizontal slots are much less than $\lambda/2$ in length, impedance-matching becomes difficult. A vertically polarized pair of slots on a coaxial line is shown in Fig. 16.20. Vertically polarized double dipoles also can be used as shown in Fig. 16.21 for vertical polarization. For horizontal polarization, three half-wave dipoles arranged on the circumference of a circle can be used, each of the dipoles being fed by a three-wire line, the central line serving as a probe to couple the dipole to the interior of the coaxial line. This is called a tridipole radiator and is shown in Fig. 16.22.

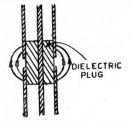

Fig. 16.20   Vertically polarized
pair of slots on co-
axial line

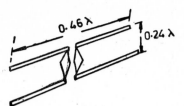

Fig 16.21   Vertically polarized
double dipole.

Fig. 16.22   Tridipole
radiator.

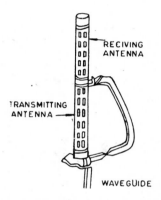

Fig. 16.23   Beacon antenna with
external feed line.

Figure 16.23 shows a beacon antenna with an external feed-line, where the receiving and transmitting antennas are located one above the other.

16.9 Aircraft Antennas

In designing aircraft antennas, it should be remembered that they are capable of withstanding severe static and dynamic mechanical stresses and the size and shape of the airframe plays an important role in determining the important characteristics of the antenna. Flat ground-plane T antennas as shown in Fig. 16.24 are used at low frequencies and these are known as flush antennas. Low frequency loop antennas are also used. For high-frequency communication, wire antennas supported between the vertical fin and an insulated mast are sometimes used. Folded dipoles, annular slots, loops,

$L \gg d$
and
$L \gg h$

Fig. 16.24   Flat ground plane
T-antennas (Flush
antennas).

turnstile antennas and also longitudinal slots are also used for different purposes.

16.10   Direction-Finding Antennas

Antennas used for radio direction finding (DF) are rotatable antennas like rotatable loop, rotatable Adcock, and an antenna with rotatable reflector as shown in Fig. 16.25. In manually-operated systems, the antenna is positioned to produce a null output from the receiver and the direction is read

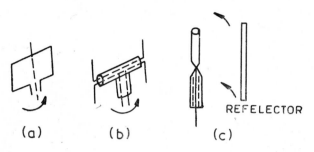

Fig. 16.25   Rotatable DF Antennas: (a) Loop, (b) Adcock, (c) Rotatable reflector.

on a scale on the antenna-positioning shaft. Automatic versions employ a servo to position the antenna system to a null or other usable position. In continuously rotated motor-driven systems, the receiver output may be displayed on a cathode-ray oscilloscope and either null-points or maximum-response points may be read. In the 2–30 MHz region, the loop is unsatisfactory because the sky-wave signals contain some horizontally polarized energy. In this region, the horizontal elevated Adcock antenna with crossed horizontal dipoles, shown in Fig. 16.26, is useful. Further,

Fig. 16.26   Elevated-H-Adcock antenna

horizontally-spaced loops are used in this region. Arrays of 4 or 8 separate Adcock antennas as shown in Fig. 16.27 are also used. When the wavelength of interest is about 3 meters, the signal received by a fixed vertical antenna as shown in Fig. 16.25(c) may be modulated with a rotated reflector.

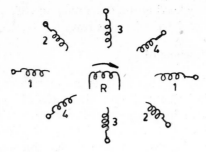

Fig. 16.27   Adcock array. Goniometer, stator coils are connected diametrically across centre as 1 to 1, 2 to 2, 3 to 3, 4 to 4.

16.11   Radomes

Many microwave antennas and some UHF antennas require a housing called a radome to protect against weather, for proper incorporation into the stream lined structure of an aircraft or missile, to provide for pressurization, to reduce scanning-power requirements, or to match feed horns. Radomes must be strong, stiff, must not get heated very much, should not absorb water, and should be resistant to abrasion and erosion. Radomes can be of many shapes, curved or conical as shown in Fig. 16.28 and they

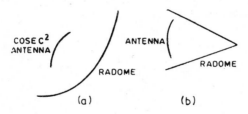

Fig. 16.28   Radomes.

usually change the radiation characteristics of the antennas, sometimes they may distort so that the signal disappears. Most solid-wall rigid radomes are made of plastic base with glass fibres as a reinforcing agent to give strength. Plexiglas, polystyrene, teflon or rexolite are used above 3 cm wavelength. If a half-wave wall of suitable dielectric material is too heavy and if the thin wall is structurally unsatisfactory, a three-layered sandwich construction is used, in which the outer layers or skins are thin compared with λ and with dielectric constant $4\epsilon_0$, and the middle layer is of low dielectric constant (about $1.2\epsilon_0$).

If Fraunhofer diffraction is neglected, the far-field pattern as distorted by the radome is given by

$$P_R(\theta', \phi') = |\tau(\theta', \phi')|^2 P_O(\theta', \phi') \qquad (16.12)$$

where $P_O(\theta', \phi')$ is the far-field pattern of the antenna in direction (θ', ϕ'), the direction of the ray leaving the antenna, and $\tau(\theta', \phi')$ is the local transmission coefficient associated with this ray.

16.12   Impedance-Matching of Antennas

Impedance-matching of antennas is usually done by using lumped reactance elements in impedance-matching networks at lower frequencies. At higher frequencies, distributed-impedance networks like transmission lines are used; stub-matching and tapered transmission lines are also used. A balun is an impedance transformer designed to couple a balanced transmission line and an unbalanced transmission line. This is usually done by introducing a high impedance of a resonant coaxial-choke structure between the outer conductor of the unbalanced coaxial circuit and ground, or by a half-wave delay line, or by helically-wound two-wire transmission lines or by a hybrid ring.

SUGGESTED READING

1. Jasik, H., *Antenna Engineering Handbook*, McGraw-Hill Book Co., New York, 1961.
2. Jordan, E. C. and Balmain, K. G., *Electromagnetic Waves and Radiating Systems*, Prentice-Hall of India Private Ltd., New Delhi, 1969.
3. Savitski, G. A., *Calculations for Antenna Installations, Physical Principles*, Oxonian Press Pvt. Ltd., New Delhi, 1978.

17

ANTENNA MEASUREMENTS

In this chapter we discuss some of the important measurements on antenna, such as, input and mutual impedance, radiation patterns, gain, phase front and polarization, which determine the characteristics of the antenna.

17.1 Input and Mutual Impedances

For radio frequencies below 30 MC, it is usual to use bridge measurements. The fundamental Wheatstone-bridge shown in Fig. 17.1 is quite useful for this measurement. This bridge utilizes a null method, and is useful for measurements of impedance, resistive or reactive from dc to the

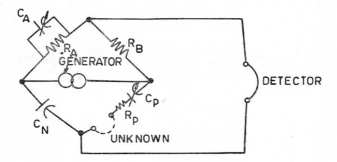

Fig. 17.1 Wheatstone-bridge.

lower VHF band. The measurements are usually preceded by a calibration of the bridge in which the latter is balanced with the unknown impedance terminals short-circuited or open-circuited. There are many bridges, derived from Fig. 17.1, with many fixed known resistors, inductances and capacitances and with one or more variable calibrated elements. The generator signal source should give at least 1 mv output, and the detector should be a well-shielded receiver having at least a sensitivity of $5\mu v$.

At higher UHF frequencies and microwave frequencies, slotted-line measurements are more convenient. Figure 17.2 shows the set-up for slot-

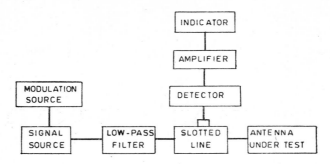

Fig. 17.2   Set-up for slotted-line impedance measurement.

ted-line impedance measurement. Slotted-lines may be coaxial line, slab lines or waveguide lines. The characteristic impedance of coaxial or slab lines is usually 50 ohms, and waveguide slotted-lines are available in different sizes corresponding to different waveguide sizes for different bands. The standing wave patterns with the slotted-line shorted, and with the antenna as load are drawn as shown in Fig. 17.3.

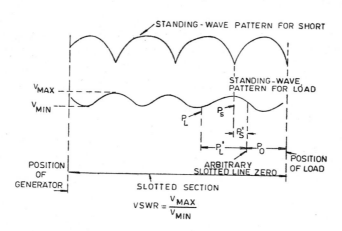

Fig. 17.3   Standing-wave patterns.

$$\text{SWR} = 20 \log \text{VSWR} = 20 \log \frac{V_{\text{MAX}}}{V_{\text{MIN}}} \tag{17.1}$$

The input impedance of the antenna is given by

$$Z_L = Z_0 \left\{ \frac{S}{\cos^2{(\beta l)} + S^2 \sin^2{(\beta l)}} + j \left[\frac{(S^2 - 1) \sin{(\beta l)} \cos{(\beta l)}}{\cos^2{(\beta l)} + S^2 \sin^2{(\beta l)}} \right] \right\} \tag{17.2}$$

where

$S = \text{VSWR}$

$\beta = 2\pi/\lambda_g$

λ_g = guide wavelength

Z_0 = characteristic impedance of line

Z_L = load or antenna impedance

and

$$|(P_l' - P_s')| = |l|$$ (17.3)

If P_l is closer to the generator than P_s, then l is taken as positive, and if P_s is closer to the generator than P_l, then l is taken as negative.

The Smith Chart (Fig. 17.4) can be easily used to determine the impedance. For example, if $P_s = 25$ cm = short position, $P_l = 17$ cm = load position, and VSWR = 4, and $\lambda_g = 26$ cm, then

$$|l| = P_l - P_s = 8$$

and

$$\frac{|l|}{\lambda_g} = \frac{8}{26} = 0.308 \text{ wavelength.}$$

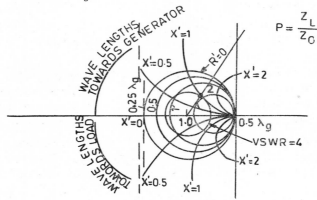

Fig. 17.4 Smith chart.

Assuming that P_s is closer to the load-end of the line than P_l, the value $|l|/\lambda_g$ is plotted using the position scale on the Smith Chart which reads 'wavelengths toward load'. If P_s is closer to the generator than P_l, then $|l|/\lambda_g$ is plotted using the scale which reads 'wavelength toward generator'. The resulting impedance (normalized value) is given by the point P of intersection VSWR = 4 circle and the radius vector representing $l/\lambda_g = 0.308$.

To measure the mutual impedance between two antennas 1 and 2, 2 is short-circuited and the terminal impedance Z_1 of 1 is measured. Then,

$$Z_1 = Z_{\text{self}} + \frac{I_2}{I_1} Z_{\text{mutual}}$$ (17.4)

$$0 = Z_{\text{self}} + \frac{I_1}{I_2} Z_{\text{mutual}}$$ (17.5)

so that

$$(Z_{\text{mutual}})^2 = Z_{\text{self}} (Z_{\text{self}} - Z_1)$$ (17.6)

where Z_{self} is the self-impedance of element 1, or element 2, if the antennas are identical.

17.2   Radiation Pattern

Figure 17.5 shows a set-up for the measurement of the radiation pattern of an antenna, which consists of a transmitting source, an antenna under

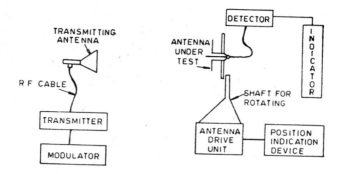

Fig. 17.5   Set-up for measuring radiation pattern.

test, a mount for rotating the antenna under test, and a detector for indicating the relative magnitude of the received field. The equipment may be either fully automatic or manual. The following precautions should be taken care of to be sure that the radiation patterns are correct:

(i)   To obtain the accurate far-field, or Fraunhofer, radiation patterns, it is necessary that the distance R between the transmitting antenna and the antenna under test be sufficiently large so that

$$R \geqslant 2d^2/\lambda \qquad (17.7)$$

where d is the maximum dimension of the aperture of the antenna and λ is the operating wavelength. Reducing R tends to give broader radiation patterns and higher minor lobes and shadower nulls.

(ii)   A uniform illumination of the antenna under test should be ensured. Ground reflections and reflections from buildings, trees etc. should be avoided.

(iii)   Antenna measurements can be done in a properly designed anechoic chamber to avoid reflections and to simulate free-space conditions.

(iv)   An automatic range equipment avoids manual errors.

(v)   The transmitting antenna should have as uniform a wavefront as possible to minimize the phase error of the antenna under test, should have as high a gain as possible, and should have very insignificant minor lobes (at least 20 dB less than the major lobe). Horns, parabolic dishes and arrays of dipoles or horns are usually used.

17.3   Gain

The gain and directivity of the antenna are defined as

$$G = \text{Gain} = \frac{\text{Maximum radiation intensity (test antenna)}}{\text{Maximum radiation intensity (reference antenna)}} \quad (17.8)$$

and

$$D = \text{Directivity} = \frac{\text{Maximum radiation intensity}}{\text{Average radiation intensity}} \quad (17.9)$$

Assuming that the test antenna and the reference antenna both have the same input power, the values of the radiation intensity are generally obtained by a measurement of the field intensity of the power density. In deriving the expression for directivity, the values of radiation intensity are usually obtained only in a relative manner from the radiation patterns.

The gain G of an antenna is always less than its directivity D, and

$$G = \alpha D \quad (17.10)$$

where α can be called the effectiveness ratio and $0 \leqslant \alpha \leqslant 1$. Also,

$$G = \gamma_0 G_0 \quad (17.11)$$

where G_0 is the gain of the test antenna over the arbitrary reference antenna, and γ_0 is the gain of the reference antenna over an isotropic antenna. For lossless antennas, $1 \leqslant \gamma_0$.

Figure 17.6 shows a set-up for measuring the gain of an antenna by a comparison method. The distance R between the test antenna and the transmitting antenna should be $\geqslant d^2/\lambda$, and reflections from nearby objects should be minimized.

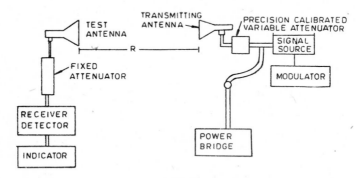

Fig. 17.6   Set-up for gain measurement by comparison method.

A standard antenna is connected to the receiver and is pointed in the direction of maximum intensity. The input to the transmitting antenna is set to a convenient level, and the reading on the receiver is noted. Let W_1 be the attenuator-dial setting and P_1 the power-bridge reading. The standard antenna is now replaced by the test antenna and let W_2 and P_2 be the attenuator dial-setting and power-bridge reading. Then the power gain G is given by

$$G = \frac{W_2 P_1}{W_1 P_2} \qquad (17.12)$$

If $P_d = 10 \log \dfrac{P_1}{P_2}$, then the

$$\text{Decibel gain} = W_2 - W_1 + P_d \qquad (17.13)$$

17.4 Phase-Front

To determine the contours of constant phase or phase fronts of an antenna, the radiated field is sampled and its phase is compared with the phase of a reference signal obtained from the signal source. By varying the amplitude and phase of either the reference or the sampled signal, it is possible to mix the two signals and produce an easily recognizable interference condition between them. The interference condition may be either a maximum or a null. A set-up to make this measurement is shown in Fig. 17.7. Modulated r-f power is fed into the transmitting antenna, and a sample of this power is fed, through a matching pad and a variable attenuator, to a crystal or bolometer detector. The transmitted energy is picked up at a convenient distance by a dipole probe, the output of which is fed into the same detector. The probe carriage is moved radially with respect to the transmitting antenna until a null is obtained and this location

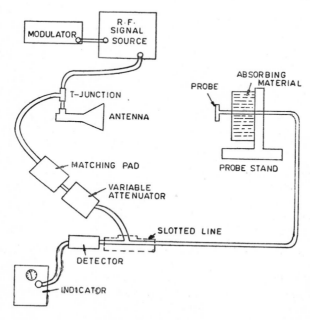

Fig. 17.7   Set-up for phase-front measurements using null-detection method.

is recorded. Moving the probe slowly and carefully, the constant phase contour can be obtained by following the course of the null. By moving one wavelength toward or away from the antenna, another null can be detected and another constant phase contour can be measured.

A modification of this measurement can be made by using a slotted line, if only a limited amount of freedom of probe is available. The slotted-line is inserted in the set-up of Fig. 17.7, as shown by the dotted lines. In this modified set-up, the field probe is allowed to move only in a circular path with the centre of the circle at the transmitting antenna. The null position is obtained at the initial position of the field probe by moving the slotted-line probe along the slotted-line. At each point the slotted-line is moved to a new null position. If ΔS is the shift of the slotted-line probe, then the corresponding phase change is

$$\Delta\phi = \frac{2\Delta S}{\lambda_g}$$

Using this, the constant-phase contour can be determined.

17.5 Polarization

The polarization of an electromagnetic wave is in general elliptic, and linear and circular polarizations are only special cases. Elliptic polarization is the resultant of two linearly polarized components of the wave. If E_x and E_y are the components and if

$$E_x = E_1 \sin (\omega t - \beta z) \tag{17.14}$$

$$E_y = E_2 \sin (\omega t - \beta z + \delta) \tag{17.15}$$

where δ is the phase difference between

$$E_x \text{ and } E_y$$

then

$$\mathbf{E} = \mathbf{u}_x E_x + \mathbf{u}_y E_y$$
$$= \mathbf{u}_x E_1 \sin (\omega t - \beta z) \, \mathbf{u}_y E_2 \sin (\omega t - \beta z + \delta) \tag{17.16}$$

The tip of $\mathbf{E}$ describes an ellipse if $E_1 \neq E_2$ and $\delta \neq 0$, will describe a circle if $E_1 = E_2$ and $\delta = 90°$, and will describe a straight line if $\delta = 0°$ or $180°$ or if E_1 or $E_2 = 0$ and the other not zero. The first case is elliptical polarization, the second case is circular polarization and the third case is linear polarization. Elliptical polarization can also be considered as being produced by two circularly polarized waves.

The polarization ellipse is tilted in space with respect to the coordinate axes as shown in Fig. 17.8, and the tilt angle is given by

$$\tan 2\tau = \frac{2E_1 E_2 \cos \delta}{E_1^2 - E_2^2} \tag{17.17}$$

Antennas for which polarization measurements are necessary are helical antennas, crossed-dipole antennas, double-ridge horns, and crossed-loop antennas. There are three methods of measurement of the polarization characteristics.

The first method is called the *polarization-pattern method*. A linearly polarized antenna, which can be rotated and is calibrated to read the field intensity, is rotated and the signal received from an elliptically-polarized test antenna is measured. This signal traces out a polarization pattern as shown in Fig. 17.8. The direction of rotation of polarization may be obtained by comparing the signals received by two circularly polarized antennas, one being polarized clockwise and the other anticlockwise. The antenna with the larger response gives the proper polarization rotation.

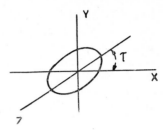

Fig. 17.8   Tilted polarization ellipse.

The second method is called the *linear component method*. One horizontally polarized antenna and the other vertically-polarized antenna receive signals from the test antenna. If E_1 and E_2 are the signals received by these two antennas, then

$$E_x = E_1 \sin (\omega t - \beta z) \tag{17.18}$$

$$E_y = E_2 \sin (\omega t - z\beta + \delta) \tag{17.19}$$

where δ is the phase difference between the two signals and can be measured using the slotted-line phase measurement of the type described in Section 17.4. The direction of rotation is clockwise for $0 < \delta < 180°$ and anticlockwise for $0 > \delta > -180°$, and the angle of tilt is given by Eq. 17.17.

The third method is the *circular component* method. Two circularly-polarization-identical antennas of opposite polarization rotation (like left- and right-hand helical antennas) are used to receive the signals E_R and E_L respectively from the test antenna. Then the axial ratio is given by

$$AR = \frac{E_R + E_L}{E_R - E_L} \tag{17.20}$$

SUGGESTED READING

1. Jasik, H., *Antenna Engineering Handbook*, McGraw-Hill Book Co., New York, 1961.

2. Kraus, J. D., *Antennas*, McGraw-Hill Book Co., Inc., New York, 1950.

3. Hollis, J. S. *et al.*, *Microwave Antenna Measurements*, Scientific Atlanta Co. Inc., 1970.

18

ELECTROMAGNETIC WAVE PROPAGATION

The electromagnetic power radiated by an antenna travels a long distance before it reaches the receiving antenna. Only a small fraction of the power radiated reaches the receiving antenna. The ratio of the power radiated to the power received is called transmission loss which may be as high as 10^{15} to 10^{20} (i.e., 150–200 dB). Thus, transmission loss determines whether the received signal will be useful; and if it exceeds a certain value, the quality and reliability of the received signal becomes poor. The transmission loss depends on the path of the electromagnetic wave which is affected by different factors like the geometry of the path, ground effects, atmospheric conditions, ionospheric conditions, etc. The ideal path lies in free-space, but in practice, this is far from the truth.

18.1 Transmission Ratio in Free Space

The free-space transmission ratio at a distance is given by

$$\frac{P_r}{P_t} = \frac{\lambda}{4\pi d} g_t g_r \qquad (18.1)$$

where P_r and P_t are the received and transmitted power in watts respectively. λ is the wavelength, and g_t and g_r are the power gains of the transmitting and receiving antennas respectively. When antenna dimensions are very large (for example, for parabolic dishes), a more convenient form of the transmission ratio is given by

$$\frac{P_r}{P_t} = \frac{A_t A_r}{(\lambda d)^2} \qquad (18.2)$$

where A_t and A_r are the effective areas of the transmitting and receiving antennas respectively.

The free-space electric field intensity is given by

$$E_0 = \frac{\sqrt{30 P_t g_t}}{d} \text{ volts/metre} \qquad (18.3)$$

where d is in metre and P_t is in watts.

The field intensity concept is more convenient than the transmission loss or transmission ratio concept at frequencies below 30 MHz, where external noise is important and the antenna dimensions are comparable to λ. The free-space field intensity E_0 is independent of frequency. The concept of free-space transmission assumes that the atmosphere is perfectly uniform and lossless and that the earth is at an infinite distance from the antennas.

18.2   Propagation Within Line of Sight

The effect of a plane earth on the propagation of electromagnetic waves is given by

$$\frac{E}{E_0} = \underset{\substack{\text{Direct} \\ \text{waves}}}{1} + \underset{\substack{\text{Reflected} \\ \text{waves}}}{Re^{j\Delta}} + \underset{\substack{\text{Surface} \\ \text{wave}}}{(1 - R)Ae^{j\Delta}} + \underset{\substack{\text{Induction field}+ \\ \text{secondary effects} \\ \text{of the ground}}}{\cdots}$$

where R is the reflection coefficient of ground, A is the surface-wave attenuation factor, and $\Delta = 4h_1h_2/\lambda d$; h_1 and h_2 are the heights measured in the same units as the wavelength and distance d of the transmitting and receiving antennas respectively. R and A depend to some extent on polarization and ground constants. However, for near grazing paths, R is approximately equal to -1, and A can be neglected if both antennas are at a height greater than a wavelength above the ground. Under these conditions, the ground effects are independent of polarisation and ground constants, and Eq. 18.3 simplifies to

$$\left|\frac{E}{E_0}\right| = \left\{\frac{P_r}{P_0}\right\}^{1/2} = 2 \sin (\Delta/2) = 2 \sin (2h_1h_2/\lambda d) \qquad (18.4)$$

where P_0 is the received power expected in free-space and P_r is the actual received power. Equation 18.4 shows that the signal oscillates around the free-space value, and it represents the sum of the direct and the ground waves. In most practical cases, $\Delta/2 < \pi/2$, and the transmission loss over plane-earth becomes

$$\frac{P_r}{P_0} = (\lambda/4\pi d)^2(4h_1h_2/\lambda d)^2 g_t g_r = (h_1h_2/d^2)g_t g_r \qquad (18.5)$$

The transmission loss given by Eq. 18.5 has been derived from optical concepts and hence is not valid for antenna heights less than a few wavelengths. However, for lower heights this formula still gives fairly approximate results if the larger value between h_1 and h_2 is used, and the error does not exceed ± 3 dB, which occurs when the antenna height is approximately equal to the minimum effective height as shown in Fig. 18.1.

The received field intensity oscillates around the free-space value as the antenna heights are increased as shown by Eq. 18.4. The first maximum occurs when the difference between the direct and ground-reflected waves is a half-wave length. The signal maximum is $(1 + |R|)$ and the signal minimum is $(1 - |R|)$.

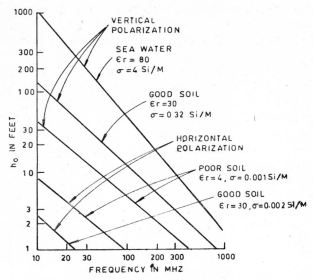

Fig. 18.1   Minimum effective antenna height.

All points from which a wave can be reflected with a path difference of $\lambda/2$ form the boundary of the first Fresnel zone, the boundary of the n-th Fresnel zone consists of all points from which the path difference is $n\lambda/2$. The n-th Fresnel zone clearance H_n at any distance d_1 is given by

$$H_n = \left[\frac{n\lambda d_1(d - d_1)}{d} \right]^{1/2} \qquad (18.6)$$

For rough terrains, R may be different from -1 even for grazing angles. The classical Rayleigh criterion for roughness says that specular reflection occurs when the phase deviations are $< \pm\pi/2$ and the reflection coefficient will be <1 when the phase deviations are $> \pm\pi/2$. This occurs when the variations in terrain are greater than $1/8$ to $1/4$ of the first Fresnel zone clearance. For microwave transmission, most of the terrains are rough and the reflection coefficient varies from 0.2 to 0.4.

Variations in signal level occur on line-of-sight paths as a result of the atmospheric conditions. This is called fading phenomenon. Fading cannot be predicted properly. There are two types of fading, (i) inverse bending and (ii) multipath effects. The latter includes the fading caused by interference between direct and ground-reflected waves as well as interference between two or more separate paths in the atmosphere. The path of an electromagnetic wave is not a straight line except in a uniform medium, and it may be bent up or down depending on the atmospheric conditions. Inverse bending may transform line-of-sight path into an obstructed one. Fading due to this may sometimes last for hours.

Extreme fading can also occur on over-water or other smooth paths, because the phase difference between the direct and reflected-rays may vary with atmospheric conditions. This type of fading may be minimised by placing one antenna at a great height and the other antenna at a low height, so that the phase difference between the direct and reflected-rays is kept steady.

On rough paths, fading is the result of interference between two or more rays travelling slightly different routes in the atmosphere. This is multipath fading and its extreme condition approaches the Rayleigh distribution in which the probability that the instantaneous value of the field is greater than E_1 is exp $[-(E/E_0)^2]$, where E_0 is the rms value.

To receive a signal in spite of fading, space diversity is used where two antennas are spaced so that one antenna receives a maximum field intensity while the other a minimum and vice versa. At frequencies from 5 kMHz to 10 kMHz, rain, snow or fog absorb the electromagnetic energy. The first absorption peak occurs at 24 kMHz for water vapour, and at about 64 kMHz for oxygen, and these frequencies are in the millimetre-wave band.

18.3 Surface Wave and Space Wave : Plane-Earth

The radiation field of an antenna vertically-oriented along the z-axis (Fig. 18.2) and located at a height $z = h$ from the surface ($z = 0$) of a finitely conducting plane-earth consists of surface and space waves. Surface waves are propagated close to the earth's surface and hence suffer attenuation depending on the complex permittivity (ϵ^*), irregularities of the terrain and wavelength of the propagating wave. The space wave consists of the direct wave and the reflected wave which appears to come from the image.

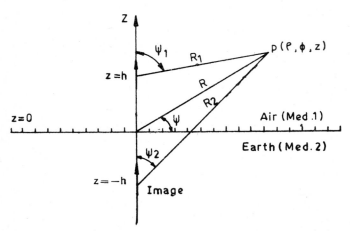

Fig. 18.2   Vertical source placing at $z = h$ over a plane earth.

The problem of finding the radiation field is formulated in terms of the Hertz vector $\boldsymbol{\pi}$ which is related to the electric field intensity vector $\mathbf{E}$ by the following relation

$$\mathbf{E} = \nabla\nabla\cdot\boldsymbol{\pi} + k^2\boldsymbol{\pi} \tag{18.7}$$

where k denotes the propagation constant.

Since the source is z-directed, $\boldsymbol{\pi}$ has only one component π_z which satisfies the following inhomogeneous equation

$$\nabla^2\pi_z^{(1)} + k_1^2\pi_z^{(1)} = -\hat{z}\mu_0\delta(x)\delta(y)\delta(z-h) \tag{18.8}$$

in the region above ($z > 0$) the surface of the earth and the following homogeneous equation

$$\nabla^2\pi_z^{(2)} + k_2^2\pi_z^{(2)} = 0 \tag{18.9}$$

below ($z < 0$) the surface of the earth. The current density $\mathbf{J}$ of the source is given by

$$\mathbf{J} = \hat{z}\delta(x)\delta(y)\delta(z-h) \tag{18.10}$$

The component π_z can be determined by solving the equation for π_z and applying the following boundary conditions.

$$k_1^2\pi_z^{(1)} = k_2^2\pi_z^{(2)} \tag{18.11}$$

and

$$\frac{\partial\pi_z^{(1)}}{\partial z} = \frac{\partial\pi_z^{(2)}}{\partial z} \tag{18.12}$$

at the interface between the two media 1 and 2. According to Norton[1]

$$\pi_z = \frac{\exp(-jk_1R_1)}{R_1} - \frac{\exp(-jk_2R_2)}{R_2} + [1 + \Gamma_v + (1-\Gamma_v)F]\frac{\exp(-jk_1R_2)}{R_2} \tag{18.13}$$

where Γ_v denotes the reflection coefficient for a vertically-polarised wave. Γ_v is given by

$$\Gamma_v = \frac{(\epsilon_r - j\sigma_2/\omega\epsilon_0)((z+h)/R_2) - \sqrt{(\epsilon_r - j\sigma_2/\omega\epsilon_0) - (1 - (z+h)^2/R_2^2)}}{(\epsilon_r - j\sigma_2/\omega\epsilon_0)((z+h)/R_2) + \sqrt{(\epsilon_r - j\sigma_2/\omega\epsilon_0) - (1 - (z+h)^2/R_2^2)}} \tag{18.13a}$$

The attenuation function

$$F = [1 - j\sqrt{\pi\Omega}\exp(-\Omega)\operatorname{erfc}(j\sqrt{\Omega})] \tag{18.13b}$$

where

$$\Omega = \frac{4p\exp(-jb)}{(1-\Gamma_v)^2} \tag{18.13c}$$

and erfc. signifies complementary error function. The Sommerfeld numerical distance p and the phase constant b are given by

$$p = \frac{k_0d}{[2\epsilon_r^2 + (\sigma_2/\omega\epsilon_0)^2]}, \qquad R \gg \lambda \text{ and } R \gg h \tag{18.13d}$$

---

[1] K. A. Norton, 'The Physical Reality of Space and Surface Waves in the Radiation Field of Radio Antennas', *Proc. I.R.E.* Vol. 25, No. 9, pp. 1192–1202, Sept. 1937.

$$b = \text{arc tan} \left\{ \frac{\omega \epsilon_0 \epsilon_r}{\sigma_2} \right\} \tag{18.13e}$$

where d denotes the horizontal distance between the transmitting and receiving antennas. k_1 and k_2 denoting the propagation constants in media 1 and 2 respectively are given by

$k_1^2 = \omega^2 \mu_0 \epsilon_0,\ k_2^2 = \omega^2 \mu_0 \epsilon_2 - j\omega \mu_0 \sigma_2$

$\epsilon_2 = \text{Re}\ (\epsilon^*),\ \epsilon_r = \epsilon_2/\epsilon_0 = $ dielectric constant of the earth

$\epsilon_0 = 10^{-9}/36\pi$ f/m, $\mu_0 = 4\pi \times 10^{-7}$ h/m

$\omega = $ angular frequency of the wave

$\sigma_2 = $ conductivity of the earth s/m.

For a current filament of elemental length d carrying a uniform current I, the z-component of the electric field intensity is given by

$$dE_z = -\frac{j\omega\mu_0}{4\pi}\ Idl\left[\sin^2 \psi_1\ \frac{\exp\ (-jk_1 R_1)}{R_1} + \Gamma_v \sin^2 \psi_2\ \frac{\exp\ (-jk_1 R_2)}{R_2} \right.$$

$$+ (1 - \Gamma_v)\left\{ 1 - \frac{k_1^2}{k_2^2} + \frac{k_1^4}{k_2^4} \sin^2 \psi_2 \right\} F\ \frac{\exp\ (-jk_1 R_2)}{R_2}$$

$$\left. + \text{electrostatic and induction field terms} \right] \tag{18.14}$$

The vertical component E_z consists of the direct wave radiated into space given by the first term, and the indirect wave representing the effect of the ground and electrostatic and induction fields which can be neglected when $R \gg \lambda$ and the distance d of the observer is large.

The radial field component E_ρ of $\mathbf{E}$ can be determined from the relation

$$E_\rho = \frac{\partial^2 \pi}{\partial \rho\ \partial z} \tag{18.15}$$

The total electric field intensity for the space-wave is

$$E(\text{space-wave}) = [E_z^2\ (\text{space-wave}) + E_\rho^2\ (\text{space-wave})]^{1/2}$$

The E-field of surface-wave can be similarly written. For antennas grounded or located very close to the earth ($h \to 0$), the surface and space-wave fields are given by

$d\mathbf{E}$ (surface-wave)

$$= -\frac{j\omega\mu_0}{4\pi}\ Idl(1 - \Gamma_v)\left[\hat{\rho}\ \frac{k_1^2}{k_2^2} \sin \psi \left(1 + \frac{\cos^2 \psi}{2} \right)\left(\frac{k_2^2}{k_1^2} - \sin^2 \psi \right)^{1/2} + \hat{z} \right]$$

$$\times F\ \frac{\exp\ (-jk_1 R)}{R} \tag{18.16}$$

$$d\mathbf{E}\ (\text{space-wave}) = \hat{\theta}\ \frac{j\omega\mu_0}{4\pi}\ Idl(1 + \Gamma_v) \sin \psi\ \frac{\exp\ (-jk_1 R)}{R} \tag{18.17}$$

where $\hat{\theta}$ denotes the unit vector normal to $\mathbf{R}$. Equations 18.16 and 18.17 hold good for large distances where the approximations $R_1 = R_2 = R$; $\psi_1 = \psi_2 = \psi$ are valid.

The Sommerfeld reduction factor $A = |F|$ reduces the surface-wave field intensity depending on f, d, ϵ_r and σ_2. If the distance d is so small that $F \to 1$, in which case, the surface-wave is unattenuated. Figure 18.3 shows the variation of A with respect to p as a function of b.

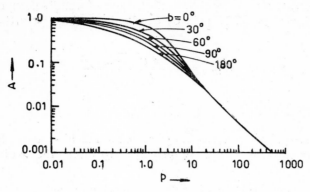

Fig. 18.3 A vs. p as $f(b)$.

A can also be calculated from the following approximate relation.

$$A \simeq \frac{2 + 0.3p}{2 + p + 0.6p^2} - \sqrt{\frac{p}{2}} \sin b \, \exp\,(-0.625p) \qquad (18.18)$$

For $b = 180°$, or $b < 5°$, the attenuation factor A becomes

$$A \simeq \frac{2 + 0.3p}{2 + p + 0.6p^2} \qquad (18.19)$$

The following conclusions can be drawn:

(i) For $b < 5°$, the earth behaves as a resistive impedance.

(ii) For $b = 90°$, the earth behaves as a capacitive impedance.

(iii) The earth which behaves as a conductor at low frequency, giving rise to Joulean heat-loss will behave like a dielectric with small loss.

(iv) For $\Gamma_v = -1$, the space-wave component vanishes and the surface-wave predominates.

(v) For $\Gamma_v = +1$, the surface-wave component vanishes and the space-wave predominates.

(vi) For any value of Γ_v other than ± 1, both the components of the ground-wave exist.

For horizontally-polarised waves, the propagation characteristics are significantly altered. If a horizontal dipole is located immediately above the earth with ($\sigma_2 = \infty$), the radiation field will be cancelled as the image dipole is of opposite polarity. For a finitely conducting earth, however, the radiation field of a horizontal dipole placed above the earth will not be completely cancelled. For horizontally-polarised wave

$$p = \frac{\pi d}{\lambda} \sqrt{(\epsilon_r - 1)^2 + (60\lambda\sigma)^2} \qquad (18.20)$$

For grounded vertically-polarised antennas, the ground-wave field intensity $E_{(rms)}$ at a distance d (km) due to the power P_T (kw) radiated by a transmitting antenna of gain G_T can be determined from the following relation

$$E_{(rms)} = \frac{173\sqrt{P_T G_T}}{d} F \frac{mV}{m} \qquad (18.21)$$

18.3.1   Wave Tilt

For a plane-polarised wave launched over the air-earth interface, the field components for the wave propagating along the x-direction are given by

$$E_{x1} = \frac{E_{z1m}}{[\epsilon_r^2 + (60\sigma\lambda)^2]^{1/4}} \cdot \exp(-j\phi/2)$$

$$E_{z1} = E_{z1m} \qquad (18.22)$$

$$H_{y1} = -\frac{E_{z1m}}{120\pi}$$

in medium 1. In medium 2, the z-component is given by

$$E_{z2} = \frac{E_{z1m} \exp(-j\phi)}{[\epsilon_r^2 + (60\lambda\sigma)^2]^{1/2}} \qquad (18.23)$$

where the phase shift ϕ is given by

$$\phi = \text{arc tan}(60\lambda/\epsilon_r)$$

The other components are given by

$$E_{x2} = E_{x1}, \qquad H_{y1} = H_{y2}$$

There is a phase shift between E_x and E_z depending on σ, ϵ_r of the earth and λ. Hence, the plane-polarised wave becomes elliptically-polarised (Fig. 18.4) as it progresses in the x-direction. Thus, there are both conduction as well as displacement currents. If $\epsilon_r \ll 60\sigma\lambda$, the conduction current predominates, and $\phi \to \pi/2$, whereas, if $\epsilon_r \gg 60\sigma\lambda$, the displacement current becomes more predominant and $\phi \to 0$.

For the usually encountered values of ϵ_r and σ of the earth, the eccentricity of the ellipse is large and hence the wave is almost plane-polarised. The vertical component E_{z1} tips forward towards the ground. Hence, the wavefront is tilted and consequently the Poynting vector $\mathbf{E} \times \mathbf{H}$ points towards the ground, thus maintaining the earth losses. The angle of tilt τ with respect to the normal is given by

$$\tau = \text{arc tan} \frac{E_{x1m}}{E_{z1m}} = [\epsilon_r^2 + 60\lambda\sigma)^2]^{-1/4} \qquad (18.24)$$

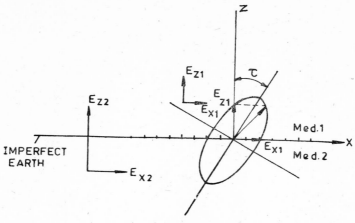

Fig. 18.4   Wave-tilt τ.

For example, if $\epsilon_r = 10$, $\lambda = 150$ m, the wave-tilt $\tau \cong 4°8'$.

The wave-tilt measurement provides a good method for determining soil conductivity.

18.3.2   Spherical Earth

The assumption of plane-earth yields only approximate results for the propagation characteristics of the surface-wave. The results discussed before hold good only for a short distance up to $d = 50/\sqrt[3]{f}$ miles, where f is the MHz. Due to this restriction, broadcast transmission at MW and LW is satisfactory only up to a short distance.

Beyond this distance d, the ground-wave field intensity is affected in several ways due to the curvature of the earth, viz., (i) the plane-wave reflection coefficient of the ground-reflected wave is different for the curved earth than for a plane earth, (ii) since the energy of the ground-reflected wave reflected from the curved earth's surface diverges more than indicated by the inverse-distance law, the ground-reflected wave needs to be multiplied by a divergence factor, (iii) the heights of the transmitting and receiving antennas above the plane tangent to the surface of the spherical earth at the point of reflection of the ground-reflected wave, are less than the heights of the respective antennas above a plane earth. Norton has summarised the results reported by several authors about ground-wave field intensity and has presented equations and curves which simplify the calculation of ground-wave field intensity over a finitely conducting spherical earth for transmitting and receiving antennas of arbitrary heights and polarisation.[1] Figures 18.5 and 18.6 show field intensity variation with distance for poor soil and sea, respectively.

[1]K. A. Norton, 'The Calculation of Ground-wave Field Intensity Over a Finitely Conducting Spherical Earth', *Proc. I.R.E.*, vol. 29, No. 12, p. 623, 1941.

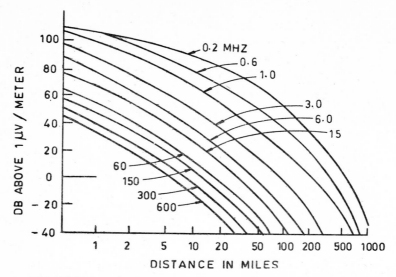

Fig. 18.5   Field intensity for vertical polarisation over poor soil for 1 kW radiated power from a grounded λ/4 vertical (whip) antenna.

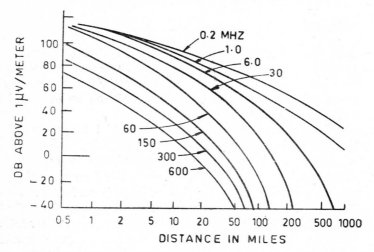

Fig. 18.6   Field intensity for vertical polarization over sea water for 1 kW radiated power from a grounded whip antenna.

18.4   Tropospheric Propagation

The troposphere is the region of the atmosphere extending from the surface of the earth up to a height of 9 to 10 kms at polar latitudes, 10 to 12 kms at the moderate latitudes, and up to 16 to 18 kms at the equator. This region is characterised by decrease of temperature with height except at the region of inversion at some height, the average vertical temperature gradient being 6°C per km. The average temperature of the tropopause (an isothermal region) is $-55°C$ in the polar regions and $-80°C$ at the equator.

The refractive index n of the atmosphere is nearly unity. Consequently, the wavelength λ in the medium is practically the same as λ_0, the wavelength in vacuum. Though the changes in the refractivity $(n - 1)$ with height h are small, nevertheless, they exert significant influence in tropospheric wave propagation.

If we consider n as spherically stratified, being a function of the distance from the earth's centre, refraction or reflection can occur according to the ray theory, provided: (i) the local λ is much less than the radius of curvature of the surfaces of equal phase, and (ii) $\dfrac{dn}{d(z + a)} \dfrac{\lambda}{2\pi} \cos i \ll 1$, where a denotes the actual radius of the earth $(a = 6370$ kms$)$ and i denotes the angle between the ray path and the local vertical.

The refractivity of the atmosphere which consists mainly of air and water vapour mixture is given in terms of the total atmospheric pressure P (mb), where 1 bar $= 1.019$ kgm/cm$^2 \simeq 1$ atmosphere, partial pressure of water vapour e (mb) and temperature (°K) by the following relation

$$n - 1 = \left(79\frac{P}{T} - 10\frac{e}{T} + \frac{3.8 \times 10^5 e}{T^2}\right) \times 10^{-6} \qquad (18.25)$$

For the ray to be refracted towards the earth dn/dz must be negative. The ray follows the earth's curvature when $dn/dz = -0.15 \times 10^{-6}$ per metre.

18.4.1 Modified Refractive Index

Since the earth is spherical, its curvature should be taken into account while developing the theory for propagation. It is usual to assume that the earth is plane and the ray is curved. This enables the problem of diffraction to be treated as a phenomenon of refraction. For this to be valid, the earth is regarded as bigger than its actual size, and hence the actual radius a of the earth is replaced by ka where $k = 4/3$ and $ka \simeq 8500$ km. With this concept for the effective radius of the earth, all the equations developed for propagation through a homogeneous atmosphere can be extended to an inhomogeneous atmosphere by replacing a by ka, and n by the modified index of refraction

$$N = n(1 + z/a) \simeq n + z/a \qquad (18.26)$$

As N differs from unity by about 3 parts in 10,000, it is convenient for many purposes to introduce the term refractive module M which is related to N by

$$M \equiv (N - 1) \times 10^6 \qquad (18.27)$$

18.4.2 Profiles for Modified Index

The atmospheric layers may be classified as:

(i) $dN/dz = 3.6 \times 10^{-8}$ per ft. for standard refraction.

(ii) $0 < dM/dz < 3.6$ per 100 ft and $dM/dz < 0$ (which corresponds to inversion) for superstandard refraction. These conditions are favourable for beyond the line-of-sight communication.

(iii) $dM/dz > 3.6$ per 100 ft for substandard refraction which is unfavourable for communication at UHF.

18.4.3   Formation of Ducts

Figure 18.7 shows two typical N-profiles obtained by experiment in the troposphere. These types of N-profiles help in the formation of tropospheric ducts which trap and guide UHF (300–3000 MHz) and SHF (3–30 GHz) waves to beyond the line-of-sight distance depending on the width of the ducts in the z-direction.

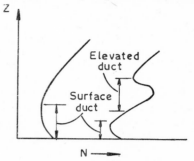

Fig. 18.7   Typical N-profile.

The formation of ducts can be explained with the help of geometrical optics according to which the inclination ρ of the ray at a height z is given by[1]

$$\beta = \pm\sqrt{2}\left[N(z) - N(z_1) + \frac{\alpha^2}{2}\right]^{1/2} \tag{18.28}$$

where α denotes the launching angle (Fig. 18.8) of the ray at a height z_1.

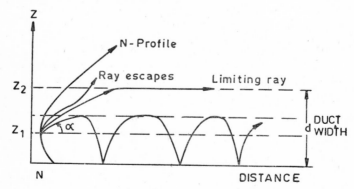

Fig. 18.8   Tropospheric duct-wave guide, d : duct width.

18.4.4   Mode Theory of Duct Propagation

The tropospheric ducts may be considered as analogous to conventional

[1]Donald E. Kerr, *Propagation of Short Radio Waves*, McGraw-Hill Book Co. Inc., New York, 1951.

metallic waveguides with the difference that in the latter case, specular reflection takes place at the walls irrespective of the inclination of the ray, whereas in the former, the boundary-walls being diffuse, the reflection is non-specular and only when α is small, the rays are treated by the mode theory. Booker and Walkinshaw's[1] rigorous treatment to tropospheric propagation reveals that the electric field E at a horizontal distance r and height z with respect to the free-space field E_0 that would exist in a completely homogeneous medium with atmospheric properties at the ground level due to a dipole located at a height z_1 in the presence of the earth is given by

$$\left|\frac{E}{E_0}\right| = \sqrt{r\lambda} \left| \sum_{n=1}^{\infty} \frac{1}{r_n} \exp\left(-j\pi\sigma_n^2 r/\lambda\right) f_n(h_T) f_n(h) \right| \tag{18.29}$$

where r_n denotes the single-hop horizontal range, α_n is the angle of elevation for a ray leaving the ground level corresponding to H_n-waves, $f_n(h_T)$ corresponds to the excitation function depending on the transmitter height h_T and $f_n(h)$ indicates the dependence of the received field on the receiver height.

18.4.5   Duct Cut-off Wavelength

In analogy with the cut-off wavelength of a metallic waveguide, there exists a maximum wavelength λ_m that can be trapped inside a tropospheric duct-guide of width d. The wavelength λ_m is given by

$$\lambda_m \text{ (cm)} = 0.014 d^{3/2} \text{ (ft)} \tag{18.30}$$

For example, a duct-guide of width $d = 80$ ft can trap a wave of maximum wavelength $\lambda_m = 0.1$ m. Similarly, a duct of width $d = 400$ ft can trap a wave of $\lambda_m = 1$ m.

Since, in nature, ducts of width $d > 500$ ft occur very rarely, tropospheric-duct propagation is practicable for UHF and SHF waves.

18.4.6   Tropospheric Scatter Propagation

The phenomenon of scattering occurs when a radio-wave propagates through a turbulent atmosphere. Turbulence in the atmosphere is associated with the velocity gradient of the wind which causes eddies. Turbulent condition is characterised by the Reynolds number $R_0 = \rho l u/\eta$ where ρ denotes the fluid density (gm/cm^2), η and u denote viscosity of the fluid and velocity of the fluid within the eddies of scale length l respectively. Turbulence is produced when $R_0 \gg 1$. $R_0 \cong 10^6$ in the case of the troposphere.

[1]H. G. Booker and W. Walkinshaw, 'The Mode Theory of Tropospheric Refraction and Its relation to Waveguides and Diffraction', in *Meteorological Factors in Radio Wave Propagation*, The Physical Society, and the Royal Meteorological Society, London, 1946.

The magnitude of scattering depends on the fluctuation Δn in n where Δn can be written as

$$\Delta n = \bar{N} + N(\mathbf{r}, t) \qquad (18.31)$$

where $\bar{N}$ represents a slowly varying function of z. N changes much faster than $\bar{N}$. The variation in n can be described by the spatial correlation function $c(R)$ which can be defined by $c(r) = f(r/l_0) = f(\rho)$ where l_0 denotes the average size of the eddies. Different forms, e.g., exponential, Bessel and Gaussian have been used for $c(R)$, i.e., $f(\rho)$ by different authors. The scale length (l), in general, is defined in terms of $c(R)$ as

$$l = \int_0^{\infty} c(r)\, dr \qquad (18.32)$$

which determines to a first approximation the radius of a sphere within which the statistical fluctuation is strong. When R is very small, the variation is synchronous and $c(0) \to 1$. But, for large R, variation at any point within the sphere is independent of the variation at any other point inside the sphere and $c(\infty) \to 0$.

The spatial variation $N(\mathbf{r})$ can also be described in terms of the spectral function $F_n(k)$, where $k = 2\pi/l$. A plot of $F_n(k)$ vs. k (Fig. 18.9) reveals three distinct zones.

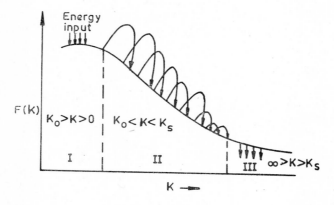

Fig. 18.9 $F_n(k)$ vs. k.

(i) Energy input range $(0 < k < k_0)$ which corresponds to large size eddies $\simeq 100$ m where $k_0 = 2\pi/l_0$ and l_0 denotes the largest size eddies. This is an anisotropic region.

(ii) $k_0 < k < k_s$. This region is the inertial, subrange where $k_s = 2\pi/l_s$, and l_s denotes small-scale eddies $(\simeq 1$ mm$)$. The propagation depends mostly on this region. The energy in this region is conserved. The eddy size decreases to a limiting value.

(iii) $k_s < k < \infty$. In this region, energy of the eddies is dissipated due to viscous force.

18.4.7 Scattered Power

In the turbulent medium each eddy comprises a zone where there is a discontinuity δn in the refractive index or in the gradient of refractive index. All the elementary eddies, each of volume dv, generate elementary radiations due to the incident-energy emanating from a transmitting antenna. These elementary radiations scattered in all directions are not coherent. The power of these incoherent radiations are integrated to get the resultant scattered power.

DuCastel[1] has used the following wave equation for a turbulent medium

$$\nabla^2 \pi_0 + k^2 \pi_0 + \nabla^2 \delta \pi + k^2 \delta \pi + k^2 2 \delta n(\pi_0 + \delta \pi) = 0 \qquad (18.33)$$

of refractive index

$$n(\mathbf{r}) = 1 + \delta n(\mathbf{r}) \qquad (18.34)$$

where π_0 denotes the mean value of the Hertz vector $\pi = \pi_0 + \delta \pi$.

The scattered power P_R received by a receiving antenna of gain G_R due to the power P_T radiated by a transmitting antenna of gain G_T is given by

$$P_R = P_T \frac{\lambda^2}{16\pi^2} \int_V \frac{G_T G_R}{d_T^2 d_R^2} \sigma(\theta) \, dv \qquad (18.35)$$

where d_T and d_R indicate distances of the transmitting and receiving antennas respectively from the centre of scatterer of volume V and scattering cross-section $\sigma(\theta)$ which is given by

$$\sigma(\theta) = \frac{\pi^2}{\lambda^4} \overline{\delta n^2} F\left(\frac{2\pi}{\lambda}, \theta\right), \qquad (18.36)$$

where the spectrum function is $F\left(\dfrac{2\pi}{\lambda}, \theta\right)$ and θ denotes the direction in which the energy is scattered by each elemental volume of the scatterer. The spectrum function depends on the form of the correlation function $c(r)$. Booker and Gordon[2] have used the exponential form for $c(R) = \exp\{-(R/l_0)\}$ for which the turbulence spectrum function $F(k)$ is given by $[8\pi < N^2 > l_0^3]/[1 + k^2 l^2]^2$.

18.4.8 Attenuation

Electromagnetic waves undergo attenuation while propagating through the troposphere due to scattering and absorption by (i) gas molecules, (ii) solids such as dust smoke, etc. suspended in the atmosphere, and (iii) precipitation particles, such as, fog, hail, snow and rain.

[1]Francois DuCastel, *Tropospheric Radio Wave Propagation Beyond the Horizon*, Pergamon Press, London, 1966.
[2]H. G. Booker, W. E. Gordon, "A Theory of Radio Scattering in the Troposphere", *Proc. I.R.E.*, Vol. 38, p. 401, 1950.

Molecular absorption occurs since electromagnetic energy is dissipated to ionise or excite atoms and molecules. The excited atoms and molecules pass from a lower energy to a higher energy state. Since, the energy states are discrete, these transitions are resonant in character, which gives rise to resonant absorption.

The main constituents of dry atmosphere are oxygen and nitrogen which are electrically non-polar and hence no absorption occurs by dipole resonance. However, the oxygen molecule has a permanent magnetic moment which gives rise to absorption by resonance.

The major atmospheric gases that absorb energy are oxygen and water vapour. The total absorption A_1 due to oxygen and water vapour in the atmosphere over a path length l (km) is given by

$$A_1 = \int_0^l [\gamma_{O_2}(r) + \gamma_{H_2O}(r)] \, dr \text{ dB} \tag{18.37}$$

where γ_{O_2} and γ_{H_2O} denote specific attenuation (dB/km) due to O_2 molecules and H_2O vapour respectively depending on temperature, pressure and humidity at a distance r (km) along the path of propagation of the wave. Absorption by O_2 molecules has a broad peak at 60 GHz and a narrow peak at 119 GHz, whereas, the absorption peaks for water vapour occur at 22.5, 183 and 320 GHz. All these resonant peaks are pressure-broadened.

The loss due to molecular scattering can be evaluated in terms of Rayleigh scattering as follows

$$A_2 = \frac{4.34 \times 32\pi^3 \times 10^3}{3N\lambda^4} \times (n_0 - 1)^2 \text{ dB/km} \tag{18.38}$$

where N (Avogadro number) $= 2.69 \times 10^{25}$ per cubic metre. λ is in metres and n_0 normally $= 1.000325$ is the refractive index at the earth's surface. Equation 18.38 shows a rapid decrease in attenuation with increasing wavelength.

The solid particles suspended in the atmosphere may act as nuclei of condensation if their sizes are not greater than 1 micron and, thus, produce attenuation.

18.5 Ionospheric Propagation

At sea-level the atmosphere consists of N_2 (78%), O_2 (20.95%) and Ar, CO_2, Ne, He, H_2 comprising the rest of the percentage by volume. The atmosphere maintains the same percentage of the constituent gases up to a height of approximately 90 km from the surface of the earth. At heights greater than 90 km, the atmosphere is stratified.

The atmospheric pressure p at a height h from the surface of the earth is given by

$$p = p_0 \exp\left(-\frac{hMgm}{RT}\right) \tag{18.39}$$

where

p_0 = pressure at the sea-level

R = universal gas constant = 8.316×10^3 Joules/kg-mole °C

g = acceleration due to gravity = 978.05 cm/sec²

M = mass of 1 kg-mole = 29.0 within the constant constituent region. At h = 130–250 kms, M = 25 and at $h > 250$ kms, M = 15.

m = mass of a molecule

T = temperature °C

In a rarefied atmosphere, energy $(h\nu)$ of solar radiation dissociates nitrogen (N_2) and oxygen (O_2) molecules according to

$$N_2 + h\nu \rightarrow N + N$$
$$O_2 + h\nu \rightarrow O + O \tag{18.40}$$

where

h (Planck's constant) = 6.616×10^{-27} erg-sec

ν = radiation frequency in Hz.

The photon energy $(h\nu)$ causing dissociation of N_2 and O_2 molecules corresponds to λ (ion) = 0.25 and 0.168 microns respectively. In the region of visible spectrum λ (ion) ranges from 0.75 to 0.4 microns. Consequently, X-rays and UV-radiation are responsible for the dissociation of N_2 and O_2 molecules. Ionisation can occur when the radiation frequency $\nu > \nu$ (ionisation), where ν (ion) = ϕ/h Hz, where ϕ denotes the ionisation work-function. For example

$$\phi(O_2) = 12.2 \text{ eV}$$

and

$$\phi(N_2) = 15.51 \text{ eV}.$$

The ionosphere is divided into several layers, viz., (i) D(60–90 km during day-time and disappearing at night, (ii) E(100–140 km), (iii) F_1(180–240 km by day and disappearing at night), and (iv) F_2(230–400 km). The ionisation is caused not only by UV and X-rays, but also by corpuscular radiation emanating from the Sun. The causes of radiation in the D-region are soft X-rays and Lyman-alpha radiation, whereas, in the E-region ionisation is by X-rays, and in the F-region by UV, X-rays and probably corpuscular radiation.

The electron or ion densities are $100–10^3$ for electrons and $10^6–10^8$ for ions in the D-region, up to 10^5 to 4.5×10^5 (day-time), 5×10^3 to 10^4 (night-time) in the E-region, $2 \times 10^5–4.5 \times 10^5$ in the F_1-region and maximum 2×10^6 (daytime in winter)–maximum (daytime in summer)–3×10^5 (nighttime in winter) for the F_2-layer. Figure 18.10 shows electron density profiles.

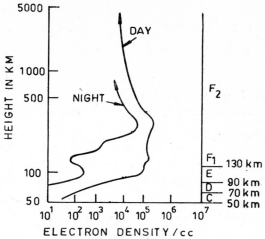

Fig. 18.10   Electron density profiles of ionosphere.

18.5.1   Dielectric Constant and Conductivity

When an electromagnetic wave of angular frequency ω is incident on a homogeneous, isotropic ionised region of density N, the charged particles are set in motion which can be described by the following equation

$$eE = m\frac{d^2z}{dt^2} + \nu mN\frac{dz}{dt} \tag{18.41}$$

where E is the electric field intensity in V/m, ν denotes the collision frequency in Hz. The velocity dz/dt and position z at any time t of the particle can be determined by solving Eq. 18.41.

$$\frac{dz}{dt} = \left[-\frac{e}{m(\omega^2 + \nu^2)} - j\frac{e\nu}{m\omega(\omega^2 + \nu^2)}\right]\frac{dE}{dt}\frac{m}{s} \tag{18.42}$$

which yields z. The total current density J in the ionised region consists of the conduction current density $(=Ne\ dz/dt)$ and the displacement current density $(=\epsilon_0\ dE/dt)$. Hence,

$$J = \left[\epsilon_0 - \frac{Ne^2}{m(\omega^2 + \nu^2)} - j\frac{Ne^2\nu}{m\omega(\omega^2 + \nu^2)}\right]\frac{dE}{dt} = \epsilon^*\frac{dE}{dt} \tag{18.43}$$

which yields

$$\mathrm{Re}\ (\epsilon^*) = \epsilon_0\left[1 - \frac{Ne^2}{m\epsilon_0(\omega^2 + \nu^2)}\right] \tag{18.44}$$

or dielectric constant

$$\epsilon_r = \left[1 - \frac{Ne^2}{m\epsilon_0(\omega^2 + \nu^2)}\right]\frac{F}{m} \tag{18.45}$$

and

$$\mathrm{Im}\ (\epsilon^*) = \sigma\ (\text{conductivity}) = \frac{Ne^2\nu}{m(\omega^2 + \nu^2)}\frac{S}{m} \tag{18.46}$$

Hence, we may conclude that

(i) The ionised region behaves like an imperfectly conducting medium.

(ii) If $\omega^2 \gg \nu^2$ and in the limit $\nu \to 0$, then the conductivity $\sigma \to 0$ and the ionized medium behaves like a dielectric of dielectric constant

$$\epsilon_r = [1 - Ne^2/m\omega^2\epsilon_0] \tag{18.47}$$

(iii) If

$$\nu^2 \gg \omega^2, \qquad \epsilon_r = 1 - \frac{Ne^2}{m\epsilon_0\nu^2} \frac{F}{m} \tag{18.48}$$

and

$$\sigma = \frac{Ne^2}{m\nu} \frac{S}{m} \tag{18.49}$$

18.5.2   The Gyromagnetic Frequency

In the presence of the earth's constant magnetic field (H_e) in the iono-sphere, the electrons describe a complicated trajectory under the influence of the combined electric and magnetic field.

The Lorentz force is given by

$$\mathbf{F} = e\mathbf{v} \times \mathbf{B}_e$$

or

$$F = evB_e \sin \theta, \tag{18.50}$$

where $B_e = \mu_0 H_e$, where θ is the angle between v and B_e. The perpendicular component (v_T) of v is perpendicular to B_e and F. The trajectory of electrons becomes a circle of radius R_H and the angular velocity is ω_H. Equating Lorentz force with the centrifugal force, we obtain,

$$ev_T B_e = mR_H\omega_H^2$$

or

$$eR_H\omega_H B_e = mR_H\omega_H^2 \tag{18.51}$$

which yields

$$\omega_H = \frac{|e|}{m} B_e \tag{18.52}$$

Since

$e = 1.60 \times 10^{-19}$ Coulomb

$m = 9 \times 10^{-31}$ kg

$B_e = 0.5 \times 10^{-4}$ Weber/m^2

therefore

$$f_H = \omega_H/2\pi = 1.40 \text{ MHz} \tag{18.53}$$

f_H is called the gyro-magnetic frequency. Since for a constant value of B_e, f_H is a constant, the rotation of the electron in the magneto-ionic region shows resonance characteristic. Hence, the phenomenon is called gyro-magnetic resonance.

18.5.3   Magneto-Ionic Theory

The equation of motion of electrons in the magneto-ionic ionosphere can be written as

$$e\mathbf{E} + e\ \frac{\partial \mathbf{r}}{\partial t} \times \mathbf{B}_e = m\ \frac{\partial^2 r}{\partial t^2} + m\nu\ \frac{\partial \mathbf{r}}{\partial t} \tag{18.54}$$

(Force due   (Lorentz   (Centrifugal   (Collision
to **E**-field)   Force)   Force)   damping)

Assuming the time factor as $\exp(-j\omega t)$ and introducing the following notations:

$$X = \frac{Ne^2}{\omega^2 m \epsilon_0}, \qquad Y = \frac{e}{m\omega} B_e, \qquad Z = \nu/\omega$$

$$\mathbf{P} \text{ (Polarisation vector)} = -\frac{Ne^2}{m} \frac{1}{\omega^2 + j\nu}$$

(18.55)

the equation of electron motion (18.54) is reduced to

$$-\epsilon_0 X \mathbf{E} = \mathbf{P}(1 + jZ) - j\mathbf{P} \times \mathbf{Y}$$

(18.56)

which can be written in a 3×3 matrix form by resolving the vector $\mathbf{P}$ and $\mathbf{E}$ into Cartesian coordinates.

Choosing the spatial variation of field components as $\exp(jknz)$ for a plane wave $\left(\frac{\partial}{\partial x} = 0, \frac{\partial}{\partial y} = 0, \frac{\partial}{\partial z} = jkn \right)$ progressing in the z direction and choosing the x, y axes such that B_e is parallel to the x-z plane and if l, m, n are the direction cosines of Y, the constitutive relation given by 3×3 matrix reduces to

$$\begin{pmatrix} P_x \\ P_y \\ P_z \end{pmatrix} \begin{pmatrix} U & -jnY & 0 \\ +jnY & U & -jlY \\ 0 & +jlY & U \end{pmatrix} = -\epsilon_0 X \begin{pmatrix} E_x \\ E_y \\ E_z \end{pmatrix}$$

(18.57)

where $m = 0$, $U = 1 + jZ$.

Combining the matrix Eq. 18.57 with Maxwell's equations, the polarisation quadratic Eq. 18.58 of the magneto-ionic theory is obtained as

$$\rho^2 + \frac{jY_T^2}{Y_L(1 - X + jZ)} \rho + 1 = 0$$

(18.58)

where ρ is defined as the ratio $\rho = P_y/P_x = E_y/E_x$. The transverse and longitudinal components of Y are $Y_T(=lY)$ and $Y_L(=nY)$.

Since

$$P_x = \epsilon_0(n^2 - 1)E_x$$

(18.59)

which can be obtained from Maxwell's equations, where n denotes the refractive index of the ionosphere, the relation between the parameters X, Y, Z of the ionosphere and n is given by

$$n^2 = 1 - \frac{X}{\left[1 + jZ - \frac{Y_T^2/2}{1 - X + jZ} \pm \left\{ \frac{Y_T^4/4}{(1 - X + jZ)^2} + Y_L^2 \right\}^{1/2} \right]}$$

(18.60)

which is the *Appleton-Hartree relation*.

The following conclusions can be drawn from this analysis:

(i) In the presence of the earth's magnetic field, the ionosphere behaves as an anisotropic medium. This manifests itself in that the permittivity of the ionosphere becomes a tensor.

(ii) The quadratic nature of the refractive index n indicates that the wave travelling in the ionosphere undergoes double refraction.

(iii) The $\pm$ signs in Eq. 18.60 signify the splitting of the wave into two components with different values for n. These waves are called ordinary (with +ve sign) and extraordinary (with −ve sign).

18.5.4   Secant Law

Figure 18.11 shows the stratified ionosphere such that the ion density N_i ($i = 1, 2, 3, \ldots$) satisfies

$$0 < N_1 < N_2 \ldots < N_n < N_{n+1} \tag{18.61}$$

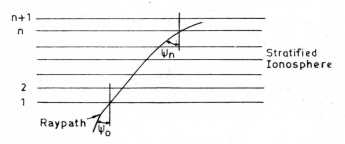

Fig. 18.11   Reflection of waves in a stratified ionosphere.

It is easy to prove that

$$n = [1 - 80.8N/f^2]^{1/2} \tag{18.62}$$

Therefore,

$$1 > n_1 > n_2 \ldots > n_n > n_{n+1} \tag{18.63}$$

From Snell's law we obtain

$$\sin \psi_0 = n_1 \sin \psi_1 = \ldots = n_n \sin \psi_n \tag{18.64}$$

where n_n and ψ_n denote respectively the refractive index at the nth layer and angle of incidence with respect to the local normal at the nth layer.

(i) If the angle of incidence $\psi_n = 90°$, the ray becomes horizontal at the nth layer and $\sin \psi_0 = n_n$, i.e.,

$$\psi_0 = \text{arc} \sin [1 - 80.8N/f^2]^{1/2} \tag{18.65}$$

(ii) If $\psi_0 = 0°$, i.e., the ray incident at the ionosphere is vertically upwards,

$$f_v^2 = 80.8 \, N \tag{18.66}$$

This is the *condition for a vertically-directed ray to return to the earth.*

(iii) From Eqs. 18.62, 18.64 and 18.66 we obtain

$$\sin \psi_0 = \left(1 - \frac{f_v^2}{f^2}\right)^{1/2} \quad \text{or} \quad \cos^2 \psi_0 = f_v^2/f^2$$

which yields the *secant law*

$$f = f_v \sec \psi_0 \tag{18.67}$$

which states that a wave of frequency f projected obliquely into the iono-sphere will return to the earth from the same layer as that of a vertically projected beam.

18.5.5 MUF and Critical Frequencies

The equation $\sin \psi_0 = n_n$ indicates that if N for a layer remains cons-tant, the wave can be made to return to the earth by adjusting ψ_0. But due to the earth's curvature as shown in Fig. 18.12, there is a limiting

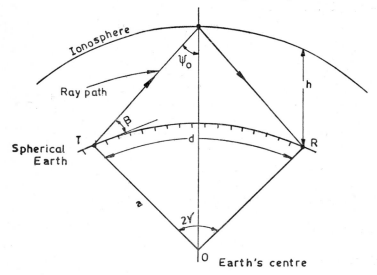

Fig. 18.12   Spherical earth of radius 'a' effect of earth's curvature.

maximum value of ψ_0 which is given by

$$\sin \psi_{0(max)} = a \,|\, (a + h) \qquad (18.68)$$

where h denotes the height of the ionosphere from the surface of the earth where the receiver is located. Since

$$\sin \psi_0 = \cos \beta / (1 + h/a) \qquad (18.69)$$

we obtain

$$\cos^2 \beta \cong \left(1 + \frac{2h}{a}\right)\left(1 - \frac{80.8 N_n}{f^2}\right)$$

which yields

$$f = \left[\frac{80.8 N_n (1 + 2h/a)}{\sin^2 \beta + 2h/a}\right]^{1/2} \text{ Hz} \qquad (18.70)$$

For $N_n = N_{max}$, $f \to f_{max}$ which occurs for $\beta = 0°$, i.e., when the ray is tan-gential to the earth's surface. The maximum usable frequency (MUF) is given by

$$f_{max} (\beta = 0°) = \left[\frac{80.8 N_{max}(a + 2h)}{2h}\right]^{1/2} \qquad (18.71)$$

If the ray is projected vertically upwards, i.e., if $\beta = \pi/2$, the critical frequency is given by

$$f_{\text{crit}}(\beta = \pi/2) = (80.8 N_{\max})^{1/2} \qquad (18.72)$$

Critical frequency is the limiting frequency for vertically-projected waves to be returned to earth by the ionosphere. Figure 18.13 shows a typical variation of f_c. The relation between the MUF and critical frequency is given by

$$f_c = f_{\text{MUF}}(2h/a + h)^{1/2} \qquad (18.73)$$

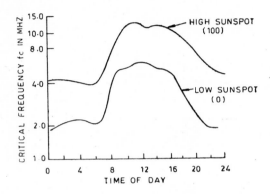

Fig. 18.13   Typical diurnal variation of critical frequency for January at Latitude 40°.

18.5.6   Skip Distance

Since (Fig. 18.12) the angle

$$2\gamma = \pi - 2\psi_0 - 2\beta \qquad (18.74)$$

the horizontal distance d between the transmitter and the receiver is given by

$$d = a(\pi - 2\psi_0 - 2\beta) \qquad (18.75)$$

since

$$\cos \psi_0 = \left[1 - \frac{a^2 \cos^2 \beta}{(a+h)^2} \right]^{1/2} \cong \left(\frac{a \sin^2 \beta + 2h}{a + 2h} \right)^{1/2} \quad \text{for } h \leqslant a \quad (18.76)$$

As β decreases (Fig. 18.12), d increases and hence in the limit $\beta \to 0°$, $d \to d_{\max}$. Equation 18.76 yields for $\beta = 0°$

$$\cos_{\substack{\psi_0 \\ (\text{limit } \beta \to 0°)}} = [2h/(a + 2h)]^{1/2} \qquad (18.77)$$

Therefore, from Eq. 18.75, we obtain

$$d_{\max}(\beta = 0°) = 2a \left(\frac{\pi}{2} - \psi_0 \right) = 2a \left[\frac{\pi}{2} - \text{arc cos} \left(\frac{2h}{a + 2h} \right)^{1/2} \right] \cong \sqrt{8ah} \qquad (18.78)$$

as the *skip distance*. For example, the skip distances for the

E-region ($h = 120$ km) : d_E(skip) $\simeq 1200$ km

and

F-region ($h = 400$ km) : d_F(skip) $\simeq 3000$ km

18.5.7   Ionospheric Phenomena

(i) *The virtual height*:   The virtual height h' (Fig. 18.14) of reflection of a pulse of radio-waves from the ionosphere is an important parameter involved in ionospheric soundings. The virtual height h' is related to the real height h by the following relation

$$h' = \int_0^h \mu' \, dh \tag{18.79}$$

where the group index $\mu' = c/v_g$ where v_g denotes the group velocity.

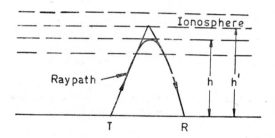

Fig. 18.14   Actual and virtual heights.

(ii) *Absorption*:   Due to collisions between electrons and other particles such as molecules, the ionosphere absorbs energy from the passing electromagnetic wave. If the complex refractive index of the ionosphere is denoted by n^*, the absorption index k is given by

$$k = \frac{\omega}{c} \, \text{Im} \, (n^*) = \frac{e^2}{2m\epsilon_0 c \, \text{Re} \, (n^*)} \frac{N\nu}{\omega^2 + \nu^2} \tag{18.80}$$

(a) For Re $(n^*) \approx 1$ which is the case for no refraction

$$k \cong 4.6 \times 10^{-2} \frac{N\nu}{\omega^2 + \nu^2} \text{ dB/km} \tag{18.81}$$

(b) For $\omega^2 \gg \nu^2$

$$k \cong 1.15 \times 10^{-3} \frac{N\nu}{f^2} \text{ dB/km} \tag{18.82}$$

(c) For $\nu^2 \gg \omega^2$

$$k \cong 4.6 \times 10^{-2} \frac{N}{\nu} \text{ dB/km} \tag{18.83}$$

(d) The total absorption (L) is obtained by integrating k over the ray-path s, and is given by

$$L = 8.68 \int_s k \; ds \; \text{dB} \tag{18.84}$$

where k is in Neper per unit length. Figure 18.15 shows ionospheric absorption with frequency.

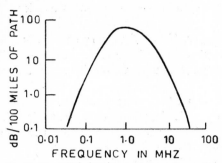

Fig. 18.15   Mid-day ionospheric absorption.

(iii) *Group and phase velocities*: In a non-dispersive medium, electromagnetic waves travel with phase velocity v_p i.e., the velocity with which an equiphase surface travels in the direction of the wave normal. In this case, $v_p = c \mid \sqrt{\epsilon_r}$ m/s. But in the case of the ionosphere $\epsilon_r < 1$, consequently $v_p > c$.

In a dispersive medium, the velocity is a function of frequency. Such a signal can be composed of an infinite sum of harmonic components which travel with different velocities. The velocity of the composite wave is defined by introducing the concept of group velocity v_g i.e., the velocity of propagation of the crest of a group of interfering waves where the components have slightly different frequencies and v_p. v_g is defined by

$$v_g = \frac{c}{n + f\dfrac{dn}{df}} \; \text{m/s} \tag{18.89}$$

where $n = \text{Re} \; (n^*) = \sqrt{\epsilon_r} = [1 - 80.8N/f^2]^{1/2}$.

Therefore,

$$v_p v_g = \frac{c^2}{1 - 80.8\dfrac{N}{f^2} + \dfrac{f}{2} \, 80.8 \times \dfrac{2N}{f^3}} \cong c^2 \, \frac{m^2}{s^2} \tag{18.90}$$

A knowledge of v_p enables the treatment of reflection and refraction phenomena, whereas v_g is useful for measuring the delay of waves reflected from the ionosphere.

18.5.8 *Experimental Observations*

Some of the experimental observations of ionospheric phenomena are summarised as follows:

(i) During the day and seasons of the year the virtual height h' for the E-layer remains between 110 to 120 kms. The critical frequency f_c of the E-layer varies during the day and seasons of the year. f_c (E-layer) $\propto$ (cos ψ)$^{1/4}$, where ψ is the Zenith angle of the Sun.

(ii) For the F_1-layer $h' \simeq 225$ kms irrespective of the seasonal variations, but f_c (F_1-layer) undergoes seasonal variations in a similar way as the f_c (E-layer).

(iii) For the F_2-layer, h' and f_c undergo variations during the day and seasons of the year. f_c (during the summer season) $< f_c$ (during the winter season). h' (during the summer) varies between 300 to 400 kms but in winter $h' \simeq 225$ kms.

(iv) Sometimes during the summer season, the waves get refracted from the E-layer even for $f > f_c$. This is called a sporadic E-layer phenomenon.

(v) Due to solar bursts, ionisation below the E-layer increases to a high value giving rise to a signal fade-out.

(vi) Sunspots appearing as vortices in the visible surface of the sun and corpuscular radiation emanated from the sun, influence ionospheric propagation significantly. The intensity of sunspot activity is measured by 'Wolf Sunspot number'. The increase in the sunspot number increases N for the E- and F-layers and thus improves the propagation characteristics as $f_c(E)$ and $f_c(F)$ are increased.

(vii) Disturbances in the earth's magnetic field create ionospheric storms which lower $f_c(F_2)$, thus impairing long-distance communication.

(viii) Rapid multipath fading occurs on ionospheric transmission, and the amplitude of the fast fading follows the Rayleigh distribution.

(ix) As the frequency increases above the MUF, the signal level decreases rapidly but does not completely disappear. Reliable transmission can be obtained at frequencies up to 50 MHz or higher and to distance of 1200 to 1500 miles. This is due to ionospheric scatter which is the result of turbulence giving rise to many patches of ionisation in the E-layer.

18.5 Noise

The usefulness of a received radio signal is limited by the noise in the receiver. This noise may be due to either unwanted external interference or the circuit noise in the receiver itself.

At frequencies below a few MHz, atmospheric static-noise is important while the circuit noise is the primary noise between 200 to 500 MHz. There is also man-made noise and cosmic noise. The theoretical minimum circuit noise is caused by the thermal agitation of the electrons and is 204 dB below one watt/cycle of bandwidth at usual atmospheric temperatures. The first circuit noise is usually higher than the theoretical minimum by a factor known as the noise figure. For example, the set noise in a receiver with a 6 kHz noise bandwidth and an 8 dB noise figure is 158 dB below one watt which is equivalent to 0.12 μV across 100 ohms.

Atmospheric static noise is caused by lightening and other natural electrical disturbances and is propagated over the earth by ionospheric transmission.

Figure 18.16 shows the typical average values of different types of noises in a 6-kHz band.

Man-made noise is primarily ignition noise while cosmic noise is thermal-type-interference of extraterrestrial origin.

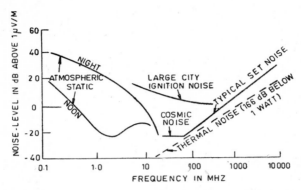

Fig. 18.16.   Typical average noise levels in a 6-kHz band.

PROBLEMS

1. An electromagnetic wave is propagated in dry soil of dielectric constant 5 and conductivity 10^{-3} s/m(mhos/m). Calculate the distance over which the amplitude of the field is decreased to 10^{-7} of its original value.

2. A transmitter radiates 100 watts of power at a wavelength 15 cm from a transmitting antenna of height 20 m and gain 50. The receiving antenna of height 5 m is placed at a distance of 15 km. The electromagnetic wave is vertically polarised and travels over a moist soil. Determine the (i) electric field intensity at the point of reception, and (ii) the attenuation function.

3. An electromagnetic wave of frequency 2 MHz is propagated over a moist soil of dielectric constant 12 and conductivity 10^{-1} s/m. The value of the vertical component of the electric field is specified. Calculate the other components for air and ground. Calculate also the wave tilt.

4. Assuming that the earth's diameter is 12,740 km, find the variation of refractive index which gives rise to a ray travelling parallel to the earth's surface.

5. If two aircrafts are 6 km apart and are flying low over the sea, their height above the water being 9 m, calculate the field strength at the second aircraft if the first one radiates 600 watts of power at a frequency of 400 MHz. The transmitting antenna has a gain of 3 over a half-wave dipole.

6. An electromagnetic wave of power 25 kW and wavelength 1000 m is radiated from an antenna of gain 1.5 and is travelling over an earth of dielectric constant 15 and conductivity 10^{-2} s/m. Calculate the field strength at a distance of 200 km from the transmitting antenna.

7. An electromagnetic wave of wavelength 200 m is propagated over a ground of dielectric constant 3 and conductivity 2×10^{-3} s/m for the first 50 km and then over sea of conductivity 4 s/m. Plot the attenuation function over a distance of 20 to 150 km.

8. Prove that the refractive modulus M gradient with respect to height in the troposphere of refractive index n is given by the relation

$$\frac{dM}{dh} = \left(\frac{dn}{dh} + \frac{1}{a} \right) \times 10^6$$

    where a denotes the radius of the earth.

9. If the permittivity of the troposphere is given by $\epsilon = \epsilon_0 + \delta\epsilon$, prove that the wave equation in the medium due to irregularities in the refractive index assumes the form given by

$$\nabla^2 \mathbf{E} + \omega^2 \mu_0 (\epsilon_0 + \delta\epsilon) \, \mathbf{E} = \operatorname{grad} \left(-\frac{E}{\epsilon} \operatorname{grad} \epsilon \right)$$

    Use Maxwell's equation.

10. Prove that the effective earth's radius factor k is given by

$$k = \frac{1}{1 + a\dfrac{dN}{dh} \times 10^{-6}}$$

11. In a turbulent tropospheric medium if the radiated power is P_1 watts from a transmitting antenna of gain G_T and the wavelength is λ, prove that the scattered power received by an antenna of gain G_R located at a distance d is given by

$$P_{SR} = \frac{P_1 G_T G_R \lambda^2 F^2}{(4\pi d)^2} \text{ watts}$$

    where F denotes the spectrum function.

12. Discuss the different causes of attenuation and fading in tropospheric propagation.

13. Use Maxwell's equations to deduce the main features of a plane linearly-polarised electromagnetic wave in a homogeneous medium. Explain the significance of 'refractive index' and 'characteristic impedance' for such a medium. Explain the significance of the complex values of these quantities.

14. The refractive index n in the ionosphere is given by

$$n^2 = 1 - \beta(z - h) \quad \text{for } z \geqslant h$$
$$n^2 = 1 \qquad\qquad\quad \text{for } z \leqslant h$$

    where β is a real constant. A transmitter sends up a wave packet to the ionosphere at an angle ψ to the vertical. Neglecting the earth's magnetic field, find the distance at which the wave returns to the earth's surface.

15. If the earth's magnetic field is vertical and electron-collisions are neglected, prove that for waves travelling vertically, the refractive index n and group refractive index n' are related as follows

$$nn' = 1 - \frac{XY}{2(1 + Y)^2}$$

16. A linearly-polarised plane electromagnetic wave of angular frequency ω travels in a homogeneous medium containing N free electrons per unit volume with its wave-normal and its electric vector both perpendicular to the magnetic field which is constant and superimposed in the medium. Neglecting collision-damping, prove that

$$n^2 = 1 - \frac{X(1 - X)}{1 - X - Y^2}$$

where n denotes the refractive index of the medium.

17. The electron densities of D, E and F layers in the ionosphere are 400, 5×10^5 and 2×10^6 electrons/cc respectively. Calculate the dielectric constant of the three layers, if an electromagnetic wave of frequency 30 kHz is propagated through the layers.

18. An electromagnetic wave of frequency 12.73 kHz is incident in the F-layer at an angle of 45° which is equal to the critical frequency of the layer. Calculate the (i) MUF and (ii) skip distance.

19. Prove that the product $v_p v_g$ of an electromagnetic wave of frequency f in an ionised medium of ion density N is given by

$$v_p v_g = \frac{c^2}{1 - 80.8 \frac{N}{f^2} + \frac{f}{2} 80.8 \times \frac{2N}{f^3}} \frac{m^2}{s^2}$$

20. Prove that the relative permittivity of the ionosphere treated as a perfect dielectric is given by

$$\epsilon_r = 1 - 80.8 N / f^2$$

where N denotes the electron density. Neglect the collisions between electrons and neutral molecules.

SUGGESTED READING

1. Francois Du Castel, Translated and Edited by E. Sofaer, *Tropospheric Radiowave Propagation Beyond the Horizon*, Pergamon Press, 1966.

2. Kenneth Davies, *Ionospheric Radio Propagation*, National Bureau of Standards Monograph 80, United States Department of Commerce, 1965.

3. K. G. Budden, *Radiowaves in the Ionosphere*, Cambridge University Press, 1961.

PROBLEMS IN ANTENNA DESIGN

4.5 A Vertical Broadcast Antenna

Problem

Design a grounded vertical aerial of height h carrying a loop current $I = 10$ amp at 1000 kHz to produce a field intensity $E=10$ mV/m at a distance of $d=115$ km. The major lobe should point in a direction $\theta=85°$ measured from the aerial axis. Assume the ground to be perfectly conducting.

Solution

For a grounded vertical antenna, the radiation pattern is given by the relation

$$E = \frac{60I}{d} \left[\frac{\cos (\beta h \cos \theta) - \cos (\beta h)}{\sin \theta} \right] \text{ volts/m} \qquad \text{(A.4.43)}$$

where $\beta = 2\pi/\lambda$, $\lambda = 300$ m.

Since all the data except h are given, the design requires the determination of height h of the aerial.

Given $d = 115 \times 10^3$ m, $E = 10^{-2}$ V/m, $I = 10$ amps, $\theta = 85°$, height h is determined by solving

$$\cos (\beta h \cos \theta) - \cos (\beta h) = \frac{Ed}{60I} \sin \theta \qquad \text{(A.4.44)}$$

$$\sin 85° = 0.9962, \qquad \cos 85° = 0.0872$$

For $h/\lambda = 0.55$ Eq. A.4.44 yields

$$\text{LHS} = \cos (0.301) - \cos (3.454) = 1.9066 \cong 1.91$$

$$\text{RHS} = 1.909 \cong 1.91$$

Therefore, $h = 300 \times 0.55 = 165$ m satisfies Eq. A.4.44.

In designing a broadcast antenna, the aim should be to avoid the effect of fading which results from the interaction between the ground wave and the wave reflected from the ionosphere. Fading can be minimised by choosing the height of the antenna such that the radiation pattern lies almost parallel to the ground. The optimum height in wavelength h/λ should be between 0.5 to 0.6 in order to minimise the upward radiation. Hence, the design calculation giving $h/\lambda = 0.55$ conforms to the condition of ground-wave propagation and reception without fading.

4.6 Capacitive-Top-Loaded Vertical Grounded Antenna

The radiation resistance of a vertical antenna can be increased by increasing the effective height which can be done either by increasing the actual height or by changing the current distribution in the antenna so as to make it uniform. The current distribution can be made uniform either totally or partially by loading the antenna at the top. It is usual to provide a capacity top to a vertical grounded antenna in order to increase the effective height. The capacity top may be in the form of a ring (Fig. A.4.8). The field intensity for a vertical grounded antenna with top

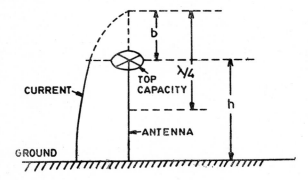

Fig. A.4.8   Capacitive top-loaded vertical antenna
$\lambda/2 < (h + b) > \lambda/4.$

capacitive loading is given by

$$E = \frac{60I}{d}\left[\cos\frac{2\pi b}{\lambda}\cos\left(\frac{2\pi h}{\lambda}\cos\theta\right) - \cos\theta\sin\frac{2\pi b}{\lambda}\sin\left(\frac{2\pi h}{\lambda}\cos\theta\right)\right.$$
$$\left. - \cos\frac{2\pi}{\lambda}(h+b)\right]\frac{1}{\sin\theta} \text{ volts/m} \qquad\qquad (A.4.45)$$

where

$$I \text{ (loop current)} = \text{base current}\left/\left[\sin\left(2\pi\frac{h+b}{\lambda}\right)\right]\right.$$

θ = angle measured from the axis of the antenna

b = reduction in the height of the antenna due to loading

Problem

Design a vertical grounded antenna with a capacitive top carrying a loop current = 10 amps at 1000 kHz to produce a field intensity 10 mV/m at

a distance   115 km. The major lobe should point in a direction 85° mes-
ured from the aerial axis. Assume that the earth is perfectly conducting.
The capacitive loading is such that $b/\lambda = 0.068$.

Solution

Given $I = 10$ amp, $d = 115$ km, $\lambda = 300$ m, $b/\lambda = 0.068$, $\theta = 85°$.
Hence the design involves the determination of h.
Equation (A.4.45) can be written as

$$\frac{Ed \sin \theta}{60I} = \left[\cos \frac{2\pi b}{\lambda} \cos \left(\frac{2\pi h}{\lambda} \cos \theta\right) - \cos \theta \sin \frac{2\pi b}{\lambda} \sin \left(\frac{2\pi h}{\lambda} \cos \theta\right)\right.$$

$$\left. - \cos \frac{2\pi}{\lambda} (h + b)\right] \qquad \text{(A.4.46)}$$

Substituting appropriate values, we obtain

$$\text{LHS} = 1.909.$$

If $h/\lambda = 0.44$, we obtain RHS $= 1.9087 \cong 1.909$.
Hence Eq. A.4.46 is satisfied for $h/\lambda = 0.44$, i.e., $h = 132$ m. Note that
the antenna height without capacitive loading (previous example) is
165 m. Hence, we conclude that top loading an antenna permits the
reduction of the actual antenna height.

6.14   Linear Array

Problem

Design a short-wave antenna array consisting of eight vertical elements
spaced half-wavelength apart to yield resultant amplitude of the field
$R = 8$ relative to that of an individual element of the array. The distant
point at which the amplitude is measured lies on a line making an angle θ
with the array-axis.

Solution

The design involves the determination of the phase difference between
any two adjacent elements of the array which is given by

$$\phi = \frac{2\pi d \cos \theta}{\lambda} \qquad \text{(A.6.87)}$$

where d denotes the spacing ($d = \lambda/2$) between any two adjacent elements
of the array.
Since for n elements in an array

$$R = \sin \left(\frac{n\pi d \cos \theta}{\lambda}\right)\bigg/ \sin \left(\frac{\pi d \cos \theta}{\lambda}\right), \qquad \text{(A.6.88)}$$

for $n = 8$, R reduces to

$$R = \sin (4\phi)/\sin (\phi/2) \qquad \text{(A.6.89)}$$

For $R = 8\phi \to 0°$, i.e., $\theta = 90°$, i.e., the maximum of radiation points at right angle to the array axis. Hence it acts as a broadside array.

7.8 Loop Aerial

Problem

Design a loop aerial of 0.25 sq. meter so that a series emf 1.5 mV is induced in it when it is pointed in the direction of maximum response and is illuminated by an incoming wave of wavelength 400 metres producing field intensity 19.1 mV/m. Comment on the efficiency of the aerial. The loop is placed well above the ground so that its radiation-pattern characteristic is not affected by the ground reflection.

Solution

When the loop is pointed in the direction of maximum response, the emf (E) induced in the loop is given by

$$E = \frac{2\pi}{\lambda} \text{ FNA volts} \tag{A.7.84}$$

where

$\lambda = 400$ metres

$A = 0.25$ m^2

$F = 19.1$ mV/m $= 19.1 \times 10^{-3}$ V/m

$E = 1.5$ mV $= 1.5 \times 10^{-3}$ V

Hence the design involves the determination of the number of turns of the wire (N) which is given by

$$N = E\lambda/2\pi \text{ FA} \tag{A.7.85}$$

Substituting the given data into Eq. A.7.85 we obtain

$$N = 20 \text{ turns of wire}$$

The remark about the efficiency of the antenna requires a knowledge of the radiation resistance (R_r) of the loop which is given by

$$R_r = 31200 \times \frac{N^2 A^2}{\lambda^4} = 0.3 \times 10^{-4} \ \Omega.$$

Hence the antenna is very inefficient.

8.9 Design Data for Helical Antenna Operating in Axial Mode

The axial mode in helical antennas is generated by a coaxial feed (Fig. A.8.16), the inner conductor of which is connected to one end of the antenna and the outer conductor is connected to the ground plane.
The symbols used in describing the geometry of the helical antenna are:

L = length of one turn of the helix

D_h = diameter of the helix

S = spacing (centre to centre) between any two adjacent turns

N = total number of turns in the helix

α = pitch angle

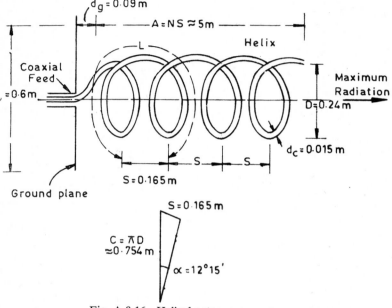

Fig. A.8.16 Helical antenna geometry.

$A = NS$ (axial length of the helix)

d_c = diameter of the helix conductor

D_G = diameter of the ground plane

d_g = distance of the helix proper from the ground plane

The design parameters for the helix operating between 300 MHz to 500 MHz in axial mode are given by[1]

$d = 0.02\lambda$ $d_g = 0.12\lambda$

$D = 0.32\lambda$ $c = \pi D = 1.005\lambda$

$D_G \geqslant 0.8\lambda$ $\alpha = \arctan \dfrac{S}{c}$ = pitch angle

$S = 0.22\lambda$ $R = 140\dfrac{c}{\lambda}$ ohms = terminal resistance

Axial ratio for maximum directivity of the polarisation in the axial direction of the helix $(AR) = 2N + 1/2N$.

---

[1]John D. Kraus, 'Helical Beam Antenna Techniques', *Communications*, Sept. 1949, p. 6.

N = total number of turns of the helix. The performance characteristics of the helix are

$$\text{Gain} \cong 15(c/\lambda)^2 \frac{NS}{\lambda}$$

Half power beamwidth of the radiation pattern $\beta = \dfrac{52}{\dfrac{c}{\lambda}\sqrt{\dfrac{A}{\lambda}}}$ degrees

Beamwidth between the two nulls of the major lobe $= \delta = \dfrac{115}{\dfrac{c}{\lambda}\sqrt{\dfrac{A}{\lambda}}}$ degrees

Problem

To construct a broadband (300 MHz to 500 MHz) helical antenna operating in the axial mode and having a gain of 20 dB. Find the helix parameters and also its performance characteristics.

Solution

Centre frequency = 400 MHz. The corresponding wavelength is $\lambda = 0.75$ m.

$$\alpha = 12°15'$$

$d_c = 0.015$ m	$N \cong 30.3$ since $G = 100$ (gain)
$D = 0.24$ m	$A = 5$ m
$D_G = 0.6$ m	$AR = 1.0165$
$S = 0.165$ m	$R \cong 140\ \Omega$
$d_g = 0.09$ m	$\beta = 20.28°$
$c \cong 0.754$	$\delta = 44.82°$

10.8   Pyramidal Horn

Problem

Design two pyramidal horns each of gain 20 dB to operate at (a) 8 GHz and fed from RG-51/U rectangular waveguide in TE_{01}-mode and (b) 25 GHz fed from RG-66/U rectangular waveguide in the same mode as above.

Solution

We will follow the method outlined by Stutzman and Thiele[1]. Let

a = broad dimension (width) of the rectangular waveguide in cms
b = height of the rectangular waveguide in cms
A = dimension of the pyramidal horn in the X-direction (cms)
B = dimension of the horn in the Y-direction (cms)

---

[1]Warren L. Stutzman and Gary A. Thiele, *Antenna Theory and Design*, John Wiley & Sons Inc., 1981.

l_E, l_H = slant lengths (cms) in the E and H planes of the horn respectively

R_E, R_H = the lengths directly from the apex to the mouth of the horn such that $R_E = R_H$ (see Fig. A.10.11)

G = gain of the horn

The procedure is as follows:

(i) Given a, b, G, determine σ from the following equation:

$$\left| \sqrt{2\sigma} - \frac{b}{\lambda} \right|^2 (2\sigma - 1) = \left(\frac{G}{2\sqrt{2\pi}} \frac{1}{\sqrt{\sigma}} - \frac{a}{\lambda} \right)^2 \left(\frac{G^2}{18\pi^2} \frac{1}{\sigma} - 1 \right)$$

Before solving this transcendental equation, it is convenient to make a trial solution for σ using the relation

$$\sigma_1 = G/2\pi\sqrt{6}$$

(ii) Find the slant length $l_E = \sigma\lambda$ cms.

(iii) Find the horn dimension $B = \sqrt{2l_E}$ cms.

(iv) Find the horn dimension $A = 2\lambda^2 G/4\pi B$ cms.

(v) Find the slant length $l_H = A^2/3\lambda$.

(vi) Find $R_E = (B - b)[(l_E/B)^2 - 0.25]^{1/2}$.

(a) Solution for $\sigma = 6.1644$ obtained by using this procedure.

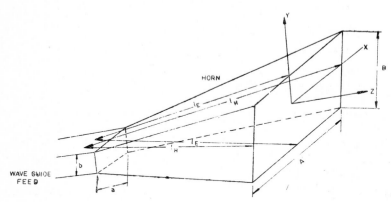

Fig. A.10.11. Pyramidal horn (X and K bands).

Nomenclature of the standard waveguide: RG-51/U, (Brass)

Inside dimensions: a (width) = 2.85 cms b (height) = 1.26 cms

Outside dimensions: a = 3.175 cms b = 1.588 cms

Recommended frequency range (TE$_{01}$-mode):   6.57 to 9.99 (GHz) (X-band)

l_E = 23.116 cms, l_H = 25.65 cms, A = 16.99 cms,
B = 13.167 cms, R_H = 20 cms,

(b) For RG-66/U waveguide (Brass), the

Inside dimensions: $a = 1.067$ cms, $b = 0.432$ cms

Outside dimensions: $a = 1.27$ cms, $b = 0.635$ cms

Recommended frequency range 17.6 to 26.7 GHz (K-band)

Using this procedure for design, we obtain

$\sigma = 6.06$, $l_E = 7.272$ cms, $l_H = 8.34$ cms,

$A = 5.48$ cms, $B = 4.177$ cms, $R_E = R_H = 6.067$ cms

For dimensions of the RG-51/U and RG-66/U rectangular waveguides, refer to the *Microwave Engineering Handbook*[1].

11.7 Cassegrain Reflector Antennas

The Cassegrain reflector antenna (Fig. A.11.15) design involves the design of:

(i) a main parabolic dish A whose parabolic contour is determined by using the following equation:[2]

$$\rho = \frac{2F}{1 + \cos \theta} = F \sec^2 \frac{\theta}{2} \qquad \text{(A.11.31)}$$

(ii) a hyperbolic subdish B whose shape can be found with the aid of

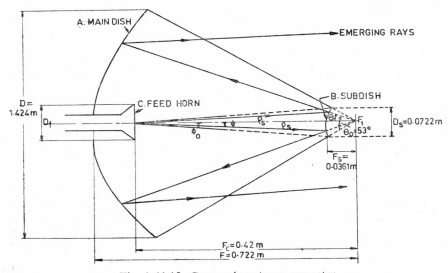

Fig. A.11.15 Cassegrain antenna geometry.

[1]Vol. 1, published by Horizon House Microwave Inc., USA, 1971.
[2]Warren L. Stutzman and Gary A. Thiele, *Antenna Theory and Design*, John Wiley & Sons, New York, 1981.

the following relation

$$\rho_s = -\frac{ep}{1 - \cos \phi} \tag{A.11.32}$$

where

$$p = \frac{F_c}{2}\left(1 - \frac{1}{e^2}\right)$$

F_c = distance between $F_1 F_2$

and e (eccentricity) $= \dfrac{\sin\left[\frac{1}{2}(\phi_0 + \theta_0)\right]}{\sin\left[\frac{1}{2}(\theta_0 - \phi_0)\right]} > 1$ $\qquad$ (A.11.33)

(iii) a feed horn c whose diameter D_f is such that

$$F_c/F = D_f/D_s \tag{A.11.34}$$

for minimum blocking condition, where F and D_s denote the focal length of the parabolic dish and diameter of the subdish respectively. For an efficient Cassegrain system, the following relation should be satisfied

$$F/D = F_s/D_s$$

The antenna gain is given in terms of the aperture area $(= \pi D^2/4)$ and aperture efficiency η_a of the parabolic dish as follows:

$$G = \eta_a(\pi^2 D^2/\lambda^2)$$

where η_a is the resultant of the product of the radiation efficiency η_r, efficiency due to illumination η_t, efficiency due to spillover η_1, efficiency due to surface tolerance η_2, efficiency due to subdish blockage η_3, etc., i.e.,

$$\eta_a = \eta_r\eta_t\eta_1\eta_2\eta_3\cdots$$

Problem

Determine the parameters of a Cassegrain reflector antenna of gain 50 dB at $\lambda = 1.25$ cms (K-band).

Solution

In practice, η_r can be considered ≈ 1. The feed horn and the subdish is designed to produce uniform illumination of the main parabolic reflector so as to maximise gain. Usually $\eta_t \approx 0.85$ and $\eta_1 \approx 0.94$. Since $\eta_2 = \exp(-4\pi\delta^2/\lambda^2)$ and since the rms phase deviation δ of the wavefront at the aperture of the main parabolic dish, from a plane wavefront is, in practice, $\ll 1$, $\eta_2 \approx 1$.

In order to cause minimum blockage of the main reflector, the feed-horn diameter D_f should be small. If $D_f/D \simeq 0.03$, $\eta_3 \approx 0.99$. Hence, the aperture efficiency $\eta_a \cong 0.79$.

Since G, η_a are known, the diameter of the main parabolic dish is given by

$$D = \frac{\lambda}{\pi}\sqrt{\frac{G}{\eta_a}} = 1.424 \text{ m where } G = 10^5$$

Considering the ratio $F/D = 0.5$, $F = 0.722$ m and since for an efficient Cassegrain system $F/D = F_s/D_s$, therefore $F_s = 0.5D_s$. If $D_s = 0.05D$, $D_s = 0.0722$ m and $F_s = 0.0361$ m.

The half-angle θ_0 between the axis of the antenna and the line joining the focus F_1 and the rms of the main dish and subdish is given by the relation

$$\theta_0 = 2 \text{ arc tan } (D/4F) = 53°$$

From the minimum blocking condition, we obtain

$$F_c = FD_f/D_s$$

If $D_f = 0.0418$ m, $F_c = \dfrac{0.722 \times 0.0418}{0.0722} \cong 0.42$ m. Hence the distance of the feed horn aperture from the vertex of the main parabolic dish $= F - F_c = 0.722 - 0.42 = 0.302$ m.

The use of Cassegrain antenna for satellite communication is well illustrated by Tsuyoshi Takahashi et al.[1] Computer-aided analysis of reflector antennas has been elegantly treated by Rusch and Potter.[2]

14.6 Log-Periodic Dipole Antennas

This antenna consists of a number of dipoles (Fig. A.14.10) of successively increasing lengths outward from the feed-point near the apex forming a cone of angle α. A two-wire transmission line transposed between adjacent pair of dipoles acts as the feeder. The design[3] of this antenna

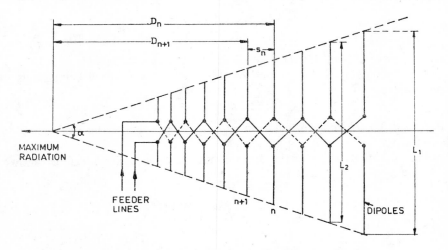

Fig. A.14.10   Geometry of log-periodic dipole antenna.

[1]Tsuyoshi Takahashi et. al. 'Electrical. Performance of the 30M Cassegrain Antennas at the Kashima Earth Station', J. Radio Res. Lab. Japan, Vol. 15, No 77, Jan. 1958, 1041.
[2]W. V. T. Rusch and P.D. Potter, Analysis of Reflector Antennas, Academic Press, 1970.
[3]D. E. Isbell, 'Log-Periodic Dipole Arrays', IEEE-Trans., vol. AP–8, p. 260, May 1960.

involves the determination of the spacing factor σ and the scale factor τ. The spacing factor yields the successive dipole spacing and the scale factor determines the lengths of the successive dipoles forming a wedge-shaped cone.

The scale factor τ and the spacing factor σ are given respectively by

$$\tau = \frac{D_{n+1}}{D_n} = \frac{L_{n+1}}{L_n}, \qquad n = 1, 2, 3, \ldots \qquad \text{(A.14.11)}$$

and

$$\sigma = \frac{S_n}{2L_n} \qquad \text{where } S_n = D_n - D_{n+1} \qquad \text{(A.14.12)}$$

The relation between σ, τ and the wedge-angle α is given by

$$\sigma = \frac{1 - \tau}{4 \tan \dfrac{\alpha}{2}} \qquad \text{(A.14.13)}$$

Isbell has given a set of design curves σ and τ with the gain of the antenna in dB. Carrel[1] has also treated the design of this type of antenna.

Problem

Determine the design parameters of a log-periodic, dipole array of gain 9 dB to operate over a frequency range of 125 MHz to 500 MHz.

Solution

The wavelength range is

$$\lambda \text{ (longest)} = 2.4 \text{ m}$$

$$\lambda \text{ (shortest)} = 0.6 \text{ m}$$

Hence the longest dipole $(L_1) = 1.2$ m and the shortest dipole $= 0.3$ m.

For 9 dB gain, the optimum values of σ and τ (from Isbell's design curves) are

$$\sigma = 0.162 \quad \text{and} \quad \tau = 0.861$$

Hence, using the relation between τ, σ and α, we obtain the wedge-angle $\alpha \cong 24°$.

Since the dimensions of the dipoles and their successive spacing follow the scaling law (Eq. A.14.14)

$$\frac{D_{n+1}}{D_n} = \frac{L_{n+1}}{L_n} = \frac{S_{n+1}}{S_n} \qquad \text{(A.14.14)}$$

which yields

$$L_2 = \tau L_1, \quad L_3 = \tau L_2 \ldots$$

and

$$s_1 = 2\sigma L_1, \quad s_2 = 2\sigma L_2 \ldots$$

---

[1] R. Carrel, 'The Design of Log-Periodic Dipole Antennas', *IRE International Convention Record*, Pt. 1, 1961, p. 61.

we obtain the following design values for the complete array:

$L_1 = 1.2$ m, $L_2 = 1.033$ m, $L_3 = 0.8894$ m,

$L_4 = 0.767$ m, $L_5 = 0.66$ m, $L_6 = 0.568$ m,

$L_7 = 0.489$ m, $L_8 = 0.421$ m, $L_9 = 0.362$ m,

$L_{10} = 0.312$ m, $L_{11} = 0.269$

and

$s_1 = 0.389$ m, $s_2 = 0.335$ m, $s_3 = 0.288$ m,

$s_4 = 0.248$ m, $s_5 = 0.214$ m, $s_6 = 0.184$ m,

$s_7 = 0.158$ m, $s_8 = 0.136$ m, $s_9 = 0.117$ m,

$s_{10} = 0.101$ m, $s_{11} = 0.087$ m.

It may be remarked that for any given value of α, there is a minimum possible value of τ which lies between 0 and 1. Larger values of α and smaller values of τ go together and result in more compact design for a given bandwidth. But larger values of τ and smaller values of α result in improved performance such as higher gain at the expense of a larger antenna structure.

16.13   Horizonal Rhombic Antenna

Problem

Design a horizontal rhombic receiving antenna to yield maximum output in the direction of its principal axis which is directed towards the incoming wave which makes an angle $\varDelta = 30°$ with respect to the ground. The ground conductivity is assumed to be infinite. The frequency of the incoming wave is 30 MHz.

Solution

The horizontal rhombic antenna placed at a height H above the ground

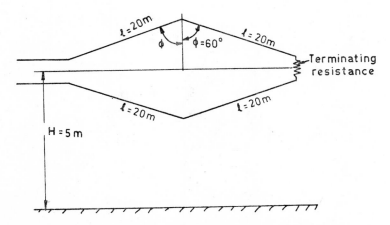

Fig. A.16.29   Plan view of the horizontal rhombic antenna.

and parallel to it is shown in Fig. A.16.29. The design involves the determination of the height H of the antenna above the ground, the length l of each arm of the rhomboid and the semi-angle ϕ between any two adjacent arms[1]. Since the principal axis of the rhomboid is to be directed towards the incoming wave, for obtaining maximum output, the following optimum design conditions are of interest.

$$H = \lambda/4 \sin \varDelta$$

$$\phi = \text{arc sin } (\cos \varDelta) \qquad\qquad (A.16.13)$$

$$l = \lambda/2(1 - \sin \phi \cos \varDelta)$$

since the angle ϕ is the complement of the angle $\varDelta$ ($\sin \phi = \cos \varDelta$), the length of each arm is given by

$$l = \lambda/2 \sin^2 \varDelta$$

Since $\lambda = 10$ m and $\varDelta = 30°$, we obtain

$$H = 5 \text{ m}, \qquad l = 20 \text{ m and } \phi = 60°$$

as the dimensions of the receiving rhombic antenna. Since the antenna is non-resonant, it can be used as a wideband device.

16.14 Yagi-Uda Array

The Yagi-Uda antenna array (Fig. A.16.30) consists of a driven element (usually a folded dipole) of dimension L, a reflector of dimension L_R and a number of directors of dimension L_D. The spacing between the reflector

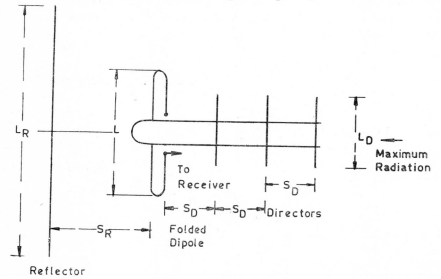

Fig. A.16.30   Yagi-Uda array.

---

[1]E. Bruce, A. C. Beck and L. R. Lowry, 'Horizontal Rhombic Antennas', *Proc. I.R. E.*, vol. 23, No. 1, p. 24, 1935.

and the driven element is S_R. The distance between the driven element and the first director is B_D. The spacing between successive directors is S_D. For optimum performance of the array, separate adjustments of L, L_R, L_D, S_R and S_D are required.

It is known that if $L_R > 0.5\lambda$, it behaves as an inductive element and screens the back radiation. Hence it is called a wave reflector. If $L_D < 0.5$, the element is capacitive and reinforces the radiation in the end-fire direction. Hence it is called a wave director. L is 0.5λ, i.e., the folded dipole is tuned to the incoming wave.

The antenna array which was first studied by Uda[1], was later analysed rigorously[2]. Green[3] and Viezbiekie[4] have studied the design of the array. Exhaustive design data are available from this literature for all TV channels. Typical design parameters and performance characteristics of UHF and VHF channels are given as follows:

Design parameters

No. of elements	L_R	L	L_D	S_R	S_D	Conductor diameter AWG8
5	0.51λ	0.5λ	0.456λ	0.25λ	0.15λ	0.3264 cm

Performance characteristics

Gain (dB)	Front to back ratio	Z_{input} (Ω)	Half-power beamwidth	
10	13.1 dB	9.6 + j13.0	76° (*H*-plane)	62° (*E*-plane)

For UHF TV channel (500–506 MHz), $\lambda_{(mean)} = 0.5964$ m. Hence

$$L = 0.2982 \text{ m}, \qquad L_D = 0.272 \text{ m}, \qquad L_R = 0.3042 \text{ m},$$

$$S_R = 0.1491 \text{ m}, \qquad S_D = 0.0895 \text{ m}$$

For VHF TV channel (204–210 MHz), $\lambda_{(mean)} = 1.449$ m. Hence

$$L = 0.7245 \text{ m}, \qquad L_R = 0.739 \text{ m}, \qquad L_D = 0.661 \text{ m},$$

$$S_R = 0.3622 \text{ m}, \qquad S_D = 0.2174 \text{ m}$$

[1]Shintaro Uda, *Shortwave Projector*, 1974.
[2]Ronold W. P. King and Richard B. Mack, Sheldon's Sandler, *Arrays of Cylindrical Dipoles*, Cambridge University Press, London, 1968.
[3]H. E. Green, 'Design data for short and medium length Yagi-Uda Array', Institution of Engineers (Australia), *Electrical Engineering Transaction*, p. 1–8, March 1966.
[4]P. Viezbiekie, 'Yagi Antenna Design', *NBS Technical Note* 688, US Govt. Printing Office, Dec. 1976.

INDEX